Advances in Intelligent Systems and Computing

Volume 862

Series editor

Janusz Kacprzyk, Systems Research Institute, Polish Academy of Sciences, Warsaw, Poland
e-mail: kacprzyk@ibspan.waw.pl

The series "Advances in Intelligent Systems and Computing" contains publications on theory, applications, and design methods of Intelligent Systems and Intelligent Computing. Virtually all disciplines such as engineering, natural sciences, computer and information science, ICT, economics, business, e-commerce, environment, healthcare, life science are covered. The list of topics spans all the areas of modern intelligent systems and computing such as: computational intelligence, soft computing including neural networks, fuzzy systems, evolutionary computing and the fusion of these paradigms, social intelligence, ambient intelligence, computational neuroscience, artificial life, virtual worlds and society, cognitive science and systems, Perception and Vision, DNA and immune based systems, self-organizing and adaptive systems, e-Learning and teaching, human-centered and human-centric computing, recommender systems, intelligent control, robotics and mechatronics including human-machine teaming, knowledge-based paradigms, learning paradigms, machine ethics, intelligent data analysis, knowledge management, intelligent agents, intelligent decision making and support, intelligent network security, trust management, interactive entertainment, Web intelligence and multimedia.

The publications within "Advances in Intelligent Systems and Computing" are primarily proceedings of important conferences, symposia and congresses. They cover significant recent developments in the field, both of a foundational and applicable character. An important characteristic feature of the series is the short publication time and world-wide distribution. This permits a rapid and broad dissemination of research results.

More information about this series at http://www.springer.com/series/11156

Suresh Chandra Satapathy
Vikrant Bhateja · Radhakhrishna Somanah
Xin-She Yang · Roman Senkerik
Editors

Information Systems Design and Intelligent Applications

Proceedings of Fifth International
Conference INDIA 2018 Volume 1

 Springer

Editors
Suresh Chandra Satapathy
School of Computer Engineering
KIIT University
Bhubaneswar, Odisha, India

Xin-She Yang
School of Science and Technology
Middlesex University
London, UK

Vikrant Bhateja
Department of Electronics and
Communication Engineering
Shri Ramswaroop Memorial Group
of Professional Colleges (SRMGPC)
Lucknow, Uttar Pradesh, India

Roman Senkerik
Faculty of Applied Informatics
Tomas Bata University in Zlín
Zlín, Czech Republic

Radhakhrishna Somanah
Universite des Mascareignes
Beau Bassin-Rose Hill, Mauritius

ISSN 2194-5357   ISSN 2194-5365 (electronic)
Advances in Intelligent Systems and Computing
ISBN 978-981-13-3328-6   ISBN 978-981-13-3329-3 (eBook)
https://doi.org/10.1007/978-981-13-3329-3

Library of Congress Control Number: 2018961687

This Springer imprint is published by the registered company Springer Nature Singapore Pte Ltd.
The registered company address is: 152 Beach Road, #21-01/04 Gateway East, Singapore 189721, Singapore

Preface

The Fifth International Conference on Information System Design and Intelligent Applications (INDIA 2018) was successfully hosted by the Université des Mascareignes (UDM) from 19th July 2018 to 20th July 2018 at the InterContinental Resort Balaclava in Mauritius.

The Université des Mascareignes has strategically partnered with Université de Limoges, France, to offer a double degree with a strong emphasis on industrial placement, design its curriculum, introduce industry-based electives and facilitate students' internship at the industry for skill development. It has established a Centre of Excellence in collaboration with industry for various research and training purposes. The value addition training and career augmentation services prepare the students to meet expectations of industry demands.

Research focuses of the faculty of information and communication technology include algorithms and theory of computation, artificial intelligence, bioinformatics, cloud computing, database and data mining, data analytics, human–computer interaction, information and network security, Internet technology, image processing, mobile computing, pattern recognition, program analysis and testing, parallel and distributed computing, real-time systems, service-oriented architecture, soft computing, software engineering and wireless sensor networks.

The objective of this conference was to provide opportunities for the researchers, academicians and industry persons to interact and exchange the ideas, experience and gain expertise in the cutting-edge technologies pertaining to soft computing and signal processing. Research papers in the above-mentioned technology areas were received and subjected to a rigorous peer-reviewed process with the help of programme committee members and external reviewers. The INDIA 2018 conference received research articles in various domains, and after the rigorous review, only quality articles were accepted for publication in Springer AISC series. The hosting of the conference and full sponsorship approved by the director were the key factors in generating quality papers from UDM staff.

Special thanks to keynote speakers Mr. Aninda Bose, Senior Editor, Hard Sciences Publishing, Springer Nature; Dr B. Annappa, Professor, Department of Computer Science and Engineering, National Institute of Technology Karnataka,

Surathkal, Mangalore, India; and Dr. Jagdish Chand Bansal, Professor, South Asian University, New Delhi, and Visiting Research Fellow, Maths and Computer Science, Liverpool Hope University, UK.

We would like to express our gratitude to all session chairs and reviewers for extending their support and cooperation.

We would like to express our special gratitude to Organizing Committee Co-chair, Prof. Binod Kumar Pattanayak, Siksha 'O' Anushandhan University, Bhubaneswar, Odisha, and Publication Chair, Dr. Suresh Chandra Satapathy, KIIT, Bhubaneswar, for initiating this conference and for their valuable support and encouragement till the successful conclusion of the conference.

We are indebted to all the committee members for organizing the conference, namely Mohammad Reza Hosenkhan, Organizing Committee Chair, Founding Dean Faculty of Information and Communication Technology; Mrs Nundini Akaloo, Programme Committee Chair, Founding Dean Faculty of Business Management; Mrs Rubina Fareed and Dr. Swaleha Peeroo, Lecturers from the Faculty of Business Management; Mr Seeven Amic, Programme Committee Co-chair; and Mr Sanjeev Cowlessur, Mrs Shameera Lotun, Mrs Mahejabeen Peeermamode and Mrs Dosheela Ramlowat, Lecturers from the Department of Software Engineering.

Special thanks to administrative staff, namely Mrs Farzana Antoaroo, Mr Ravi Langur, Mr Deojeet Nohur, Mrs Vimi Lockmun-Bissessur and Mrs Yasoda Benoit.

We express our heartfelt thanks to our Chief Patron, the Hon. (Mrs) Leela Devi Dookhun-Luchoomun, Minister of Education and Human Resources, Tertiary Education and Scientific Research, who supported and inaugurated the conference.

We are first and foremost indebted to the General Chair Dr. Radhakhrishna Somanah, Outstanding Scientist, Director General of UDM, Commander of the Star and Key of the Indian Ocean (CSK)—one of the highest national awards received on the Republic Day of Mauritius on 12 March 2011, Chevalier dans l'Ordre des Palmes Academiques—Award by the Republic of France on 4 September 2009, Award by International Astronomical Union (IAU): minor planet Somanah 19318 in our solar system named after his name "Somanah" in 2007 for his contribution to research in astronomy award by NASA to my research group in June 2004 for excellence in Education and Discovery. We express our deep gratitude to Dr. R. Somanah for having believed in the importance and value of the conference and for having found solutions to many challenges and obstacles in Mauritius at the initial stage (and even after) to ensure that the conference materializes and becomes reality and has finally been a success.

Last, but certainly not least, our special thanks to all the authors without whom the conference would not have taken place. Their technical contributions have made our proceedings rich and praiseworthy.

Bhubaneswar, India   Suresh Chandra Satapathy
Lucknow, India   Vikrant Bhateja
Beau Bassin-Rose Hill, Mauritius   Radhakhrishna Somanah
London, UK   Xin-She Yang
Zlín, Czech Republic   Roman Senkerik

Contents

About the Editors

Suresh Chandra Satapathy is currently working as Professor, School of Computer Engg, KIIT Deemed to be University, Bhubaneswar, Odisha, India. He obtained his PhD in Computer Science Engineering from JNTUH, Hyderabad and Master degree in Computer Science and Engineering from National Institute of Technology (NIT), Rourkela, Odisha. He has more than 28 years of teaching and research experience. His research interest includes machine learning, data mining, swarm intelligence studies and their applications to engineering. He has more than 98 publications to his credit in various reputed international journals and conference proceedings. He has edited many volumes from Springer AISC, LNEE, SIST and LNCS in past and he is also the editorial board member in few international journals. He is a senior member of IEEE and Life Member of Computer Society of India.

Vikrant Bhateja is Associate Professor, Department of Electronics & Communication Engineering, Shri Ramswaroop Memorial Group of Professional Colleges (SRMGPC), Lucknow and also the Head (Academics & Quality Control) in the same college. His areas of research include digital image and video processing, computer vision, medical imaging, machine learning, pattern analysis and recognition. He has around 100 quality publications in various international journals and conference proceedings. Prof. Vikrant has been on TPC and chaired various sessions from the above domain in international conferences of IEEE and Springer. He has been the track chair and served in the core-technical/editorial teams for international conferences: FICTA 2014, CSI 2014, INDIA 2015, ICICT-2015 and ICTIS-2015 under Springer-ASIC Series and INDIACom-2015, ICACCI-2015 under IEEE. He is associate editor in International Journal of Synthetic Emotions (IJSE) and International Journal of Ambient Computing and Intelligence (IJACI) under IGI Global. He is also serving in the editorial board of International Journal of Image Mining (IJIM) and International Journal of Convergence Computing (IJConvC) under Inderscience Publishers. He has been editor of four published volumes with Springer (IC3T-2015, INDIA-2016, ICIC2-2016, ICDECT-2016) and few other are under press (FICTA-2016, IC3T-2016, ICMEET-2016).

Radhakhrishna Somanah is Director of Universite des Mascareignes. He is one of the three pioneers of professional astronomy in Mauritius (together with Dr Nalini Issur and Dr Kumar Golap who is now working for the VLA, NRAO). He is fully involved with the design, construction and research at the Mauritius Radio Telescope (MRT) project since 1989. MRT is one of the biggest scientific research projects at the University of Mauritius. The main aim of the project was to produce a map of the southern sky at 151.5 MHz (unique in the world), perform astrophysical interpretation of the map, develop new algorithms and software etc. He has been involved with the supervision of the civil, mechanical, electrical and electronics work. Because of its immensity, it took nearly 5 years (1989–1993) to build. He was also part of the team which developed the software to convert the raw data into deconvolved radio images, and astrophysical interpretation of the latter. He has many publications in reputed journals.

Xin-She Yang obtained his DPhil in Applied Mathematics from the University of Oxford. He then worked at Cambridge University and National Physical Laboratory (UK) as a Senior Research Scientist. Now he is Reader in Modelling and Optimization at Middlesex University, an elected Bye-Fellow at Cambridge University and Adjunct Professor at Reykjavik University (Iceland). He is the chair of the IEEE CIS Task Force on Business Intelligence and Knowledge Management. He is on the list of both 2016 and 2017 Clarivate Analytics/Thomson Reuters Highly Cited Researchers.

Roman Senkerik was born in the Czech Republic, and went to the Tomas Bata University in Zlin, Faculty of Applied Informatics, where he studied Technical Cybernetics and obtained his MSc degree in 2004, PhD degree in 2008 (Technical Cybernetics) and Assoc. Prof. degree in 2013 (VSB – Technical University of Ostrava – degree in Informatics). He is now a researcher and lecturer at the same university. His research interests are: Evolutionary Computation, Theory of chaos, Complex systems, Soft-computing methods and their interdisciplinary applications, Optimization, Neural Networks, Data analysis, Information retrieval, and Cyber-security. He is Recognized Reviewer for many Elsevier journals as well as many other leading journals in computer science/computational intelligence. He was a part of the organizing teams for several conferences, and special sessions/symposiums at IEEE CEC and IEEE SSCI events. He was a guest editor of several special issues in journals, editor of Springer proceedings for several conferences.

New Strategy for Mobile Robot Navigation Using Fuzzy Logic

B. B. V. L. Deepak and D. R. Parhi

Abstract The current research work aims to develop an efficient motion planner for a differential vehicular system inspired from the Fuzzy inference system. In this strategy, rolling and sliding kinematic constraints have been considered while implementation. The proposed fuzzy model requires two inputs: (1) the distance between the robot and the obstacles in the environment and (2) position of the target, i.e., the robot heading angle towards the destination. Once the system receives information from its search space, the robot obtains the suitable steering angle for an intelligent system. Experimental analysis has been conducted to a differential robot in order to represent its effectiveness.

Keywords Design for assembly · Assembly sequence planning
Assembly constraints · Firefly algorithm · Computer aided design (CAD)

1 Introduction

Motion planning of a mobile robot is nothing but the robot should avoid the static/dynamic obstacles while reaching its target position. In the last decades, a lot amount of research work has been devoted to solve this mobile robot motion planning problem [1–3]. Past studies [4, 5] showed that implementation of Fuzzy is right idea for robot motion planning problems.

Deepak and Parhi [6] treated the motion planning of mobile robot as an optimization problem and then they solved it. The path planning of an autonomous robot is solved in [7] subjected to the multilayer feed forward fuzzy inference controllers. Past research found that fuzzy inference system has been integrated with other techniques for obtaining better results in robot motion planning problem [8, 9].

B. B. V. L. Deepak (✉) · D. R. Parhi
National Institute of Technology, Rourkela 769008, Odisha, India
e-mail: bbv@nitrkl.ac.in

D. R. Parhi
e-mail: drkparhi@nitrkl.ac.in

© Springer Nature Singapore Pte Ltd. 2019
S. C. Satapathy et al. (eds.), *Information Systems Design and Intelligent Applications*,
Advances in Intelligent Systems and Computing 862,
https://doi.org/10.1007/978-981-13-3329-3_1

The current research work aims to develop an efficient motion planner for a differential mobile robot based on the Fuzzy logic. The current strategy, rolling and sliding kinematic constraints have been considered while implementation. The proposed fuzzy model requires two inputs: (1) the distance between the robot and the obstacles in the environment and (2) position of the target, i.e., the robot heading angle towards the destination. Finally, the replication values have been compared with real-time values to test the adeptness of the addressed strategy.

2   Kinematic Model of a Differential Robot

To know the robot position with respect to a global reference frame as shown in Fig. 1, a mapping matrix is required as represented in Eqs. (1)–(3). Robot's position can be represented as a point P with respect to a global reference frame as function of its x-coordinate, y-coordinate and orientation of robot reference frame angle θ.

$$\xi_I = \begin{bmatrix} x & y & \theta \end{bmatrix}^{\mathrm{T}} \tag{1}$$

$$R(\theta) = \begin{bmatrix} \cos\theta & \sin\theta & 0 \\ -\sin\theta & \cos\theta & 0 \\ 0 & 0 & 1 \end{bmatrix} \tag{2}$$

$$\xi_R = R(\theta)\xi_I \quad \text{and} \quad \dot{\xi}_R = R(\theta)\dot{\xi}_I \tag{3}$$

Fig. 1   Robot reference with respect to world reference frame

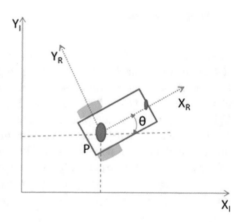

2.1 Kinematic Constraints of Wheel Configuration

A differential robot comprises one omnidirectional wheel and two fixed wheels. These two fixed wheels are motorized and the omnidirectional wheel is non-motorized. For any wheeled robot, there are two motion constraints related to the wheel rolling and sliding. The motion constraints of the robot are expressed as a function of three parameters: (i) Robot chassis distance l, (ii) wheel alignment β and (iii) chassis orientation α. The free body diagram of the fixed standard wheel and the representation of various influencing parameters are represented in Fig. 2.

The fixed standard wheel has two motion constraints as follows (Fig. 3):

(i) Rolling motion constraints is shown in Eq. (4)

$$[\sin(\alpha + \beta) \quad - \cos(\alpha + \beta) \quad (-l)\cos \beta] * R(\theta)\dot{\xi}_I - r\dot{\varphi} = 0 \tag{4}$$

(ii) The sliding constraint for this wheel which must be zero is shown in Eq. (5)

$$[\cos(\alpha + \beta) \quad \sin(\alpha + \beta) \quad l \sin \beta]R(\theta)\dot{\xi}_I = 0 \tag{5}$$

The configuration of the robot can be expressed as below

- Position coordinates: $\xi_I = \begin{bmatrix} x(t) & y(t) & \theta(t) \end{bmatrix}^T$
- Angular coordinates: β_{f1} and β_{f2}
- Rotational coordinates: $\begin{bmatrix} \varphi_{f1}(t)\varphi_{f2}(t) \end{bmatrix}^T$

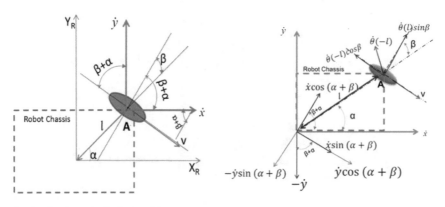

Fig. 2 Fixed standard wheel and its parameters

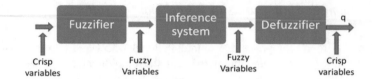

Fig. 3 The structure of the fuzzy logic control

3 Sugeno Fuzzy Model

The Fuzzy system works on the basis of If-then rules provide by the users. The fuzzy inference system can receive either fuzzy inputs or crisp inputs but the outputs it obtains are almost fuzzy sets as shown in Fig. 4.

The rules for Sugeno fuzzy model are considering as follows:

If input-1 is A and input-2 is B then $z = f(x, y)$

Defuzzification means the way of a crisp value is taken out from a fuzzy set as a representative value. For Sugeno fuzzy system, defuzzification approach is not necessary and the weighted average can be calculated as follows:

$$z = \frac{W_1 * Z_1 + W_2 * Z_2}{W_1 + W_2},$$ (9)

where Z is the weight aggregated output membership function

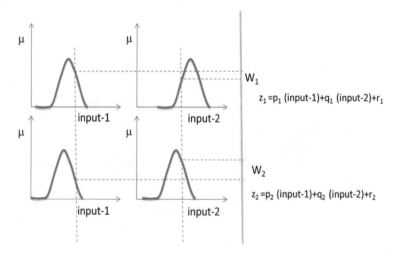

Fig. 4 Sugeno Fuzzzy inference system with two inputs and one output

4 Generalized Bell Membership Functions

This type of membership functions (MF) are indicated by three variables a, b, c

$$\text{bell}(a, b, c) = \frac{1}{1 + \left|\frac{x-c}{a}\right|^{2b}} \tag{10}$$

The parameters "a" for width of the MF, and b is a positive value for representing slope of the curve and "c" for center of MF. In this paper, two inputs and one output are considered; the membership values for these are the bell-MF.

4.1 Member Ship for First Input

The fuzzy system is taking the sensory information as the inputs from the environment and the suitable steering angle gives as the output. The first input parameter is the distance between the nearest obstacle and the robot. Let us assume that the maximum possible distance that can be sensed by the robot is 180 cm. So the MF values are varying from 0 to 180 cm and are represented by "ROB". Five linguistics are considered for this variable varying from V_high to V_low to form the rules.

4.2 Member Ship for Second Input

Let the system be working in first quadrant so the target angle is varying from $0°$ to $90°$ ($\Pi/2$) and the membership functions are represented by T_angle. Five linguistics are considered for this variable varying from V_high to V_low to form the rules.

4.3 Member Ship for Output

Like the same condition output as the steering angle also varying from $0°$ to $90°$ ($\Pi/2$) and the membership functions are represented by S_angle. Five linguistics are considered for this variable varying from V_high to V_low to form the rules.

5 Simulation Results

The proposed Fuzzy model is implemented through MATLAB 2008 version. Figures 5, 6 represent collision free trajectories of the considered differential robot.

6 Experimental Results

Experimental analysis has been performed to check the efficiency of the proposed. The necessary equipment used for modeling an autonomous mobile robot

Fig. 5 Single obstacle—single target environment

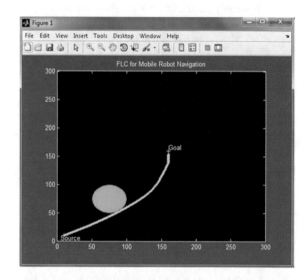

Fig. 6 Multi-obstacle—single target environment

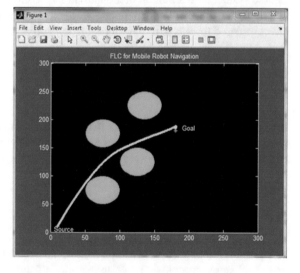

- *Microcontroller*—ATMEGA 328 is used for PWM generation for servo motor control, data acquisition, sensor reading, and data transferring.
- *Obstacle Sensor*—GP2DY Infrared Distance sensor is used, which can detect the obstacles based on the triangulation method.
- *Position Encoder*—to monitor the speed of the motor shaft.
- *Magnetic Compass*—to find out the direction movement of the robot in the environment.
- *L2938*—to control the velocity of the motorized wheels.

The developed differential drive robot and its operation in real-time environment is represented in Fig. 7. The robot reached its target position in 100 s. while moving with a velocity of 10 cm/sec.

(a) Developed real time robotic system and (b) its initial, final and object positions

(c) Detection of objects & avoidance

(d) Approaching to the target position

Fig. 7 Robot movement at different positions

7 Conclusion

New efficient system architecture has been modeled based on fuzzy inference system for solving mobile robot navigation. The proposed system architecture works on the Sugeno fuzzy type and bell-shaped membership functions has been considered. To obtain feasible path within the robotic work space, fuzzy inference system has modeled with two input parameters and one output parameter. By obtaining the proper steering angle, the mobile robot reaches its goal by avoiding obstacles within its free environment. From the simulation results it has been concluded that the robot can generate optimal collision free paths using the proposed methodology. Although the developed algorithm is suitable for generating collision free trajectories within its environments, it is required to apply more number of rules and fine tune of membership values in order to obtain better results.

References

1. B.B.V.L. Deepak, M.R. Bahubalendruni, Development of a path follower in real-time environment. World J. Eng. **14**(4), 297–306 (2017)
2. B.B.V.L. Deepak, D.R. Parhi, Control of an automated mobile manipulator using artificial immune system. J. Exp. Theor. Artif. Intell. **28**(1–2), 417–439 (2016)
3. B.B.V.L. Deepak, D. Parhi, Intelligent adaptive immune-based motion planner of a mobile robot in cluttered environment. Intell. Serv. Robot. **6**(3), 155–162 (2013)
4. A.T. Azar, H.H. Ammar, H. Mliki, Fuzzy logic controller ith color vision system tracking for mobile manipulator robot, in *International Conference on Advanced Machine Learning Technologies and Applications* (Springer, Cham, 2018), pp. 138–146
5. T.C. Lin, C.C. Chen, C.J. Lin, Navigation control of mobile robot using interval type-2 neural fuzzy controller optimized by dynamic group differential evolution. Adv. Mech. Eng. **10**(1), 1687814017752483 (2018)
6. B.B.V.L. Deepak, D.R. Parhi, B.M.V.A. Raju, Advance particle swarm optimization-based navigational controller for mobile robot. Arab. J. Sci. Eng. **39**(8), 6477–6487 (2014)
7. O.S. Syed, U.A. Muhammad, A.J. Muhammad, M. Hassam, Design and implementation of neural network based controller for mobile robot navigation, in 26th IEEEP Students' Seminar 2011 (2011)
8. M.P. Garcia, O. Montiel, O. Castillo, R. Sepúlveda, P. Melin, Path planning for autonomous mobile robot navigation with ant colony optimization and fuzzy cost function evaluation. Appl. Soft Comput. **9**(3), 1102–1110 (2009)
9. C.F. Juang, Y.C. Chang, Evolutionary-group-based particle-swarm-optimized fuzzy controller with application to mobile-robot navigation in unknown environments. IEEE Trans. Fuzzy Syst. **19**(2), 379–392 (2011)

EEG Monitoring: Performance Comparison of Compressive Sensing Reconstruction Algorithms

Meenu Rani, S. B. Dhok and R. B. Deshmukh

Abstract EEG represents the electrical activity across brain. This activity is monitored to diagnose the diseases due to brain disorders, like epilepsy, coma, sleep disorders, etc. To record EEG signal, a minimum of 21 electrodes are placed across the scalp, which generates huge amount of data. To handle this data, compressive sensing (CS) proves itself to be a better candidate than the traditional sampling. CS generates far fewer samples than that suggested by Nyquist rate and still allows faithful reconstruction. This paper compares the performance of CS reconstruction algorithms in reconstructing the EEG signal back from compressive measurements. The algorithms compared from convex optimization are basis pursuit (BP) and basis pursuit denoising (BPDN) and from greedy algorithms are orthogonal matching pursuit (OMP) and compressive sampling matching pursuit (CoSaMP). The performance of these algorithms is compared on the basis of speed and reconstruction efficiency.

Keywords Compressive sensing · EEG-monitoring · Random demodulator
Basis pursuit · OMP · CoSaMP

1 Introduction

EEG signal is recorded to diagnose the brain diseases like epilepsy, coma, sleep disorders, etc. These diseases cause abnormal brain activity. The precise diagnosis of this activity is important [1–4]. For this, the signals are oversampled and a huge amount of the data is collected for processing. This data needs to be compressed before storage and/or transmission. The compression stage consumes a lot of power.

M. Rani (✉) · S. B. Dhok · R. B. Deshmukh
Visvesvaraya National Institute of Technology, Nagpur 440010, Maharashtra, India
e-mail: meenubanait@gmail.com

S. B. Dhok
e-mail: sanjaydhok@gmail.com

R. B. Deshmukh
e-mail: mona1810@yahoo.com

© Springer Nature Singapore Pte Ltd. 2019
S. C. Satapathy et al. (eds.), *Information Systems Design and Intelligent Applications*,
Advances in Intelligent Systems and Computing 862,
https://doi.org/10.1007/978-981-13-3329-3_2

This problem becomes more severe in case of remote health monitoring, where, the devices are power constrained. At the receiver end, original is reconstructed and relevant clinical states are extracted. In this situation, compressive sensing (CS) seems to perform better than traditional method. It lowers the power consumption by sampling at much lower rate compared to the Nyquist rate and thus avoids the need for further compression [5–7].

CS is a newer signal processing technique, introduced by Candès et al. and Donoho [8, 9]. CS is applicable to the signals which are sparse or compressible, either in original domain or in some transformed domain. A signal is said to be sparse, if it can be represented using fewer significant components, compared to the number of Nyquist samples. For compressible signals, the sorted components decay rapidly obeying power law, refer Fig. 3. The Nyquist sampling rate is estimated by highest frequency content of signal, on the other hand, CS sampling rate depends upon the sparsity of underlying signal. This generates few number of CS measurements and compresses the signal during sensing only. From CS measurements, original signal can be recovered faithfully by complex nonlinear techniques [10–12].

CS works efficiently in areas, where the performance of conventional techniques is limited due to the factors like cost, speed, power, etc. The most popular hardware implementation demonstrating the CS mechanism is the 'single pixel camera', which performs compressive imaging using single pixel/photodiode [13]. Other implementations are: random demodulator (RD) [14], modulated wideband converter (MWC) [15], compressive multiplexer (CMUX) [16], etc. [17]. In this paper, the EEG signal acquisition has been done using RD technique of CS. The organization of the paper is as: Sect. 2 presents the mathematical model of CS. Section 3, describes the EEG signal acquisition using RD. Section 4, discusses the reconstruction methodology used for recovering the EEG signal. Section 5, presents the performance comparison of CS reconstruction algorithms.

2 Mathematical Model of CS

2.1 Acquisition Model

CS randomly sub-samples the input signal, which directly generates the compressed measurements, thus relaxing the requirement of a separate compression stage. In addition, the requisite CS measurements can be minimized by taking incoherent measurements. For this, the signal acquisition domain should be incoherent from the signal sparsifying basis. For example, if the sparsifying basis is frequency domain then the acquisition domain should be time, because a signal having sparse representation in frequency domain, spreads out in the time domain [10–12]. Mathematically, the CS acquisition model is given by (1) and can be represented by Fig. 1a.

$$y = \varphi x, \tag{1}$$

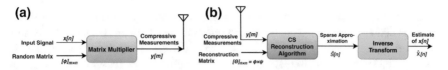

Fig. 1 Mathematical representation of **a** CS Acquisition Model and **b** Reconstruction Model

where, $x \in \mathbb{R}^n$ is the input signal, $\varphi \in \mathbb{R}^{m \times n}$ or $\mathbb{C}^{m \times n}$ is the CS measurement matrix and $y \in \mathbb{R}^m$ or $\mathbb{C}^m$ is the output vector. Here, n represents number of samples in input signal, m represents number of measurements and $m \ll n$.

2.2 Reconstruction Model

For obtaining original signal back from CS measurements, reconstruction matrix, Θ and output measurement vector, y, are needed, as shown in Fig. 1b. Here, matrix $\Theta = \varphi \times \psi \in \mathbb{R}^{m \times n}$ or $\mathbb{C}^{m \times n}$ and ψ represents the domain in which input signal, x, has sparse representation. In this sparse domain, the input signal, x, can be written as in (2).

$$x = \sum_{i=1}^{n} s_i \psi_i = \psi s, \tag{2}$$

where, $s \in \mathbb{R}^n$ is the sparse representation of x. The output generated by first stage is vector $\hat{s}$, an approximation to s. CS reconstruction basically solves an underdetermined system of linear Eq. (1), having infinitely many solutions. An exact solution to this can be obtained by posing it as a convex optimization problem (3), which finds a solution having minimum ℓ_1-norm. The solution to this convex problem can be obtained by algorithms from linear programming [18–20].

$$\hat{s} = \arg\min_{s} \|s\|_1 \quad \text{subject to} \quad \Theta s = y, \tag{3}$$

where, $\|s\|_1$ represents ℓ_1-norm of s. From this output, $\hat{s}$, an estimate of original signal, $\hat{x}$, can be achieved by computing the inverse transform of $\hat{s}$ [21].

2.3 Necessary and Sufficient Conditions

Restricted Isometry Property (RIP) The necessary condition for perfect reconstruction is that Θ should satisfy the RIP for a k-sparse vector, (4).

$$1 - \delta \leq \frac{\|\Theta v\|_2}{\|v\|_2} \leq 1 + \delta, \tag{4}$$

where, v is k-sparse vector and $\delta > 0$ is restricted isometry constant. This property ensures that Θ preserves distance among two vectors which are k-sparse. For stable solution, Θ must satisfy the RIP for $3k$-sparse vectors. Here, the difficulty lies in the calculation of δ. An alternate and simpler condition is there, which also ensures stable reconstruction is described below [11, 12, 22].

Incoherence As described earlier, incoherence between acquisition and sparsifying domains is necessary for minimizing the number of measurements necessary for faithful recovery. Incoherence ensures that each measurement carries some information about the input signal. The coherence measure (5) is used to determine incoherence between two matrices.

$$\mu(\varphi, \psi) = \sqrt{n} . \max_{1 \leq i, j \leq n} | \langle \varphi_i, \psi_j \rangle |, \tag{5}$$

where, $| \langle \cdot, \cdot \rangle |$ represents inner product operator. The value of coherence lies in the range $\mu(\varphi, \psi) \in [1, \sqrt{n}]$. Lower value of coherence indicates higher value of incoherence. A general relationship between number of measurements required for faithful reconstruction and coherence is given by (6), where c is a constant [23].

$$m \geq c\mu^2 k \log n. \tag{6}$$

3 EEG Signal Acquisition Using CS

In this paper, the EEG data set is taken from the CHB-MIT database and is shown in Fig. 2a [24]. This signal is sampled using RD as shown in Fig. 2b. The input signal is multiplied with a pseudorandom chipping sequence of +/-1s. This step is used to introduce randomness in the input signal and to smear its frequencies in the lower frequency band. The multiplied signal is then given as input a low-pass filter to retain only the lower frequency components. Finally, this low frequency signal is sampled at much lower rate [14].

The operation of RD in matrix form is described by (7) and (8), where, P is a diagonal matrix having diagonal elements as pseudorandom sequence, $p_c(t)$ and H is the accumulate and dump matrix. The number of terms that needs to be accumulated

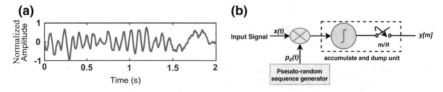

Fig. 2 EEG signal and its acquisition method **a** EEG signal taken from CHB-MIT database, **b** EEG signal acquisition via random demodulator

to generate single measurement is given by $R = \lfloor (n/m) \rfloor$.

$$P = \begin{bmatrix} p_1 & & \\ & \ddots & \\ & & p_n \end{bmatrix} ; \quad H = \begin{bmatrix} 111\cdots & & \\ & 111\cdots & \\ & & 111\cdots \end{bmatrix} \tag{7}$$

$$\left. \begin{aligned} \tilde{x} &= Px \\ y &= H\tilde{x} = \varphi x \\ \varphi &= HP \end{aligned} \right\}. \tag{8}$$

4 Reconstruction of EEG Signal Using CS

4.1 Sparsifying Basis Determination

As discussed earlier, the reconstruction matrix, $\Theta = \varphi \times \psi$. So, we need to determine the sparsifying basis, ψ, for the EEG signal. This is also a very tough task as it requires a lot of experimentation for testing each and every basis. To identify a sparse basis for EEG signal, several transform bases have been tested. The coefficients generated by these transforms are then sorted for comparison purpose, as shown in Fig. 3. From these tests it has been identified that compared to the other transforms under test, the FFT and DCT generates the very less number of coefficients for EEG signal. Based on these experiments, FFT has been selected as sparsifying basis, ψ, for EEG signal and the reconstruction matrix, Θ, is then generated. Now, the next step is to

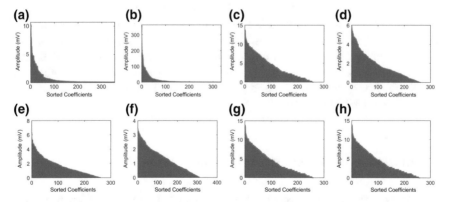

Fig. 3 Sparsity test for EEG signal using **a** Fourier transform, **b** DCT, **c** Haar wavelet trransform, **d** Coiflet wavelet transform, **e** Symlet wavelet transform, **f** discrete Meyer wavelet transform, **g** biorthogonal wavelet transform, **h** reverse biorthogonal wavelet transform

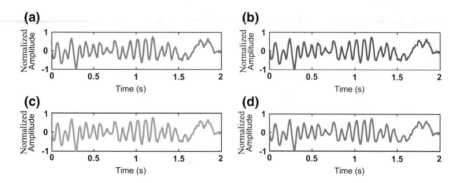

Fig. 4 EEG signal reconstruction using **a** BP, **b** BPDN, **c** OMP and **d** CoSaMP

fix a good CS reconstruction algorithm for reconstructing the EEG signal back from compressive measurements.

4.2 Reconstruction Algorithms

For identifying a better CS algorithm for EEG signal, algorithms from two popular CS approaches, namely convex optimization and greedy algorithms, have been selected. The first two methods, BP and BPDN, are convex optimization problems and their solution can be obtained by algorithms like simplex method and interior point method [20]. The two more algorithms, OMP and CoSaMP, belongs to the greedy algorithms and are relatively simpler and faster compared to the convex optimization [25, 26]. All the simulations have been done in MATLAB 2015a. The simulation settings for an undersampling factor of 2 are as: $n = 512$, $m = n/2 = 256$, x is of size 512×1, φ is of 256×512, y is of 256×1 and Θ is of 256×512.

The output of these algorithms is a sparse vector $\hat{s}$, i.e., the frequency spectrum of reconstructed EEG signal in this case. From $\hat{s}$, the time domain signal is obtained by taking the inverse Fourier transform. The EEG signal reconstructed by BP, BPDN, OMP, and CoSaMP are shown in Fig. 4a–d, respectively. For implementing BP and BPDN, CVX package has been used [27], while, MATLAB functions for OMP and CoSaMP are available at [28].

5 Reconstruction Performance Comparison

To compare the performance of these algorithms, the reconstruction process is repeated for several undersampling ratios (m/n). For each undersampling ratio the corresponding output SNR is computed as per (9). The variation of output SNR with undersampling ratio corresponding to all the algorithms is compared in Fig. 5. It

Fig. 5 Comparison of CS reconstruction algorithms based on output SNR

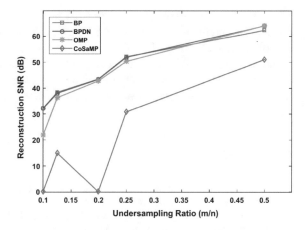

Table 1 Performance comparison of CS reconstruction algorithms

Algorithm	Speed (s)	No. of iterations	Max. output SNR (dB)
BP	4.65	22	62.35
BPDN	2.83	10	64.04
OMP	0.58	256	64.15
CoSaMP	22.45	1000	51.06

has been observed that in highly undersampled cases, the convex optimization based methods performs better than greedy algorithms. The output SNR increases as the undersampling ratio is increased.

$$\text{Output SNR} = 20 \log_{10} \frac{\|s\|_2}{\|s - \hat{s}\|_2} \tag{9}$$

The performance comparison of these algorithms is summarized in Table 1. A comparison of computational time or recovery speed indicates that OMP is the fastest of all the algorithms, despite more number of iterations and has a satisfactory output SNR. The methods based on convex optimization are slower but maintains good reconstruction quality even at higher undersampling factors. The performance of CoSaMP is not satisfactory at higher undersampling factors.

6 Conclusion

Compressive sensing is an attractive signal processing paradigm that performs compression at the time of sensing. CS works by randomly sampling at a rate proportional to the sparsity of underlying signal. There are wide variety of signal processing areas where CS has been used. In this paper, CS has been used for acquisition of EEG signal. For determining a better CS reconstruction algorithm for EEG signal, the algorithms namely BP, BPDN, OMP, and CoSaMP have been compared. A performance comparison of these algorithms shows that if reconstruction speed is a prime concern then OMP is a better choice, while, for achieving reconstruction at higher undersampling factors, the convex optimization based methods perform satisfactory.

References

1. A. Lay-Ekuakille et al., Entropy index in quantitative EEG measurement for diagnosis accuracy. IEEE Trans. Inst. Meas. **63**, 1440–1450 (2014)
2. A. Lay-Ekuakille, et al., Multidimensional analysis of EEG features using advanced spectral estimates for diagnosis accuracy. in *IEEE International Symposium on Medical Measurements and Applications (MeMeA)*, (Gatineau, QC, 2013) pp. 237–240. https://doi.org/10.1109/MeMeA.2013.6549743
3. Vergallo, P., et al.: Identification of Visual Evoked Potentials in EEG detection by emprical mode decomposition, in *IEEE 11th International Multi-Conference on Systems Signals and Devices (SSD14)* (Barcelona, 2014) pp. 1–5. https://doi.org/10.1109/SSD.2014.6808848
4. P. Vergallo, et al., Spatial filtering to detect brain sources from EEG measurements, in *IEEE national Symposium on Medical Measurements and Applications (MeMeA)* (Lisboa, 2014), pp. 1–5
5. S. Aviyente, Compressed Sensing Framework for EEG Compression, in *IEEE/SP 14th Workshop on statistical signal processing* (Madison, WI, USA, 2007) pp. 181–184
6. Z. Zhang et al., Compressed Sensing of EEG for wireless telemonitoring with low energy consumption and inexpensive hardware. IEEE Trans. on Biomed. Engg. **60**(1), 221–224 (2013). https://doi.org/10.1109/TBME.2012.2217959
7. A.M. Abdulghani et al., Compressive sensing scalp EEG signals: implementations and practical performance. E. Med. Biol. Eng. Comp. **50**, 1137–1145 (2012)
8. E.J. Candès et al., Robust uncertainty principles: exact signal reconstruction from highly incomplete frequency information. IEEE Trans. on Inf. Theory **52**(2), 489–509 (2006). https://doi.org/10.1109/TIT.2005.862083
9. D.L. Donoho, Compressed sensing. IEEE Trans. on Inf. Theory **52**(4), 1289–1306 (2006). https://doi.org/10.1109/TIT.2006.871582
10. R.G. Baraniuk, Compressive sensing [lecture notes]. IEEE Sig. Process. Mag. **24**(4), 118–121 (2007). https://doi.org/10.1109/MSP.2007.4286571
11. E.J. Candès, M.B. Wakin, An Introduction to compressive sampling. IEEE Sig. Process. Mag. **25**(2), 21–30 (2008). https://doi.org/10.1109/MSP.2007.914731
12. R. Baraniuk et al., An introduction to compressive sensing. OpenStax-CNX. April 2, 2011. http://legacy.cnx.org/content/col11133/1.5/
13. M.F. Duarte et al., Single-pixel imaging via compressive sampling. IEEE Sig. Process. Mag. **25**(2), 83–91 (2008). https://doi.org/10.1109/MSP.2007.914730
14. J.A. Tropp et al., Beyond nyquist: efficient sampling of sparse bandlimited signals. IEEE Trans. Inf. Theory **56**(1), 520–544 (2010)
15. Mishali, M., Eldar, Y.C.: From theory to practice: sub-nyquist sampling of sparse wideband analog signals. IEEE J. Sel. Top. Sig. Process. **4**(2), 375–391 (2010). https://doi.org/10.1109/JSTSP.2010.2042414
16. J.P. Slavinsky, et al., The compressive multiplexer for multi-channel compressive sensing, in *IEEE International Conference on Acoustics, Speech and Signal Processing (ICASSP)*, Prague, Czech Republic, pp. 3980–3983 (2011). https://doi.org/10.1109/ICASSP.2011.5947224

17. M. Rani et al., A systematic review of compressive sensing: concepts. Implementations Appl. IEEE Access **6**, 4875–4894 (2018)
18. D.L. Donoho, For most large underdetermined systems of linear equations the minimal L1-norm solution is also the sparsest solution. Commun. Pure Appl. Math. **59**(6), 797–829 (2006). https://doi.org/10.1002/cpa.20132
19. E.J. Candès, T. Tao, Decoding by linear programming. IEEE. Trans. Inform. Theory **51**(12), 4203–4215 (2005). https://doi.org/10.1109/TIT.2005.858979
20. S. Chen et al., Atomic decomposition by basis pursuit. SIAM J. Sci Comp. **20**(1), 33–61 (1999). https://doi.org/10.1137/S1064827596304010
21. G.H. Golub, C.F. Van Loan, *Matrix Analysis. Matrix Computations*, 4th ed. (The Johns Hopkins University Press, Baltimore, MD, 2013) ch. 2, pp. 68–73
22. G. Strang, *Linear Algebra and Its Applications*, 4th edn. (Thomson on Learning Inc, USA, 2006)
23. E.J. Candès, J. Romberg, Sparsity and incoherence in compressive sampling. Inv. Prob. **23**(3), 969–985 (2007). https://doi.org/10.1088/0266-5611/23/3/008
24. Goldberger, A.L, et al.: PhysioBank, physioToolkit, and physioNet: components of a new research resource for complex physiologic signals. Circulation **101**(23), e215–e220 [Circulation Electronic Pages; http://circ.ahajournals.org/cgi/content/full/101/23/e215]; 2000 (June 13)
25. Pati, Y.C., et al., Orthogonal matching pursuit: recursive function approximation with applications to wavelet decomposition, in *Proceedings 27th Asilomar Conference Signals, Systems, and Computers* (Pacific Grove, CA, 1993), vol. 1, pp. 40–44
26. D. Needell, J. Tropp, CoSaMP: iterative signal recovery from incomplete and inaccurate samples. Appl. Comput. Harmon. Anal., **26**(3), 301–321 (2009). https://doi.org/10.1016/j.acha.2008.07.002
27. M. Grant, S. Boyd, CVX: Matlab software for disciplined convex programming, version 2.0 beta (September 2013). http://cvxr.com/cvx
28. Becker, S., A matlab function: CoSaMP and OMP for sparse recovery, version 1.7, Aug 2016. Available online at:https://in.mathworks.com/matlabcentral/fileexchange/32402-cosamp-and-omp-for-sparse-recovery

Text Detection Through Hidden Markov Random Field and EM-Algorithm

H. T. Basavaraju, V. N. Manjunath Aradhya and D. S. Guru

Abstract The text is a dominant source and delivers semantic information about a particular content of the respective image or video. Human often gives importance to the text than any other objects in an image or a video frame. Text detection is one of the prime part of the text information extraction process. Text detection process is an exciting and emerging research area in the zone of pattern recognition, and computer vision due to the complex background, illumination, and arbitrary orientation. In this paper, the Hidden Markov Random Field (HMRF) method and Expectation-Maximization (EM) algorithm are employed to detect the arbitrarily oriented multilingual text present in an image or a video frame. The proposed method calculates the max-min cluster to maximize the discrimination between textual and non-textual region. HMRF separates the textual region. EM algorithm maximizes the likelihood of the parameters. Laplacian of Gaussian process is used to identify the potential text information. The double line structure concept is employed to extract the true text region. The proposed method is evaluated on Hua's dataset, arbitrarily oriented dataset, and horizontal dataset with performance measures recall, precision, and f-measure. The outcome shows that the approach is promising and encouraging.

Keywords Hidden Markov Random Field · EM-algorithm · Multilingual text
Arbitrary oriented · Laplacian of Gaussian

H. T. Basavaraju (✉) · V. N. M. Aradhya
Department of Master of Computer Applications,
JSS Science and Technology University (Sri Jayachamarajendra College of Engineering),
Mysore, India
e-mail: basavaraju.com@gmail.com

V. N. M. Aradhya
e-mail: aradhya.mysore@gmail.com

D. S. Guru
Department of Studies in Computer Science, University of Mysore, Mysore, India
e-mail: dsg@compsci.uni-mysore.ac.in

© Springer Nature Singapore Pte Ltd. 2019
S. C. Satapathy et al. (eds.), *Information Systems Design and Intelligent Applications*,
Advances in Intelligent Systems and Computing 862,
https://doi.org/10.1007/978-981-13-3329-3_3

1 Introduction

Nowadays, the digital information is extensively increasing in the multimedia database. Hence, a massive amount of images and videos are stored in the multimedia database. Textual information present in an image or a video delivers a crucial supplemental source and semantic information of the content to understand unambiguously. A text detection method identifies the text region from its background. The main challenges are illumination, complex background, and arbitrary orientation of the text. Arbitrarily oriented text detection is much challenging than horizontal text detection process because the text is present in the form of various directions. Arbitrary-oriented multilingual language is much more difficult than any other text detection process, because the multilingual textual information consists multiples of languages, with different geometrical shapes, multi-color, multi-size, and multi-fonts. Hence, there is a demand for a new technique, which can take care of all the above-discussed problems. The text detection process plays a prominent role in applications like, automatic annotation, indexing, tracking, retrieval and event understanding.

Markov Random Field (MRF) is countable random variable, and it is a graphical model of a joint probability distribution. MRF extensively used for computer vision and pattern recognition problems, such as surface reconstruction, depth inference, and image segmentation. MRF consists of an undirected graph in which, the nodes represent random variables. The random field becomes a Markov Random Field, when it satisfied Markov properties. The Markov property deals with that the future state conditional probability dispersal is based on the present situation. EM algorithm is an iterative model, which maximizes estimation of the parameters. It is first of is kind in the literature to propose the HMRF method and EM algorithm is employed to detect the arbitrarily oriented multilingual text present in an image or a video frame.

2 Related Works

In the past years, ample of methods are developed to detect the text from the images or video frames. Nowadays the technology becomes more advanced, hence images or videos have high resolution. Wu et al. [1] introduced a symmetry method for text identification with the help of neural network. The extremal regions are extracted to isolate the text candidates and the multi-domain strokes symmetry histogram is proposed to detect the true text regions. Skeleton cut detector has employed to detect scene text by He et al. [2] The candidate text blocks are extracted using skeleton junction detection and elimination method. Two stage classifier is used to identify the exact text region. The text candidates extraction methods and region filtering methods are discussed in Zarechensky [3], and then empirical analysis had done on those methods using own created synthetic dataset. Sharma et al. [4] extracts four directional gradient features. The text candidate image has obtained by fusing the four-pixel candidate images. Then the two-stage method is proposed to segment the words. The tensor voting method is introduced by Lim et al. [5] to identify the text layers. Initially, the region is decomposed into the chromatic or achromatic region. Finally, K-means algorithm and density estimation algorithm are employed to detect the text region.

Wavelet, Gabor and K-means concepts are used by Pavithra and Aradhya [6] to detect the text region. Sharpened information and texture properties of given frame is found by processing wavelet transform and Gabor concept and then morphological operation is performed on k-means clustered pixel of Gabor result to obtain the text region. Shivakumara et al. [7] used a color and gradient features are used to segment text. K-means technique is applied to distinguish the textual cluster and non-textual cluster. The text candidates are extracted by computing median deviation of co-efficient, K-means algorithm and morphological functions in Minemura et al. [8]. Matko Saric [9] introduced a novel approach on the basis of extremal regions to generate a character, candidates. Finally, SVM is used to classify the text region. Shivakumara et al. [10] introduced a quad tree approach to extract the curved text region using max-min cluster and region growing method. Ma et al. [11] presents rotation proposals to detect the arbitrary-oriented scene text. the rotation region proposal networks and rotation region of interest methods are employed to identify the text region.

3   Proposed Methodology

The proposed method initially, separates a given input color frame into R band, G band, and B band. The Max-Min cluster technique is used to cluster the R band, G band and B band on the basis of maximum and minimum values. Hidden Markov Random Field obtains the text detected regions through map estimation and EM algorithm. Hidden Markov Random Field preserves the edge information of the text region by removing non-text region edges. The HMRF method efficiently groups the similar regions using Markov property and EM algorithm. Therefor, in this paper HMRF has used to identify the textual region and non-textual region. Laplacian of Gaussian process is used to identify the potential text information. Finally, the double line structure analysis is applied to extract the true text region. Figure 1 illustrates the block diagram of the proposed method.

3.1   Preprocessing

The preprocessing stage is an essential process to extract the textual region from the non-textual region. In this stage, initially the given color input frame is separated into R, G, and B band respectively. The Max-Min cluster [7] is employed to maximize the discrimination between textual and non-textual region.

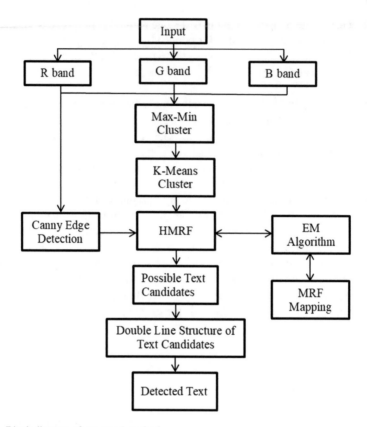

Fig. 1 Block diagram of proposed method

3.2 Hidden Markov Random Field and EM-Algorithm

The Hidden Markov Random Field (HMRF) method was first proposed for segmentation of brain MRI images. The HMRF technique is obtained from the Hidden Markov Models (HMM), where a Markov chain process generates HMM. In Markov chains, the state is obtained through a sequence of observations. Expectation Maximization algorithm is an iterative model. It estimates the maximum likelihood or maximum a posteriori parameters. Every iteration of an EM algorithm have two steps: one is M-step, and another one is E-step. The missing data are estimated through E-step from the given experimental data and present data of the method. M-step increases the likelihood task under the postulates of missing data. In this paper HMRF and EM algorithm are applied to detect the textual regions.

HMRF and EM algorithms are processed with initial parameters like the gray frame, canny edge frame, clustered frame, mean, standard deviation of the cluster and number of iterations. The HMRF-EM algorithm estimates the likelihood distribution using Eqs. 1 and 2. The Maximum a posterior (MAP) algorithm is employed to com-

(a) Input   (b) Max-Min Cluster   (c) Canny edge

Fig. 2 Intermediate Results of the Proposed Method

pute the posterior distribution for all pixels using Eq. 3. Finally, all initial parameters are updated to retain the textual region with eliminating non-textual region. Canny edge detector preserves the edge information of the text region to get better detection. Mean and standard deviation values keep on updating using Eqs. 4 and 5 on the basis of a number of EM and MAP iterations. The intermediate results are as shown in Fig. 2.

$$x^* = \text{argmax } P(y/x, \theta)P(x) \tag{1}$$

where: y is the intensity of a pixel, and x is the label.

$$\begin{aligned} P(y/x, \theta) &= \pi_i P(y_i/x, \theta) \\ &= \pi_i P(y_i/x_i, \theta_{xi}) \end{aligned} \tag{2}$$

$$x^t = \text{argmax}_{(x \in X)} P(y/x, \theta^t)P(x) \tag{3}$$

$$\mu_l^{(t+1)} = \frac{\sum_i P^{(t)}(l|y_i)y_i}{\sum_i P^{(t)}(l|y_i)} \tag{4}$$

$$(\sigma^{(t+1)})^2 = \frac{\sum_i P^{(t)}(l|y_i)(y_i - \mu_l^{(t+1)})^2}{\sum_i P^{(t)}(l|y_i)} \tag{5}$$

where $P(y_i|x_i, \theta_{xi})$ is a likelihood distribution with parameters $\theta_{xi} = (\mu_{xi}, \sigma_{xi})$. $\theta = \theta l | l \in L$ is the set of all probable labels, which was gained by an EM algorithm.

3.3 Laplacian of Gaussian

The Laplacian of Gaussian detects edges as well as noise from resultant image of HMRF model. The second order spatial derivatives are extracted by Laplacian of Gaussian. The Laplacian of Gaussian responses zeroes at a long distance from the image, negative at one side of the edge and positive at alternative side of the border. The Laplacian of Gaussian technique (refer Eq. 6) is processed on the outcome of HMRF model to extract the two line structure of the text information.

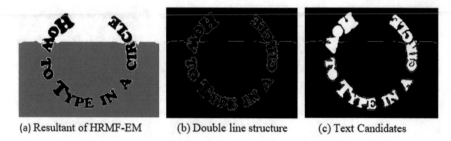

(a) Resultant of HRMF-EM   (b) Double line structure   (c) Text Candidates

Fig. 3 Final Results of the Proposed Method

$$\text{LoG}(x, y) = -\frac{1}{\pi \sigma^4}\left[1 - \frac{x^2 + y^2}{2\sigma^2}\right]e^{-\frac{x^2+y^2}{2\sigma^2}},$$ (6)

where:

σ is a Gaussian standard deviation.
x and y are Spatial coordinates.

3.4 Double Line Structure of a Text Information

After the application of Laplacian of Gaussian, the double line structure of the text information is generated. These double line structures are considered as possible text candidates. Finally, true text candidates are extracted by taking the difference between edge information (i.e., Fig. 2c) and hole-filled regions. Text cluster (i.e., Fig. 3a) is extracted by employing HMRF and double line structure is depicted in Fig. 3b. Fig. 3 represents the procedure of the double line structure of the text information and outcome of the proposed method.

4 Experimental Results

The superiority of the proposed method is evaluated by conducting experimentation on standard datasets such as Hau's dataset, the arbitrarily oriented dataset, and horizontal datasets. These datasets collected from various news, sports, films, and advertisement videos and it consist both scene and graphic text of multilingual languages like Kannada, English, Hindi, and Chinese. Evaluating measures are precision (refer Eq. 7), recall (refer Eq. 8), and f-measure (refer Eq. 9). The parameters of text detection are divided into three categories: (1) Actual Text Block (ATB) which represents total text blocks present in a single image or a video frame, (2) Truly Detected text Block (TDB) is a text information identified by the proposed algorithm and (3) Falsely Detected text Block (FDB) indicates non-text information detected by the proposed algorithm. The proposed method employed an HMRF-EM algorithm to detect the text from complex frames and Laplacian of Gaussian technique effectively used to generate the double line structure as a text information. This structure detects the accurate text region from a given frame. The following subsections show the experimental results on different challenging datasets.

$$\text{Precision (P)} = \frac{\text{TDB}}{\text{TDB} + \text{FDB}}, \tag{7}$$

$$\text{Recall (R)} = \frac{\text{TDB}}{\text{ATB}}, \tag{8}$$

$$\text{F-measure (F)} = \frac{2RP}{R + P}, \tag{9}$$

4.1 Experimental Results on Hua's dataset

Hua's dataset [12] has 45 horizontal text images with complex background and low contrast collected from news sports events. The proposed method detected the textual region from non-textual region. The capability of the proposed model on Hua's dataset is as shown in Fig. 4. Table 1 depicts the quantitative analysis of the proposed technique on Hua's dataset. From Table 1, it is quite evident that the proposal HMRF and EM algorithm outperforms other existing works in terms of recall and f-measure. The quantitative analysis of the existing techniques are taken from [13].

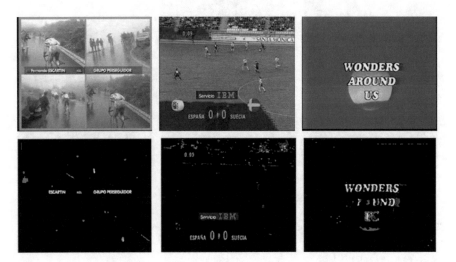

Fig. 4 Inputs and corresponding outputs of Hua's dataset

Table 1 Quantitative analysis of the proposed and existing methods on Hua's data

Methods	R	P	F
Zhou et al. [14]	72	82	77
Fourier-RGB [13]	81	73	76
Lu et al. [15]	75	54	63
Wong and Chen [16]	51	75	61
Basavaraju et al. [17]	86	62	72
Proposed method	90.33	72.66	79.56

4.2 Experimental Results on Nusdataset

The Nusdataset [7] consists of 62 images with illumination, complex background, and low resolution and multi-orientation. This dataset have both graphic and seen text. The proposed method detected the textual region from non-textual region. The performance of the proposed work on Nusdataset is as shown in Fig. 5. Table 2 depicts the comparative study between proposed and existing methods on Nusdataset. From Table 2, it is clear that precision and f-measure is quite competitive as compared to [7] and recall performs much better than [7].

Fig. 5 Inputs and corresponding outputs of Nusdataset

Table 2 Quantitative analysis of the proposed and existing methods on Nusdataset

Methods	R	P	F
Shivakumara et al. [7]	85	84	82
Proposed method	91.80	84.88	85.56

4.3 Experimental Results on Arbitrary-Oriented Dataset

An arbitrary-oriented dataset [10] consists of 142 arbitrary-oriented images with low contrast, complex background, and illumination. The proposed model accurately detects the textual region with few false positives in case of illumination effect. The outcome of the proposed model on arbitrary-oriented dataset is represented in Fig. 6. Table 3 depicts the experimental analysis of the proposed model on arbitrary-oriented dataset. From Table 3, it is observed that recall and f-measure proves the superiority of the proposed model and precision is competitive with other methods.

Fig. 6 Inputs and corresponding outputs of arbitrary-oriented dataset

Table 3 Quantitative analysis of the proposed and existing methods on arbitrary-oriented dataset

Methods	R	P	F
Zhou et al. [14]	41	60	48
Fourier-RGB [13]	52	68	58
Lu et al. [15]	47	54	50
Wong and Chen [16]	34	90	49
Basavaraju et al. [17]	72	57	64
Proposed method	93.21	74.78	81.63

5   Conclusion

In this work, the HMRF-EM algorithm is used for text detection of arbitrary multilingual text in images/video. The proposed method efficiently clusters using Max-Min cluster method. Hidden Markov Random Field (HMRF) and EM algorithm discriminates the textual region and non-textual region. Laplacian of Gaussian model generates the double line structure of the textual candidates to extract the true textual candidates. Finally, the proposed model effectively detects the textual region. The proposed model is evaluated on three challenging datasets considering all types of variations. In future work, the text detection accuracy will be increased by solving the illumination challenge.

References

1. Y. Wu, W. Wang, S. Palaiahnakote, T. Lu, A robust symmetry-based method for scene/video text detection through neural network, in *14th IAPR International Conference on Document Analysis and Recognition (ICDAR)*, pp. 1249–1254 (2017)
2. X. He, Y. Song, Y. Zhang, Scene text detection based on skeleton-cut detector, in *IEEE International Conference on Image Processing (ICIP)*, pp. 3375–3379 (2017)
3. M. Zarechensky, Text detection in natural scenes with multilingual text, in *Proceedings of the Tenth Spring Researcher's Colloquium on Database and Information Systems*, pp. 32–35 (2013)
4. N. Sharma, P. Shivakumara, U. Pal, M. Blumenstein, C.L. Tan, A new method for word segmentation from arbitrarily-oriented video text lines, in *International Conference on Digital Image Computing Techniques and Applications (DICTA)*, pp. 1–8 (2012)
5. J. Lim, J. Park, G.G. Medioni, Text segmentation in color images using tensor voting. Image Vis. Comput. **25**(5), 671–685 (2007)
6. M.S. Pavithra, V.N.M. Aradhya, A comprehensive of transforms, Gabor filter and k-means clustering for text detection in images and video. Appl. Comput. Inform. 1–15 (2014)
7. P. Shivakumara, D.S. Guru, H.T. Basavaraju, Color and Gradient Features for Text Segmentation from Video Frames, in *International Conference on Multimedia Processing, Communication and Computing Applications*, pp. 267–278 (2013)
8. K. Minemura, S. Palaiahnakote, K. Wong, Multi-oriented text detection for intra-frame in H. 264/AVC video, in *International Symposium on Intelligent Signal Processing and Communication Systems*, pp. 330–335 (2014)
9. M. Saric, Scene text segmentation using low variation extremal regions and sorting based character grouping. Neurocomputing **266**, 56–65 (2017)
10. P. Shivakumara, H.T. Basavaraju, D.S. Guru, C.L. Tan, Detection of curved text in video: quad tree based method, in *12th International Conference on Document Analysis and Recognition (ICDAR)*, pp. 594–598 (2013)
11. J. Ma, W. Shao, H. Ye, L. Wang, H. Wang, Y. Zheng, X. Xue, Arbitrary-oriented scene text detection via rotation proposals. arXiv preprint arXiv:1703.01086 (2017)
12. X.S. Hua, L. Wenyin, H.J. Zhang, An automatic performance evaluation protocol for video text detection algorithms. IEEE Trans. CSVT, 498–507 (2004)
13. P. Shivakumara, T.Q. Phan, C.L. Tan, New Fourier-Statistical Features in RGB Space for Video Text Detection, in *IEEE Transaction on CSVT*, pp. 1520–1532 (2010)
14. J. Zhou, L. Xu, B. Xiao, R. Dai, A robust system for text extraction in video. In: Proceeding ICMV, pp. 119–124 (2007)

15. C. Lu, C. Wang, R. Dai, Text detection in images based on unsupervised classification of edge-based features, in *Proceedings ICDAR*, pp. 610–614 (2005)
16. E.K. Wong, M. Chen, A new robust algorithm for video text extraction. Pattern Recognition 1397–1406 (2003)
17. H.T. Basavaraju, V.N.M. Aradhya, D.S. Guru, A novel arbitrary-oriented multilingual text detection in images/video. Inf. Decis. Sci. 519 –529 (2018)

Classification and Comparative Analysis of Control and Migraine Subjects Using EEG Signals

Abhishek Uday Patil, Amitabh Dube, Rajesh Kumar Jain, Ghanshyam Dass Jindal and Deepa Madathil

Abstract Migraine is an incapacitating neurovascular disorder that disables the brain by a severe headache and dysfunction of the autonomic nervous system. There is no perfect diagnosis of migraine till date. Migraine diagnosis if replaced by electroencephalogram (EEG) modality could help in the diagnosis of the disease. Recent advances in EEG signal processing have led to multi-resolution, processing, and methods of feature extraction. In this study, a nonlinear parametric method is used to acquire EEG features of and are used for the classification of control and migraine subjects. This EEG classification is carried out by classifiers based on supervised classification methods—backpropagation used in artificial neural network (ANN) and the results are compared with a bilinear supervised classifier support vector machine (SVM). The classification results confirm that the methodology has a potential to classify EEG and can be used to detect EEG of migraine subjects and could thus further result in improved diagnosis of migraine.

Keywords Electroencephalogram · Migraine · Classification · Entropy · SVM

A. U. Patil · D. Madathil (✉)
Department of Sensor & Biomedical Technology,
Vellore Institute of Technology (VIT), Vellore 632014, India
e-mail: deepa.m@vit.ac.in

A. Dube
S.M.S. Medical College, Jaipur, India

R. K. Jain
Electronics Division, BARC, Mumbai, India

G. D. Jindal
Bio-medical Engineering, MGM College of Engineering & Technology,
Kamothe, Mumbai, India

© Springer Nature Singapore Pte Ltd. 2019
S. C. Satapathy et al. (eds.), *Information Systems Design and Intelligent Applications*,
Advances in Intelligent Systems and Computing 862,
https://doi.org/10.1007/978-981-13-3329-3_4

1 Introduction

Electroencephalogram (EEG) is brain activity, electrical in nature, and recorded using several electrodes mounted on the scalp. EEG is usually acquired using pre-defined international standards like the 10–20 electrode system. The EEG reflects the summation of the electrical activity generated by the number of neurons in the brain. Its shows synchronized changes in the membrane potential due to the activity in particular areas of the brain. In the last five decades or so, the detection of brain abnormalities using EEG has been found significant by various researchers. The EEG consists of signals in time series of potentials which are evoked and result in neural activities. The recording of the EEG signal is acquired by the electrode placement on the scalp. The EEG is plotted as voltage magnitude against time. The dynamics of the EEG signal change with the variety of brain activities such as a subject performing a particular task would have different EEG signals compared to a subject responding to visual stimuli. EEG has been involved with various applications of clinical importance. EEG is usually considered a valuable clinical tool. It helps in the detection of various conditional neural disorders such as Alzheimer, Parkinson, and Epilepsy. Research has been conducted on various feature extraction methods, and a number of classification algorithms have been used to classify various brain disorders. Acharya et al. used higher order spectra features to classify epileptic signals from control signals [1]. They further proposed two types of entropies normalized bi-spectral and normalized bi-spectral squared entropy to classify these signals. In another study, Zaria et al. obtained better accuracy for brain–computer interface (BCI) [2]. He suggested statistical features like standard deviation, maximum, minimum, quartiles, quartile range, skewness, kurtosis, etc., to characterize and obtain more accurate results in BCI. To check the robustness of the features, obtained classifiers such as neural networks, LS-SVM, and logistic regression were used and the former two performed outstandingly well. In another study, Upadhyay et al. compared EEG signals decomposed by 16 different wavelets and obtained various features like statistical, fractal dimension, and entropy features [3]. Furthermore, the features were ranked using three ranking methods (ReliefF, Fisher Score, and Information Gain) to obtain the best possible wavelet. Obtaining the best wavelet was done to improve the overall accuracy of the classifiers used. Then, signals were classified by LS-SVM as pre-ictal, ictal, and normal.

Migraine, a chronic disease, causes outbreaks of severe headaches and further results in the dysfunction of the nervous system (autonomic). Physiologists rely on the two important factors for the diagnosis of migraine: one is the physical exam and second is the medical history. Most of the times, these factors are insufficient for the total diagnosis of migraine. For this reason, migraines are difficult to predict and can be misdiagnosed. The use of EEG for migraine detection has been very debatable. Apart from the abnormal EEG recordings in the early analysis, researchers have criticized most of the studies based on migraine for various methodological flaws [4]. Studies suggest that there are more number of spikes in migraine compared to control subjects but higher order statistics are needed [5]. In this study, three nonlinear feature extraction methods, viz., permutation entropy (PE), Katz, and Higuchi's fractal dimension, are used for the EEG classification. ANN and SVM are used for this classification.

2   Methods

2.1   Participants

Participants were 54 adult's subjects: 27 having migraine and 27 control subjects, recruited from the local region via a newspaper advertisement. Their mean age was 28 (range 25–35) years. Subjects prior to the recording were screened by a local doctor for head injuries, other neurological disorders, learning disabilities, or any kind of psychiatric conditions and were identified by the doctor for migraine.

2.2   Physiological Recordings

EEG was recorded for about 300s from standard 22 electrode sites. The electrodes were placed in accordance with the standard 10–20 international electrode placement system using an electrode cap with wet electrodes. Since research studies suggest that migraine is hyperexcitable in the occipital lobe, EEG recording was recorded via six electrodes of the 10–20 electrode system, namely, P3, P4, C3, C4, O1, and O2 [6]. Signals were recorded and amplified using BESS system and [0.5–30 Hz] band-pass filter was used. The sampling rate for the study was 256 Hz. A six-channel sample EEG signal recorded from both control and migraine subjects is shown in Fig. 1.

2.3   Procedure

When the participants entered the lab, they were asked to read and sign an information sheet and a consent form. Participants were asked to be seated in an air-conditioned sound-attenuated room and were fitted with the EEG cap and electrodes. The subjects were asked to relax at the initial stage, and the resting EEG was recorded for about 3 min. The subjects were then asked to close their eyes and be seated without any kind of movement. EEG was recorded for about approximately 3 mins. During the eyes-open state, the subjects were asked to keep the eyes open and the EEG was again recorded for about the same time period as in eyes-closed state. The baselines have not been reported here.

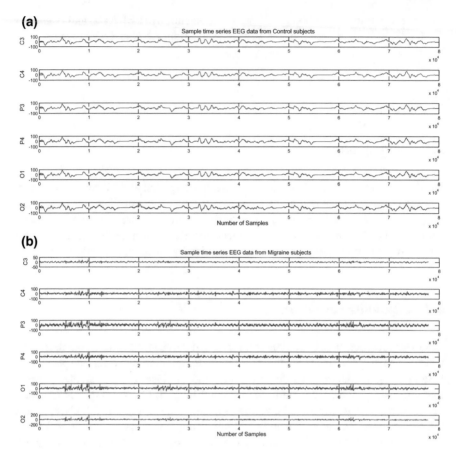

Fig. 1 Six-channel sample time series EEG activity recorded from **a** control subjects, **b** migraine subjects

3 Data Extraction

The information contained in the EEG signals are often contaminated by non-cerebral artifacts originated from eye blinks, movements of the eye, electromyogram (EMG) as well as electrocardiogram (ECG). There is also a chance of unknown random sources that lead to such disturbances in the signals. This would eventually affect the EEG signals. The presence of such strong artifacts makes it difficult to identify the use of EEG signals in feature extraction and reduces its usability. Therefore, preprocessing of an EEG signal is an important step which should be performed before extraction of useful information from the EEG. In our study, the EEG affected by blinks was eliminated by the deployed hardware and software acquisition system. The EEG is also affected by patient or head movement motion artifacts which were rejected from the study. The blinking of the subjects could also be eliminated using the software deployed with acquisition system. In our study, the technique of removal of artifacts from eye movement is illustrated in [7].

3.1 Feature Extraction

There are a various number of feature extraction methods that can be used for the EEG signals. There has been some literature that has been used to extract EEG signals. An important method includes linear analysis of 1-D signals in time–frequency or time domain. The feature extraction was done in MATLAB 2016 software using the data signal processing toolbox. The methods that we have used have been discussed in the subsections:

3.1.1 Permutation Entropy (PE)

Pompe and Bandt introduced PE, method to study the complexity of a time-domain signal, and compared them with their neighboring values [8]. The estimation of PE is done by mapping the time series into a sequence of symbols. The time series $\{y(m);$ $m = 1, 2, 3...\}$ is first embedded to perform mapping, to n-dimensional space [9]. PE is used to identify various such mappings or couplings. PE can be defined as Shannon Entropy of $m!$ patterns if $m \geq 2$ and can be expressed below:

$$H(m) = -\sum p(x)\log p(x) \tag{1}$$

where summation symbol runs over all $m!$. Also $0 < H(m) < \log(m!)$

3.1.2 Katz Fractal Dimension

Mandelbrot in 1982 suggested that the fractal dimension can be obtained for a planar curve by Mandelbrot [10],

$$FD_{\text{mandelbrot}} = \frac{\log(L_x)}{\log(s)} \tag{2}$$

where L_x is the summation of the diameter and the successive points on the curve. The distance between the first point and the farthest point which is located at a distance from the first point is the diameter and let s be the diameter. If T_j, be a point on the curve such that $T_j = (x_j, y_j)$ with $x_j < x_{j+1}, j = 1, 2, 3 \ldots N$. We also can define L and d mathematically as

$$L_x = \sum_{j=1}^{N} ||T_{j+1} - T_j|| \tag{3}$$

$$s = \max||T_{j+1} - T_j|| \tag{4}$$

where $||.||$ is called the Euclidean distance. Mandelbrot's FD depends upon the calculation of the Euclidean distance. Katz improvised fractal dimension of Mandelbrot's original contribution and he improvised the theory to calculate his own FD [11]. He calculated FD and stated that space discretization should be performed. Katz normalized distances and stated his FD as

$$FD_{katz} = \frac{\log(L_x/b)}{\log(s/b)} \tag{5}$$

The number of steps is m then m can be given as $m = L/b$. Then, the FD_{katz} changed to

$$FD_{katz} = \frac{\log(m)}{\log(m) + \log(s/m)} \tag{6}$$

3.1.3   Higuchi's Fractal Dimension

Higuchi FD is another feature-based method which a curve is length-based estimator. This division of the curve is done with samples (k) and the averaged length is tabulated [12]. $FD_{Higuchi}$, a series estimation in time domain which can be discretized functions, are analyzed in time domain, $y(1), y(2), \ldots y(N)$. The sequence of time at the start, Y_k^n, a new series in time domain can be tabulated as

$$Y_k^n : y(n), y(n+k), y(n+2k), \ldots, y\left(n + \text{int}\left[\frac{N-k}{k}\right]k\right) \tag{7}$$

For $n = 1, 2, \ldots, k$, initial time is given by m in interval k, where $k = 1, 2, \ldots, k_{max}$ and int(a) is the real number integer part of number a and k_{max} is a free parameter.

Again, $L(k) = \frac{\sum_{n=1}^{k} L_n(k)}{k}$ is the length of the curve, where

$$L_n(k) = \frac{1}{k}\left[\left(\sum_{i=1}^{\text{int}[\frac{N-n}{k}]} |y(n+ik) - y(n+(i-1)k)|\right) \frac{N-1}{\text{int}[\frac{N-n}{K}]k}\right] \tag{8}$$

Here, the original series in the time domain is of Y length and $\frac{N-1}{\text{int}[\frac{N-n}{K}]k}$ is called the factor of normalization. $L_n(k)$ is obtained by averaging all n values forming an average curve length given by $L(k)$. The total average length $L_{avg}(k)$ is proportional to k^{-D} where D is $FD_{higuchi}$. The slope of the least square linear fit $\ln(L_{avg}(k))$ v/s

$ln(1/k)$ curve is the factor of the FD_{higuchi} [10]. The work by Polychronaki et al. had an approach (2010), obtained by the k_{max} approximately [13].

$$FD_{\text{higuchi}} = \frac{\ln(L(k))}{\ln\left(\frac{1}{k}\right)} \tag{9}$$

To obtain the features using the abovementioned methods, we time windowed the signal into epochs of 3-s duration. Each of the six channels was divided into a 3-s epoch to obtain three features from the EEG. Each of the 3-s epochs annotated the PE and FD (Katz and Higuchi's) values obtaining 650 values per feature per signal. This method was performed on the six electrodes. The final matrix formed was of the size [650 × 6] values per subject.

4 Results

We performed classification for all three feature vectors in weka software (version 3.8) for detailed analysis. Tables 1 and 2 define the confusion matrices obtained using SVM and ANN, respectively. To avoid one-sidedness of data from the results of the classifiers, 10-fold cross-validation approach is performed in 10 steps and then the confusion matrix is obtained by collecting the various results of each step. The first step of 10 cross-validation contains 90% of the total feature vectors selected for training and 10% selected for validation of the feature vectors. In the next step, the training is made by 90% of feature vectors which include 10% of the feature vector used for testing and another 10% taken out from the previous training set for validation purpose. Similar steps are performed for 10 times and then after the 10th step, again the feature vector is tested. Hence, this method eliminates the possibility of eliminating one-sidedness of the data. The last column indicates the total confusion matrix for all the features, i.e., it indicates the summation of all the steps in the 10-fold validation.

Tables 1 and 2 show that SVM classifies the feature values with higher accuracy compared to ANN. We observed the prediction of control versus migraine subjects using SVM and ANN. It can be seen from Table 3 that the accuracy of classification is lesser in ANN compared to the SVM model.

Table 1 SVM classification: confusion matrix (Diseased: Migraine; Normal: Control)

PE		Katz FD		Higuchi's FD		All features		Classified as
Diseased	Normal	Diseased	Normal	Diseased	Normal	Diseased	Normal	
550	100	419	231	475	175	585	65	Migraine
286	364	389	261	370	280	87	563	Control

Table 2 ANN classification: confusion matrix (Diseased: Migraine; Normal: Control)

PE		Katz FD		Higuchi's FD		All features		Classified as
Diseased	Normal	Diseased	Normal	Diseased	Normal	Diseased	Normal	
533	117	406	244	461	189	545	105	Migraine
302	384	396	254	414	236	141	509	Control

Table 3 Comparison of classifiers computed using overall confusion matrix

Parameter	SVM	ANN
Classification accuracy (max)	88.31%	81.08%
Sensitivity	0.8705	0.7945
Specificity	0.8965	0.8290

5 Discussions

In this work, a unique diagnostic method using feature extraction which is nonlinear in nature was utilized to classify migraine and control subject's EEG. Three nonlinear feature extraction techniques (PE, Katz FD, and Higuchi's FD) were calculated from the recorded EEG signals of both migraine and control subjects. The observed classification results stated that using SVM we obtain a successful highest classification and it successfully diagnosed migraine. Similarly, ANN also showed a better accuracy and diagnosed the unknown EEG in PE but it failed to classify with respect to the Katz FD and Higuchi's FD. SVM outperforms the accuracy for classification of EEG signals from migraine and control EEG signals.

Acknowledgements We would like to show our sincere gratitude toward SMS Medical College, Jaipur where we could actually collect the data and do the processing of the same. We also thank Dr. Amitabh Dube, Mr. Rajesh Sonania, Dr. Rahul Upadhyay, Dr. R. K Jain, Dr. G. D. Jindal, Dr. Abhishek Saini, Dr. Bhupendra Patel, and Dr. Indoria for their assistance with EEG signal acquisition which has a major role in the present study.
Declaration
Our project is funded by a government organization and one of the authors from the government organization is mentioned in the author list. We have taken permissions to use the dataset/images and responsible for any kind of issues in future.

References

1. U. Rajendra Acharya, K.C. Chua, V. Chandran, C.M. Lim, Analysis of epileptic EEG signals using higher order spectra. J. Med. Eng. Technol. **33**(1), 42–50 (2009)
2. R. Zarei, J. He, S. Siuly, Y. Zhang, A PCA aided cross-covariance scheme for discriminative feature extraction from EEG signals. Comput. Methods Programs Biomed. **146**, 47–57 (2017)
3. R. Upadhyay, P.K. Padhy, P.K. Kankar, A comparative study of feature ranking techniques for epileptic seizure detection using wavelet transform. Comput. Electr. Eng. **53**, 163–176 (2106)

4. E. Niedermeyer, The EEG in migraine and other forms of headache, in *Electroencephalography*, 4th ed., ed by E. Baltimore F. Niedermeyer, L. da Silva (Williams and Wilkins, 1999), pp. 595–602; T. Sand, Funct. Neurol. 6 (1991) 7; G.S. Gronseth, M.K. Greenberg, Neurology 45 (1995) 1263; J. Schoenen, G.L. Barkely, Neurophysiology, in, The Headaches ed by J. Olesen, P. Tfelt-Hansen, K.M.A. Welch (Lippincott Raven, London, 1994), pp. 199–208

5. S.C. Schachter, M. Ito, B.B. Wannamaker, I. Rak, K. Ruggles, F. Matsuo, A. Wilner, R. Jackel, F. Gilliam, G. Morris, J. Skantz, M. Sperling, J. Buchhalter, F.W. Drislane, J. Ives, D.L. Schomer, J. Clin. Neurophysiol. **15**(3), 251 (1998)

6. T. Sprenger, D. Borsook, Migraine changes the brain—neuroimaging imaging makes its mark. Curr. Opin. Neurol. **25**(3), 252–262 (2012)

7. R.R. Vazquez, H.V. Perez, R. Ranta, V.L. Dorr, D. Maquin, L. Maillard, Blind Source Separation, wavelet denoising, and discriminant analysis for EEG artifacts and noise canceling. Biomed. Sign. Process. Control **7**, 389–400 (2012)

8. C. Bandt, B. Pompe, Permutation entropy: a natural complexity measure for time series. Phys. Rev. Lett. **88**, 174102–174104 (2002)

9. N. Nicolaou, J. Georgiou, Detection of epileptic electroencephalogram based on permutation entropy and support vector machines. Expert Syst. Appl. **39**, 202–209 (2012)

10. B.B. Mandelbrot, *The fractal geometry of nature* (Freeman, New York, 1982)

11. M. Katz, Fractals and the analysis of waveforms. Comput. Biol. Med. **18**(3), 145–156 (1988)

12. T. Higuchi, Approach to an irregular time series on the basis of the fractal theory. Physica D **31**, 277–283 (1988)

13. G.E. Polychronaki, P.Y. Ktonas, S. Gatzonis, A. Siatouni, P.A. Asvestas, H. Tsekou, D. Sakas, K.S. Nikita, Comparison of fractal dimension estimation algorithm for Epileptic Seizure onset detection. J. Neural Eng. **7**, 046007 (2010) (18 pp.)

An Empirical Analysis of Big Scholarly Data to Find the Increase in Citations

J. P. Nivash and L. D. Dhinesh Babu

Abstract The research quality and productivity of a research area are decided by the number of research articles and citations. Several factors affect the citation count of a research article. The objective of this paper is to find the influences of social media and abstract views in the increase of citations. The relationship between social media influence and abstract count on the overall citations is evaluated on the top cited research articles of cloud computing area. More research focus is needed to analyze the social media influence score. The research scholars, research organizations, funding agencies, and various communities can increase their research productivity and research impact through this analysis.

Keywords Big scholarly data · Citation network
Scientific collaboration network · Bibliometric analysis · Information science

1 Introduction

A social network containing articles and their citation information are named as citation network [1]. The impacts of citations play a major role in ranking universities [2, 3], research scholars [4, 5], journals [6], and funding agencies [7, 8]. The citation count and h-index of an individual are significant in the growth of the researchers. The objective of this paper is to find the influence of social media and abstract count on increasing the citations. The investigation of group formation and social influence of a set of people is named as social network analysis [9]. Citation network is composed of articles in the form of nodes and their citations in the form of edges [10]. The main motive behind citation network modeling is to experience the hot research topics and

---

Please note that the LNCS editorial assumes that all authors have used the western naming convention, with given names preceding surnames. This determines the structure of the names in the running heads and the author index.

J. P. Nivash (✉) · L. D. Dhinesh Babu
School of Information Technology and Engineering, Vellore Institute of Technology, Vellore, India
e-mail: jeevprabnivash@gmail.com

© Springer Nature Singapore Pte Ltd. 2019
S. C. Satapathy et al. (eds.), *Information Systems Design and Intelligent Applications*,
Advances in Intelligent Systems and Computing 862,
https://doi.org/10.1007/978-981-13-3329-3_5

bring forth academic thoughts in scientific society [11–13]. In general, according to experiment and observation, the citation networks are originated to be scale-free, assortative, and clustering. To express those properties through mechanisms, widespread attention has been involved. In general, citation networks are directed and acyclic, the association from one point to the other. Hence, the citation network is called as Directed Acyclic Graph (DAG) [14]. Citation networks can be used for identifying relevant articles, subject experts, and so on [15]. The researchers' collaboration differs significantly between different disciplines. For example, the association of authors is fewer in the works related to arts and humanities while the authors' association is higher in the works related to medicine, physics, mathematics, computer science, and engineering. Network analysis methodologies are generally used to read collaboration networks. There is an abundant growth of scholarly data since 2006, and it is continuously growing [16]. Every year, the number of articles and researchers is increasing rapidly. The research collaborations have made the scientists from different countries to work together. Many innovative and challenging issues were addressed easily by the research collaborations. Recent research works state that the growth of scholarly data is going to be an immediate challenging issue to handle. Hence, the term Big Scholarly Data (BSD) is coined to address the scholar data handling issues.

This will handle the vast data of researchers, articles, collaborations, citations, and so on [17–19]. Figure 1 shows the 5 V features of the vast scholarly network. There are various research issues in the scholarly network. This paper concentrates on the growth of article citations. There are various factors involved in the increase of citations . In general, number of citations decides the quality of an article. A separate attention is needed in future to handle the above issue. Some of the other factors which can influence the citation count are type of an article, research discipline, journal, author, affiliation, and so on. An empirical analysis can show the proof of citation connections.

2 Related Work

In 2017, Nielsen [20] explores the citation rate of gender-based classification. The paper has found that the woman scholars have an equal citation impact. Kim et al. [21] bring out the relationship in the authors' area through citation contents and proximity. Fiala et al. [22] discuss finding the influential researchers in citation networks. Previous research works claims that author collaboration plays a major role in the paper citation. Bornmann et al. [23] analyzed various factors influencing the citation impact like citation count, references, number of authors, number of pages, and compared their effects. They did not explore their abstract view counts and social media influence. Guan et al. [24] propose that knowledge elements also affect the paper citation. Patience et al. [25] show that the citation rate of top articles is increased with respect to the quality of references; the recent research articles are also cited often. They prove that the number of citations is correlated with the affiliations. Wang et al.

Fig. 1 5 V characteristics of BSD

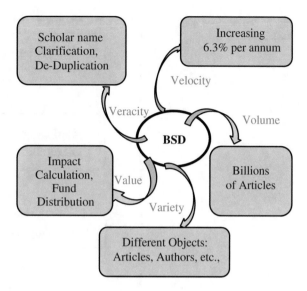

[26] conducted a citation and co-citation analysis on research articles. The citation analysis was made through degree centrality and betweenness centrality to find the important research articles. To filter and find relevant research articles, Son et al. [27] proposed a recommendation system combining citation analysis and network analysis. The above researchers explored various aspects of citation networks but they did not explore much on increasing the citation count.

3 Proposed Work

A novel Top Citation Analysis Model (TCAM) is proposed to explore the elements on promoting the citation count. To analyze the factors behind the increase in citations, top cited research articles are chosen. As displayed in Fig. 2, there are various factors behind the increase in citation count. This paper analyzes the influences of social media and abstract views in the increase of citations. The CAM model is comprised of four stages,

(i) Collecting the articles of larger citations,
(ii) Analyzing the factors involved behind the increase in citations,
(iii) Selecting the common factors, and
(iv) Analyzing the relationship between the factors and citations.

Stage-I: In the first stage, articles with largest citations were collected for analysis.

Stage-II: In the second stage, the quality of citations is analyzed. Here, LC_A is defined as the largest cited article

Fig. 2 Top citations analysis model

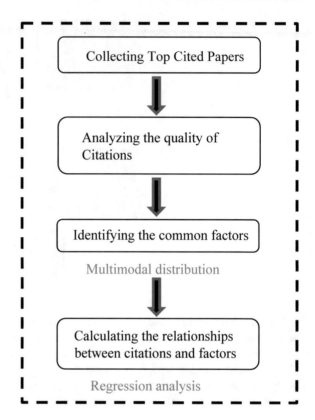

$$LC_A = \sum_{i=1}^{n}\sum_{k=1}^{5}\frac{A_i}{Y_k} \tag{1}$$

where A_i is defined as the set of articles and Y_k is defined as the year period after publication.

Stage-III: In the third stage, common factors are chosen among all the factors using multimodal classification. Now each social media influence is given a common score (10). In future research, this score will be given with the proper analytical measures of Facebook (F), Twitter (T), Blogs (B), and Wikipedia (W) separately.

$$AV_A = \sum_{i=1}^{n} plum\_abs(A_i) \tag{2}$$

where AV_A is defined as the total number of abstract views of an article and $plum\_abs(A_i)$ is defined as abstract view count obtained by the plum metrics from the Scopus database.

$$SM_A = \sum_{i=1}^{n} plum_{sm}\{F(A_i), B(A_i), T(A_i), W(A_i)\} \tag{3}$$

where SM_A is defined as the social media influence of an article, $plum\_sm(F(A_i))$ is defined as the Facebook score of an article, $plum\_sm(B(A_i))$ is defined as the blog score of an article, $plum\_sm(T(A_i))$ is defined as the Twitter score of an article, and $plum\_sm(W(A_i))$ is defined as the Wikipedia score of an article. For now each social media influence is given a common score (10).

Hence, the social media score SM_Ascore is calculated as follows:

$$SM_A score = \sum_{i=1}^{n} \{plum_{sm(F(A_i))} + plum_{sm(B(A_i))} + plum_{sm(T(A_i))} + plum_{sm(W(A_i))}\} \tag{4}$$

Stage-IV: In the final stage, the obtained factors are subjected to regression analysis. The depended variables (citations) are related to the independent variables (abstract views and social media influence) for exploring the relationships.

In this paper, top cited articles were acquired from the Scopus database along with the Plum metrics. The total number of research citations so far, number of abstract views, year-wise citations, and social media influence were collected from Plum metrics. There are many other factors involved in the increase in citations. Here, our objective is to analyze the following factors and relationships:

(i) Relationship between the overall citations and abstract view count,
(ii) Relationship between overall citations and the social media influence.

Here, the abstract views and social media influence are considered as the independent variables and the overall citations are considered as the dependent variable. The regression analysis is carried out for our proposed work. The correlation coefficient Multiple R shows about the linear relationship, the coefficient of determination R-squared shows about the regression line, and standard error estimates the mean of error deviated from the result, followed by the Sum of Squares (SS). Regression MS is defined as the ratio of regression and regression degrees of freedom. Residual is defined as the ratio of residual SS and residual degrees of freedom; the F-test is associated with the P-value.

4 Dataset Used

For our analysis, the articles of cloud computing area were chosen from the Scopus database. Scopus is considered to be the largest abstract and citation database for scientific articles. The articles with largest citations were selected and for each article, the plum metrics [28] statistics were taken for the analysis. Table 1 displays the data of top cited articles acquired from the Scopus database for evaluation. Here, a

Fig. 3 Year-wise citations
of top cited articles in cloud
computing

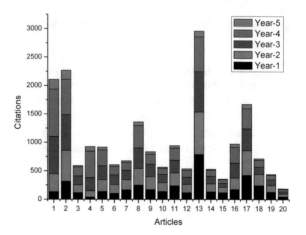

common influence score-10 is given to all the social media applications. The article
references and occurrences in the social media are considered as the influence score.
Each application has its own strength in the social media and its strength can be
analyzed in depth in future. Figure 3 displays the year-wise citations of the articles.
Depending upon the research relevance and scope, there is a steady upsurge in the
citation count every year. Our objective is to analyze the common factors behind all
top cited research articles. Hence, abstract views and social media influence scores
are taken for the analysis.

Abbreviations: [C-Citations, DT-Document type, Y-Year, AV-Abstract views, SM-
Social media influence (B-Blog, W-Wikipedia, F-Facebook, T-Twitter)]

5 Evaluations

Multiple R shows the strength of the linear relationship. In Multiple R, the value of
1 is considered to be a perfect positive relationship and a value of zero is considered
as no relationship. In Table 2, the Multiple R is achieved as 0.751687, which means
moderately positive relationship between the abstract views and the overall citations.
R-squared is defined as the coefficient of determination. It shows the total number
of points that fall on the regression line. In R-squared, 0.8 means 80% of the values
fit correctly to the model. Here, the R-squared is achieved 0.565034, which means
the average values fit the model. Similarly, adjusted R-square alters for the model
(Table 2).

By analyzing the residuals, the model suitability can be assessed. Residuals (R)
are the difference of observed (O) and predicted (E) value.

$$R = O - E \tag{5}$$

Table 1 Citation overview of top cited articles in cloud computing

Year	C	DT	Y-1	Y-2	Y-4	Y-5	Y-6	AV	SM
2010	4283	R	134	316	653	832	173	29,843	B, W
2009	2838	A	318	542	624	625	159	7737	B, F, T
2013	2315	A	117	137	159	156	23	7234	B, T
2013	2124	A	41	113	230	461	82	6347	R, W, K, T
2011	1650	A	138	213	237	278	52	3246	B
2008	1564	CA	105	154	172	148	30	4238	B, W
2010	1285	A	168	183	162	134	31	1893	W, F, T
2009	1253	A	251	295	355	394	65	3120	F, T
1984	1253	A	171	230	218	174	44	1978	W
2011	1066	R	139	159	142	107	18	3536	B, W, F, T
2012	1018	A	241	226	221	212	45	2500	B, F, T
2009	1018	CA	119	149	122	122	27	1786	B, W
2008	995	CA	790	746	708	612	103	536	W
2012	942	CA	131	145	129	111	14	476	B
2009	942	CA	117	97	69	68	7	489	B
2011	885	A	174	204	260	279	58	7749	B, W, F, T
2010	790	CA	425	435	380	357	75	2345	B
2010	704	A	244	197	131	119	25	566	W, T
2012	703	A	130	119	99	81	12	4879	W, F, T
2013	642	R	50	47	42	40	10	1900	B, W, F, T

Table 2 Regression statistics for number of citations and abstract views

Regression statistics	
Multiple R	0.751687
R-Square	0.565034
Adjusted R-Square	0.540869

Table 3 Residual output for AV

Predicted values	Residuals
19,477.88006	10,365.11994
11,195.53987	−3458.539867
8197.848571	−963.8485714
7103.089418	−756.0894177
4386.252565	−1140.252565
3893.324359	344.6756406
2294.173554	−401.1735536
2110.758407	1009.241593
2110.758407	−132.7584074
1038.926147	2497.073853
763.8034276	1736.196572
763.8034276	1022.196572
631.9737913	−95.97379129
328.1924554	147.8075446
328.1924554	160.8075446
6653	1096
1789	556
673	−107
3467	1412
2100	−200

In a graph, if the points are randomly dispersed, then a linear regression model is appropriate; else, a nonlinear model is appropriate.

In this paper, the top cited articles of cloud computing are taken and analyzed the influence of abstract views and social media. Table 3 shows the predicted and residual values of top cited research articles of cloud computing area. Same regression statistics and residuals were calculated for social media influence score of the top cited research articles. The regression analysis was made and tabulated in Figs. 4 and 5.

Fig. 4 Observed and
expected AV counts

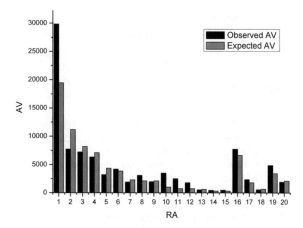

Fig. 5 Observed and
expected SMI score

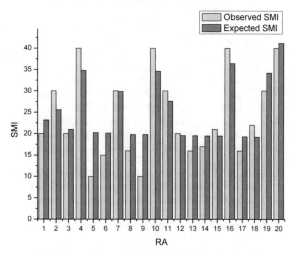

6 Conclusion

Getting an impact in the scientific community is a challenging task for any researcher who writes a research paper. A research paper is evaluated according to the number of research citations it gets. Apart from quality and originality, there are various factors involved in the increase in citations. This paper tabulates the relation between abstract view counts and social media influence on the increase in citations. The social media influence score is fixed with a common value and tabulated. In future, more attention has to be given on various social media influences like blogs, Facebook, Twitter, Wikipedia, and so on. A separate influence analysis must be done before setting the influence score. This paper analyzes the top cited research articles of cloud computing area. In future, the evaluation is planned on various other research areas, interdisciplinary areas, and so on. The research scholars, research organizations, funding agencies, and various communities can increase their research productivity and research impact through this analysis.

References

1. M.E.J. Newman, Networks: an introduction, **23**(01) (2014)
2. A.L. Kinney, National scientific facilities and their science impact on nonbiomedical research. Proc. Natl. Acad. Sci. **104**(46), 17943–17947 (2007)
3. H.F. Moed, A critical comparative analysis of five world university rankings. Scientometrics **110**(2), 967–990 (2017)
4. J.E. Hirsch, An index to quantify an individual's s scientific research output. Proc. Natl. Acad. Sci. U.S.A. **102**(46), 16569–16572 (2005)
5. I.M. Verma, Impact, not impact factor. Proc. Natl. Acad. Sci. **112**(26), 7875–7876 (2015)
6. E. Garfield, The history and meaning of the journal impact factor. J. Am. Med. Assoc. **295**(1), 90–93 (2006)
7. L. Bornmann, H. Daniel, What do citation counts measure? A review of studies on citing behavior. J. Doc. **64**(1), 45–80 (2008)
8. L. Bornmann, G. Wallon, A. Ledin, Does the committee peer review select the best applicants for funding? An investigation of the selection process for two European molecular biology organization programmes. PLoS ONE, **3**(10) (2008)
9. G. Madaan, S. Jolad, Evolution of scientific collaboration networks, in *Proceeding—2014 IEEE International Conference Big Data, IEEE Big Data 2014*, pp. 7–13, (2015)
10. Z. Xie, Z. Ouyang, Q. Liu, J. Li, A geometric graph model for citation networks of exponentially growing scientific papers. Phys. A Stat. Mech. Its Appl. **456**, 167–175 (2016)
11. D. Wang, C. Song, A.-L. Barabasi, Quantifying long-term scientific impact, Science *(80-.)* **342**(6154), pp. 127–132 (2013)
12. S. Redner, "How popular is your paper? An empirical study of the citation distribution. *Eur. Phys. J. B—Condens. Matter Complex Syst.*, **4**(2), 131–134 (1998)
13. T.A. Brooks, Evidence of complex citer motivations. J. Am. Soc. Inf. Sci. **37**(1), 34–36 (1986)
14. S. Zhang, D. Zhao, R. Cheng, J. Cheng, H. Wang, Finding Influential Papers in Citation Networks, in *2016 IEEE First International Conference on Data Science in Cyberspace*, pp. 658–662 (2016)
15. Y. Jia L. Qu, Improve the Performance of link prediction methods in citation network by using H-Index, in *2016 International Conference on Cyber-Enabled Distributed Computing Knowledge Discovery*, pp. 220–223 (2016)
16. L. Floridi, Big data and their epistemological challenge. Philos. Technol. **25**(4), 435–437 (2012)
17. C. Caragea, J. Wu, K. Williams, S. Das, G.M. Khabsa, P. Teregowda, C.L. Giles, Automatic identification of research articles from crawled documents (2014)
18. Y. Lin, H. Tong, J. Tang, K. Selc, Guest editorial: big scholar data discovery and collaboration. IEEE Trans. Big Data **2**(1), 1–2 (2016)
19. F. Xia, W. Wang, T.M. Bekele, H. Liu, Big scholarly data: a survey. IEEE Trans. Big Data **3**(1), 18–35 (2017)
20. M.W. Nielsen, Gender and citation impact in management research. J. Informetr. **11**(4), 1213–1228 (2017)
21. H.J. Kim, Y.K. Jeong, M. Song, Content- and proximity-based author co-citation analysis using citation sentences. J. Informetr. **10**(4), 954–966 (2016)
22. D. Fiala, G. Tutoky, PageRank-based prediction of award-winning researchers and the impact of citations. J. Informetr. **11**(4), 1044–1068 (2017)
23. L. Bornmann, L. Leydesdorff, Skewness of citation impact data and covariates of citation distributions: a large-scale empirical analysis based on Web of Science data. J. Informetr. **11**(1), 164–175 (2017)

24. J. Guan, Y. Yan, J.J. Zhang, The impact of collaboration and knowledge networks on citations. J. Informetr. **11**(2), 407–422 (2017)
25. G.S. Patience, C.A. Patience, B. Blais, F. Bertrand, Citation analysis of scientific categories. *Heliyon* **3**(5) (2017)
26. N. Wang, H. Liang, Y. Jia, S. Ge, Y. Xue, Z. Wang, Cloud computing research in the IS discipline: A citation/co-citation analysis. Decis. Support Syst. **86**, 35–47 (2016)
27. J. Son, S.B. Kim, Academic paper recommender system using multilevel simultaneous citation networks. Decis. Support Syst. **105**, 24–33 (2018)
28. J.M. Lindsay, PlumX from plum analytics: not just altmetrics. J. Electron. Resour. Med. Libr. **13**(1), 8–17 (2016)

Iterative Sharpening of Digital Images

B. Sravankumar, Chunduru Anilkumar, Sathishkumar Easwaramoorthy,
Somula Ramasubbareddy and K. Govinda

Abstract The fundamental point of image processing is about upgrading the quality and visual look of the picture, which can viably enhance the impression of data from pictures. Numerous pictures like medicinal pictures, satellite, flying pictures, and furthermore genuine photos experience the ill effects of poor and awful complexity and commotion. It is important to upgrade the differentiation and expel the clamor to build picture quality, thus picture improvement is an essential field to think about as it has a considerable measure of uses like in the therapeutic field, unique mark upgrade, signature protecting and so forth as appeared in this paper. Picture Enhancement is comprehensively ordered in two sections: Pixel-Based Enhancement and Enhancement based on Frequency. In Pixel-Based Enhancement, there are two sections: Sharpening and Smoothening. This report just demonstrates a Sharpening Method for Spatial Domain Enhancement.

Keywords Intrusion detection system (IDS)
Intrusion detection and prevention systems (IDPS)
Intrusion prevention system (IPS) · Network behavior analysis (NBA)

B. Sravankumar
QIS Institute of Technology, Ongole, India
e-mail: sravankumar.badithala@gmail.com

C. Anilkumar · S. Easwaramoorthy · S. Ramasubbareddy (✉) · K. Govinda
Vellore Institute of Technology, Vellore 632014, Tamil Nadu, India
e-mail: svramasubbareddy1219@gmail.com

C. Anilkumar
e-mail: chunduru.anilkumar@vit.ac.in

S. Easwaramoorthy
e-mail: srisathishkumarve@gmail.com

K. Govinda
e-mail: kgovinda@vit.ac.in

© Springer Nature Singapore Pte Ltd. 2019

53

S. C. Satapathy et al. (eds.), *Information Systems Design and Intelligent Applications*,
Advances in Intelligent Systems and Computing 862,
https://doi.org/10.1007/978-981-13-3329-3_6

1 Introduction

The fundamental point of image processing is upgrading the quality and visual look of a picture, which can viably enhance the impression of data from pictures. Picture upgrade underscores the picture highlights, for example, edge, plot and differentiate, and so on. There are numerous sorts of various pictures upgrade calculations for various pictures.

Amid catching, transmission or potentially procurement forms, the information estimation of information pictures will experience the ill effects of different sorts of Contaminations [1]. These undesirable pollutions in the picture are called Noise. Clamors are of numerous kinds. These Contaminations (clamor) for the most part as outside obstructions like air aggravations, defective instruments will make annoyances of the framework. The commotion will decrease the nature of pictures and harm the declaration of data of pictures extensively. Picture upgrade can adequately lessen the commotion and make the picture smooth. Picture improvement is a testing issue: it will probably stifle the clamor while saving the respectability of edges and detail data and subsequently we look at changed upgrade procedures. Picture Enhancement procedures are characterized into two types: Pixel-Based Enhancement and Enhancement Based on Frequency.

In pixel-based enhancement, the operations are performed on pixels whereas in enhancement based on frequency operations are done on Fourier transform of image.

Expulsion of commotion falls under Spatial Domain Enhancement and this undertaking will talk about the spatial space improvement approaches. One of the major Spatial Domain Enhancement approaches is Filtering which is of two kinds: Smoothing and Sharpening [2, 3].

Unsharp masking is a standout among the most settled sharpening algorithms. The way toward sharpening can ordinarily be repeated a few times without a recognizable picture. In the event that sharpening is repeated in excess of a predetermined number of times clear ancient rarity shows up in the picture bringing about quality debasement. To take care of the issue of iterative sharpening, a calculation that is like unsharp masking is formulated with a slight change. This calculation joins both unsharp masking and power-log transformation [4].

2 Background

2.1 Iterative Sharpening

A picture is obscured utilizing a Gaussian channel and, as a moment step, the contrast between the first picture and the obscured form is computed. The distinction picture is added straightly to the first to touch base at the coveted sharp picture. So as opposed to adding contrast to the first, we enhance the distinction and add it to the obscured picture instead of the first. We at last obscure the outcome and rehash the procedure with a similar unique high-recurrence picture. The picture is obscured at every cycle to guarantee that the outcome is constantly consistent [5, 6].

The algorithm takes the input of a grayscale image $J(x, y)$. This is then blurred by a Gaussian filter D and then difference image K is made.

$$K = J - D * J \tag{1}$$

where* is for convolution. The higher frequency image K is then raised to gamma function and the image is added back to the $D * J$ image which yields a final result,

$$J_1 = K^\gamma + D * J \tag{2}$$

Here, we see that if γ is set to 1 then the resultant image I_j is the same as I. The iteration is

$$J_{j+1} = K^\gamma + D * J_j \tag{3}$$

3   Image Quality Metrics

A.   Mean Square Error (MSE)

It calculates the difference between each pixel of the image and takes the mean of that.

B.   Peak Signal-to-Noise Ratio (PSNR)

PSNR is utilized for quantifying nature about a recreated picture of the first picture. PSNR is communicated as far as logarithmic decibel scale. Higher PSNR by and large demonstrates that the yield picture is of higher quality [7].

4   Proposed Model

As in iterative honing, the picture is obscured and subtracted from the first picture and after that, the distinction picture is intensified and added to the obscured picture. Rather than adding the intensified picture to the obscured picture, the enhanced picture is added to the first picture where gamma work utilized for intensifying is between 0 and 1 and the cycles are completed.

$$H = I - D * I \tag{4}$$

where H is the difference image.

$$I_1 = H^\gamma + I \tag{5}$$

where γ is the gamma function which lies between 0 and 1.

$$I_{j+1} = H^\gamma + I_j \tag{6}$$

This is the iterative equation.

5 Experimental Results

This paper has taken some example pictures and has got a few yields by executing the code in MATLAB. It has utilized our proposed demonstrate while figuring our code. The proposed show here incorporates its own particular variant of unsharp covering in light of the fact that in typical unsharp concealing, the greatest number of iterations before ancient rarity arrangement is less and in adjusted unsharp veiling, the honing task is not effective in any way. The proposed system works in two ways [8]:

 (i) Sharpening is done effectively
(ii) Number of iterations is generally high.

In this model, rather than adding the opened-up picture to the obscured picture, the intensified picture is added to the first picture where gamma work utilized for intensifying is between 0 and 1 and the iterations are completed. In this model, the estimation of the greatest number of cycles is contrarily relative to the gamma esteem. On the off chance that gamma diminishes, the quantity of iterations before antique development increments. For estimations of gamma near 1 even 10 iterations prompt antiquity arrangement. In this way, this model delivers a more effective yield and is more qualified for advanced picture improvement than the current models specifically typical unsharp masking and changed unsharp masking.

5.1   MATLAB Code for the Proposed Model

```
I = imread('cameraman.jpg');
a=input('Enter no.of iterations ');
fori = 1:a
    D = imgaussfilt(I,0.5);
    H = I - D;
    [M,N]=size(H);
for x = 1:M
for y = 1:N
                m=double(H(x,y));
z(x,y)=m^(0.5);
                I(x,y)=I(x,y) + z(x,y);
end
end
end
imshow(I);
```

The above code has been executed on MATLAB R2017b.

Here, this paper has taken the value of gamma as 0.5 as default. It can be easily changed by changing the value of the right-hand side function in imgaussfilt(). This is the code for cameraman picture input. D is the obscured picture adaptation of the first picture shaped by means of the use of Gaussian channel. It is then subtracted from the first picture to create a high-recurrence picture. This is the place this model veers off from the current models. Ordinary models utilize $H = I + D$ though this model uses $H = I - D$.

So, this paper has got a few outcomes after actualizing this code and has made a few correlations in light of the yield found.

Comparisons (Figs. 1, 2, 3, 4, 5, and 6).

As one can notice, there is artifact formation in normal unsharp masking method for low values of n (5 and 10 in this case). These artifacts really ruin the image and hence, make normal unsharp masking unfit for digital image enhancement.

Also, in modified unsharp masking, blurring is evident which also ruins the image and hence, modified unsharp masking is also unfit for digital image enhancement. Looking at the proposed model, there is neither artifact formation nor blurring if one knows how to intelligently adjust the values of n and gamma (Fig. 7).

Fig. 1 $n = 5$ gamma $= 0.2$ proposed model

Fig. 2 $n = 5$ gamma $= 0.2$ unsharp masking

Fig. 3 $n = 5$ gamma $= 0.2$ modified unsharp masking

Fig. 4 $n = 10$ gamma 0.2 proposed model

Fig. 5 $n = 10$ gamma $= 0.2$ unsharp masking

Fig. 6 $n = 10$ gamma $= 0.2$ modified unsharp masking

Fig. 7 $n = 15$ gamma $= 0.2$ proposed model

6 Conclusion

We developed an algorithm to sharpen the image iteratively and enhance the contrast of the image. Earlier the existing algorithm could iteratively sharpen the image but the number of iterations could not be large. Our algorithm could be used to increase the number of iterations without occurring of any artifacts and it shows that more we sharp the more effect of process spreads to a lower frequency.

From the results that we got, it can be inferred that as we increase the value of gamma, the blurring effect is more as n increases. So, for lower levels of gamma, one can achieve more iterations without considerably blurring the image or producing artifacts.

So, to get the best image, the proposed model should be used along with low levels of gamma. This would be the best choice for an optimum digital enhancement of the image. Also, the number of iterations is very high for the above situation.

References

1. B. Masschaele, M. Dierick, L. Van Hoorebeke, P. Jacobs, J. Vlassenbroeck, V. Cnudde, Neutron CT enhancement by iterative de-blurring of neutron transmission images. Nucl. Instrum. Methods Phys. Res., Sect. A **542**(1–3), 361–366 (2005)
2. P. Milanfar, H. Talebi, A new class of image filters without normalization, in *2016 IEEE International Conference on Image Processing (ICIP)*, pp. 3294–3298. IEEE (2016)

3. V.S. Patil, R.H. Havaldar, Haze removal and fuzzy based enhancement of image, in *2016 IEEE International Conference on Computational Intelligence and Computing Research (ICCIC)*, pp. 1–3. IEEE (2016)

4. A. Alsam, I. Farup, H.J. Rivertz, Iterative sharpening for image contrast enhancement, in *Colour and Visual Computing Symposium (CVCS), 2015,* pp. 1–4. IEEE (2015)

5. K. Nitta, R. Shogenji, S. Miyatake, J. Tanida, Image reconstruction for thin observation module by bound optics by using the iterative back projection method. Appl. Opt. **45**(13), 2893–2900 (2006)

6. R.J. Parada, E.B. Gindele, A.L. McCarthy, K.E. Spaulding, in *U.S. Patent No. 6,795,585*. (U.S. Patent and Trademark Office, Washington, DC, 2004)

7. S.H. Malik, T.A. Lone, Comparative study of digital image enhancement approaches, in *2014 International Conference on Computer Communication and Informatics (ICCCI)*, pp. 1–5. IEEE (2014)

8. O.N. Portniaguine, M.S. Zhdanov, U.S. Method of digital image enhancement and sharpening, in *Patent No. 6,879,735* (U.S. Patent and Trademark Office, Washington, DC, 2005)

An Effective Rumor Control Approach for Online Social Networks

S. Santhoshkumar and L. D. Dhinesh Babu

Abstract Since the rumor spreading has a negative impact on the stability of Online Social Networks (OSNs), rumor diffusion study is becoming an important research area in recent days. The main causes for rumor spreading are lack of education and lack of official information. Promotion of education and official information through social media applications against the disease is called as "Social Vaccine". In this paper, a novel social vaccine approach called Pulse Vaccination for Rumor Control (PVRC) is proposed to combat the rumor spreading. A novel algorithm is proposed to find the portion of the population to provide PVRC in regular intervals. A Spreader(S)-Educated Spreader(U)-Ignorant(I)-Educated Ignorant(V)-Recovered(R), USVIR, rumor dynamics model is proposed to study the vaccination approach. This model illustrates the impact of the vaccination on rumor propagation with respect to time in OSNs. We evaluated the proposed approach on four different datasets. Experimental results show that periodic education to the selected population will eradicate the rumor from the network and devise a rumor-free equilibrium.

Keywords Rumor control · Online social networks · Social vaccination
Education · Pulse vaccination · USVIR epidemic model

1 Introduction

With the advent of the internet-enabled devices such as smartphones and the plethora of online social media applications, a rumor spread quicker and wider in OSNs. Rumor causes panic and chaos during emergency events or it may create public

S. Santhoshkumar (✉) · L. D. Dhinesh Babu
School of Information Technology and Engineering, Vellore Institute of Technology University,
Vellore, India
e-mail: santhoshramuk@gmail.com

L. D. Dhinesh Babu
e-mail: lddhineshbabu@gmail.com

© Springer Nature Singapore Pte Ltd. 2019   63
S. C. Satapathy et al. (eds.), *Information Systems Design and Intelligent Applications*,
Advances in Intelligent Systems and Computing 862,
https://doi.org/10.1007/978-981-13-3329-3_7

awareness. Therefore, studying the rumor dynamics has been a major research interest in recent days.

Social vaccine is a health promotion movement from an official body such as government to educate the population against the medical and popular conceptions of health [1]. In the field of rumor dynamics, the social vaccine can be defined as a methodology to educate the population against the rumor. This method tries to immune the suspected or infected population by educating the people.

Pulse vaccination in the epidemic model is a widely accepted vaccination method. It is used on a portion of the population within a given period to combat the disease spread. In rumor spreading model, pulse vaccination is promoting the official and scientific knowledge constantly to combat the rumor spread. The portion of the population can be relieved from the rumors by this vaccination. Apparently, the individuals can immune themselves and stop spreading the rumors because of the vaccination. Here, vaccination is applied as an active defense to prevent the rumor spread in online social networks. Pulse vaccination emphasizes the active vaccination with a specific interval can terminate the rumor spread.

In this paper, a social vaccination approach with an optimal interval to control the rumor spread is proposed by the inspiration of pulse vaccination in epidemic spread. A new method to identify optimal interval and an optimal portion of the population is devised. Also, this paper proposes a novel rumor spread model called USVIR to study the proposed vaccination approach.

Our contributions in this paper are as follows, first, we define a novel rumor spreading model USVIR based on the classic epidemic model SIR. Second, we introduce a pulse vaccination technique with a specific interval to combat the rumor spread. We also find the high spreading ability nodes to vaccinate at every interval.

The rest of this paper is organized as follows, we briefly review related works in Sect. 2. In Sect. 3, USVIR model is derived from the inspiration of SIR model. In Sect. 4, the dynamics of the proposed model is studied to pulse vaccinate the set of high spreading ability nodes at every interval. In Sect. 5, the paper discusses the experimental setup and evaluation results. This paper concludes in Sect. 6 with some future enhancements

2   Related Works

Due to the rapid diffusion nature of information in online social networks, a rumor spread faster and wider. This causes more damage to the network and sometimes the damage is irreversible. The classical models DK and MK models were introduced to study the rumor process by Daley et al. [2] and Maki et al. [3], respectively. DK model proposed a three-state epidemic model Susceptible-Infectious-Recovered (SIR) which serves as one of the classical epidemic models. This model segregates the population into three distinct groups as follows: people who do not know the rumor and are susceptible to be affected by rumor are grouped into infectious (I), people who know the rumor and spreading it to others are grouped into susceptible

(S), people who know but not spreading the rumor are recovered group (R). In SIR model rumor spread on one-to-one communication among individuals. MK model extended this model and added another hypothesis. That is, when two individuals from susceptible group communicate, either of them will inevitably turn themselves into recovered state and stop spreading the rumor. Another classical rumor spreading model is Susceptible-Infectious-Susceptible (SIS). Here susceptible state nodes are vulnerable to infection and infectious state nodes are infected by the epidemic. In SIS, infected state nodes once recovered returns to the susceptible state. Various models and studies carried out based on these classical models through different approaches such as applied mathematical theory [4, 5], applied physics theory [6–8], and stochastic theory [9].

Most of the current rumor spreading models are derived from or inspired by the epidemic models [10, 11]. Education rate also can determine the rumor spread in social networks. Komi introduced an education rate-based model SEIR [12]. This model argues uneducated individuals have high chance to accept the rumor than educated individuals.

In rumor dynamics field, vaccination is a process to spread the official information or knowledge to the population to immunize or educate about the rumor. Huo et al. [13] introduced a modified SIR model, XWYZ with impulse vaccination and time delay. This paper argues small impulse vaccination rate and long impulse period is sufficient to eradicate the rumor permanently from the network. This is the first model to address rumor dynamics with vaccination. This approach vaccinates only ignorant population and this research evaluated the performance only with mathematical modeling.

Our method derives an approach to eradicate the rumor from the network using pulse vaccination with specific interval time. Higher spreading ability nodes are being vaccinated in our model to spread the vaccination among other individuals.

3 A Rumor Dynamic Model—USVIR

Rumor propagates through direct contact of the spreader with ignorant individuals in the network. To contribute to rumor spreading studies, this paper formulates a rumor spreading model by extending the classical SIR rumor model. This paper considers pulse vaccination with a specific time interval to remove rumor from the network. In a real-world scenario, spreaders and ignorant need to be vaccinated to eradicate the rumor quicker. To achieve this, the model distinguishes two new states to represent the vaccinated spreader and vaccinated ignorant from spreader and ignorant states, respectively.

At any time t, the total population N, divided into five different groups:

$S(t)$: A non-vaccinated spreader who know the rumor and actively spreading it among others irrespective of knowing the truthfulness of rumor,

$U(t)$: A vaccinated spreader who know the rumor and spreading it among others by knowing the information is rumor,

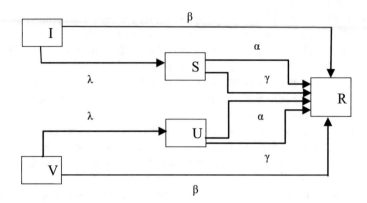

Fig. 1 Rumor dynamics model USVIR

$I(t)$: A non-vaccinated ignorant who do not know the rumor and susceptible to rumor,

$V(t)$: A vaccinated ignorant who do not know the rumor and less prone to be affected by rumor,

$R(t)$: A recovered who know the rumor and decided to not spread it.

At any time t, $S(t) + U(t) + I(t) + V(t) + R(t) = N$ satisfies.

Figure 1 shows the structure of the proposed USVIR rumor spreading process. The proposed pulse vaccination is studied on this USVIR rumor dynamics model. The rumor will be removed from the network when $S(t) \simeq 0$ where $t > 0$

4   Pulse Vaccination for Rumor Control

The proposed approach educates the individuals on how to stay away from rumor spread. Unlike other models, instead of acting on specific rumor topic in hand, this approach educates the participants on general awareness to stop spreading the rumor in OSNs. This will help the individuals to stay away from the rumor. To achieve this purpose, this paper proposes a vaccination approach called Pulse Vaccination for Rumor Control (PVRC). The rumor dynamics model described in Fig. 1 shows various types of participants and their state changing pattern.

PVRC can be defined as a repeated vaccination on a specific group of individuals to control the infection. As per our model, the vaccination is given to both ignorant and Spreader population. Assume that θ population being vaccinated at every T time. Among this, ρ is Spreader population and $\theta - \rho$ is ignorant population.

The objective of the proposed approach is to eradicate the rumor from online social networks by educating individuals about the rumor.

USVIR rumor dynamics model is studied to derive a rumor-free equilibrium. Rumor-free equilibrium can be achieved by increasing the individuals to move to recovered state faster. i.e.,

$$\beta + \beta_v > \alpha + \alpha_v + \gamma + \gamma_v \ldots \tag{2}$$

$$\beta + \beta_v + \alpha + \alpha_v + \gamma + \gamma_v > \lambda + \lambda_v \ldots . \tag{3}$$

We propose three steps to achieve the equilibrium in the system:

1. Identify vaccination population
2. Identify Max inter vaccination time
3. Apply vaccination on people identified in step 1 at time interval identified in step 2.

In our approach, we assume people who vaccinated are adhering to the rumor vaccination process and not going to ignore the vaccination.

4.1 Vaccination Population

The portion of the population that is being vaccinated to avoid rumor is called as vaccination population (pv). To improve the efficiency of vaccination, pv should be the individuals who have more impact on others in the network. So, the vaccination spreads to neighbors faster than usual. This approach helps in indirect vaccination. Impact of a person on other keeps changes over time in a dynamic network like OSNs. Hence, the proposed method finds the different set of individuals on every time interval. Therefore, the maximum number of individuals is vaccinated by direct vaccination and indirect vaccination.

Spreading ability determines whether an individual can convince the neighbor and propagates the rumor to them. i.e., the influencing ability of an individual among neighbors. We define this as follows:

$$SA_i(t) = \varphi(t) + k_i + \sum_{j \in \text{Neighbors}(i)} \frac{\omega_{ij}}{\langle \omega \rangle}$$

ω_{ij} is the acceptance probability of node j on i. Acceptance probability is a probability of accepting the rumor from its immediate neighbor. $\varphi(t)$ is the rumor attraction factor of the individual. k_i is the degree of node i.

Algorithm – Vaccination Population Finder (VPF):

Input: $G_c = \{\gamma, \delta_c\} \mid \gamma = \{1,2,3, \dots n\}$ *nodes & δ_c edges*
Input: *Rumor attraction value $\phi(t)$ for time t*
Output: $I = \{I^1, I^2, I^3, \dots, I^u \mid I \in \gamma \ \& \ |I| \leq |\gamma|\}$

FOR every node i in G_c:
    FOR every neighbors j of node i:
    // Degree Strength Co-efficient
$$di_{ij} = \frac{k_j}{k_i}$$
    // Acceptance probability j on i
    $Acc_i = Acc_i + \omega_{ij} . di_{ij}$
    ENDFOR
$$AP_i = \frac{Acc_i}{\langle \omega \rangle}$$
$$SA_i(t) = \varphi(t) + k_i + AP_i$$
$$SA(t) = SA(t) \cup SA_i(t)$$
ENDFOR
Sort nodes by $SA(t)$ in descending order
Return first u nodes $I = \{I^1, I^2, I^3, \dots, I^u \mid I \in \gamma \ \& \ |I| \leq |\gamma|\}$ as vaccination proportion for time t

    u nodes identified by VPF consists of participants from all the states S, I, and R. R nodes are ignored from vaccination as those will not return to the spreader or the ignorant state. So, the vaccination applied on S and I state nodes and those nodes vaccinated are turned to U and V states, respectively.

4.2 Maximum Inter-Pulse Interval

The period between two pulses should not be larger to avoid the rumor to grow indefinitely. To obtain "rumor free" stable solution, the inter-pulse interval should be minimum. Stone et al. derived maximum inter-pulse interval, T_{max}, for SIR epidemic model [14]. This value considers birth and death rate of the people.

$$T_{max} \simeq \frac{1}{m} \ln \left(1 + \frac{p(m+g)}{(\beta - m - g)} \right)$$

Here, m is birth and death rate, β is infection rate, people move from Ignorant to Spreader, g is recover rate, people move from Spreader to recovered. p is the proportion of the population being vaccinated.

    The maximum inter-pulse interval for USVIR model considers the vaccination applied to Spreader and Ignorant group of nodes. For simplicity, we avoid network growth rate from the equation. With vaccination, T_{max} for USVIR model can be written as

$$T_{\max} \simeq \ln\left(1 + \frac{u(\alpha + \alpha_v + \gamma + \gamma_v)}{(\beta + \beta_v - (\alpha + \alpha_v + \gamma + \gamma_v))}\right)$$

Here, u is the vaccination proportion on each pulse interval. So, the inter-pulse interval should not be greater than $T_{\max}$

4.3 Vaccination

Pulse Vaccination for Rumor Control is educating a certain number of people against the rumor propagation in regular interval. In this approach, the vaccination is applied to u nodes identified by the algorithm VPF.

Algorithm: PVRC

Input: $G_c = \{\gamma, \delta_c\} \mid \gamma = \{1,2,3, \dots n\}$ nodes & δ_c edges
Input: $I = \{I^1, I^2, I^3, \dots, I^u \mid I \in \gamma$ & $|I| \leq |\gamma|\}$
Input: $S(t) \subseteq \gamma, I(t) \subseteq \gamma, R(t) \subseteq \gamma, V(t) \subseteq \gamma,$
    $U(t) \subseteq \gamma \mid S(t) + I(t) + R(t) + V(t) + U(t) = \gamma$
Output: Updated $S(t) \subseteq \gamma, I(t) \subseteq \gamma, R(t) \subseteq \gamma, V(t) \subseteq \gamma,$
    $U(t) \subseteq \gamma \mid S(t) + I(t) + R(t) + V(t) + U(t) = \gamma$

FOR every node i in u:
    IF i $\in$ S(t) THEN:
            S(t) = S(t) − i
            U(t) = U(t) ∪ i
        ELSE IF i $\in$ I(t) THEN:
            I(t) = I(t) − i
            V(t) = V(t) ∪ i
    ENDIF
ENDFOR

This algorithm includes nodes into the new states U and V. S nodes vaccinated are vaccinated spreader grouped into U state. I nodes vaccinated are vaccinated ignorant grouped into V state.

5 Evaluation and Results

In this section, we evaluated four networks to study the effectiveness and efficiency of the proposed approach to SIR epidemic model. Throughout this evaluation, λ and λv is set to 1. i.e., the spreader will move to recovered state as soon as the spreader contacted its ignorant neighbors, irrespective of whether the ignorant turn

Table 1 Network properties of datasets for experiments

Dataset name	Network features					
	n	e	$<k>$	H	β_{th}	β
Karate club	34	78	4.5882	1.6895	0.129	0.242
RandNW_1	1000	5178	13	2.11	0.08	0.24
RandNW_2	2000	14,324	20	2.33	0.11	0.14
Ca_condmat	21,363	196,972	22	2.99	0.02	0.035

Here $<k>$ is average degree, $H = <k^2>/<k>^2$—Degree Heterogenicity index [16]
$\beta_{th} = <k>/<k^2>$—epidemic threshold [17]

into the spreader. A spreader can contact a neighbor only once in an iteration. All the following simulation results are reported by averaging the results at least fifty runs.

5.1 Datasets

To evaluate effectiveness and efficiency of our proposed approach, we apply it on two real-world social networks and two randomly generated networks. We used NetworkX [15] to generate the random network of size 1000 and 2000 nodes. Topological features of these networks are summarized in Table 1.

Karate club is a small world network, RandNW_1 and RanNW_2 are random networks, and Ca_condmat is a representative from the scale-free network. Different types of networks are considered to prove the generality of the proposed approach

5.2 Results

The evaluation first focuses on the final size of rumor to illustrate the spreading effectiveness of the proposed approach.

Rumor Final Size
Rumor final size is the number of unrecovered people remains in the network at final time step. When the final size is lesser, then the approach is effective in removing the rumor from the network. The experiment conducted with different βv.

Figure 2 depicts the rumor final size in percentage against vaccinated transmission rate (β_v) from the ignorant to the recovered state. As seen in Fig. 2, as βv increases, the rumor final size is decreasing. This reveals the pulse vaccination significantly contributes to the reduction of rumor final size. This clearly suggests that increasing the pulse vaccination rate has a great benefit in the decline of rumor spread.

Next, we evaluate the impact of vaccination population selection in rumor dissemination. The evaluation compares the rumor final size for the vaccination of

Fig. 2 Rumor % with vaccinated transmission rate

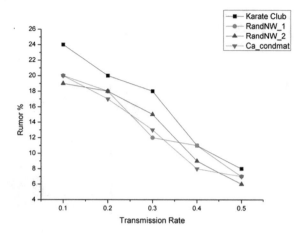

Fig. 3 Rumor % with VPF

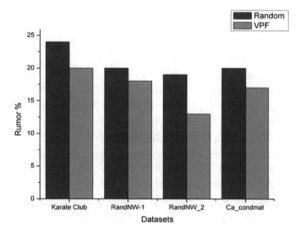

random people and the vaccination of people identified by the proposed approach. This evaluation keeps β_v constant for all the experiments.

Figure 3 depicts the rumor final size between random initial population selection and Vaccination Population Finder's initial vaccination. For all datasets, the results clearly indicate that VPF having an impact on rumor final size. Therefore, the vaccination approach that combines influential spreader selection and vaccinating those spreaders plays a key role in the control of rumor spreading in online social networks.

Average Infected Time

Next, we compare the average infected time [18] of USVIR rumor spread model with SIR epidemic model. This metric measure the average time the individual believes the rumor and keep spreading it. i.e., the average amount of time individuals remains to be spreader in the network. This measure helps to identify the efficiency of the approach in removing rumor from the network. Average Infected Time (AIT) can be defined as

Fig. 4 AIT comparison
between SIR and USVIR

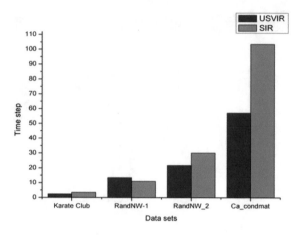

$$AIT = \frac{1}{N} \sum_{i \in N} T_i$$

In this evaluation, AIT measured for USVIR and SIR by having every transition rate between each state as constant. Since the states, U and S are spreaders in USVIR model, both the states are considered the common spreading state to compare it with SIR.

Figure 4 depicts AIT comparison between SIR and USVIR for four datasets. The results show that USVIR is performing well as AIT is less for the proposed approach in all datasets.

All these results show that the pulse vaccination has a significant impact on reducing the rumor propagation in USVIR rumor spread model. This clearly indicates pulse vaccination plays a key role to combat the rumor spread in Online social networks.

6 Conclusion

The rumor propagation in OSNs can lead to undesirable and at times devastating effects. Hence, eradicating rumors from OSNs has been a recent research with practical significance. In this paper, we proposed a social vaccination approach called pulse vaccination to control the rumor spread in OSNs. A novel rumor spread model USVIR proposed to study pulse vaccination dynamics in this paper. The evaluation is done on four different datasets and the results show that the proposed vaccination approach is effective in controlling the rumor spread.

In this paper, max inter-pulse time is defined to find the maximum interval between two consecutive vaccinations. Future work focus could be in increasing the max inter-pulse in order to reduce the cost incurred on pulse vaccination. Also, the vaccination population selection method can be optimized to increase the penetration of the education in the network to reduce the rumor spread. Overall, this research shed some lights on considering periodic vaccination will help in curbing the rumor from OSNs.

References

1. F. Baum, R. Narayan, D. Sanders, V. Patel, A. Quizhpe, Social vaccines to resist and change unhealthy social and economic structures: a useful metaphor for health promotion. Health Promot. Int. **24**(4), 428–433 (2009)
2. D.J. Daley, D.G. Kendall, Epidemics and rumors. Nature **204**(4963), 1118 (1964)
3. D.P. Maki, M. Thompson, Mathematical models and applications: with emphasis on the social life, and management sciences (1973)
4. V. Giorno, S. Spina, Rumor spreading models with random denials. Phys. A **461**, 569–576 (2016)
5. D. Li, J. Ma, Z. Tian, H. Zhu, An evolutionary game for the diffusion of rumor in complex networks. Phys. A **433**, 51–58 (2015)
6. L. Huo, L. Wang, N. Song, C. Ma, B. He, Rumor spreading model considering the activity of spreaders in the homogeneous network. Phys. A **468**, 855–865 (2017)
7. S. Han, F. Zhuang, Q. He, Z. Shi, X. Ao, Energy model for rumor propagation on social networks. Phys. A **394**, 99–109 (2014)
8. J. Ma, D. Li, Z. Tian, Rumor spreading in online social networks by considering the bipolar social reinforcement. Phys. A **447**, 108–115 (2016)
9. M.Z. Dauhoo, D. Juggurnath, N.-R.B. Adam, The stochastic evolution of rumors within a population. Math. Soc. Sci. **82**, 85–96 (2016)
10. J. Wang, L. Zhao, R. Huang, SIRaRu rumor spreading model in complex networks. Phys. A **398**, 43–55 (2014)
11. L. Zhao, H. Cui, X. Qiu, X. Wang, J. Wang, SIR rumor spreading model in the new media age. Phys. A **392**(4), 995–1003 (2013)
12. K. Afassinou, Analysis of the impact of education rate on the rumor spreading mechanism. Phys. A **414**, 43–52 (2014)
13. L. Huo, C. Ma, Dynamical analysis of rumor spreading model with impulse vaccination and time delay. Phys. A **471**, 653–665 (2017)
14. L. Stone, B. Shulgin, Z. Agur, Theoretical examination of the pulse vaccination policy in the SIR epidemic model. Math. Comput. Model. **31**(4–5), 207–215 (2000)
15. N. Developers, "NetworkX," *networkx. lanl. gov,* 2010
16. H.-B. Hu, X.-F. Wang, Unified index to quantifying heterogeneity of complex networks. Phys. A **387**(14), 3769–3780 (2008)
17. C. Castellano, R. Pastor-Satorras, Thresholds for epidemic spreading in networks. Phys. Rev. Lett. **105**(21), 218701 (2010)
18. R.M. Tripathy, A. Bagchi, S. Mehta, A study of rumor control strategies on social networks, in *Proceedings of the 19th ACM international conference on Information and knowledge management,* pp. 1817–1820 (2010)

A Novel Approach for Data Security in Cloud Environment Using Image Segmentation and Image Steganography

R. Kiran Kumar and D. Suneetha

Abstract Now days storage of data in cloud plays a vital role in any place but data security for cloud environment is a major problem now a days because the data is maintained and its is organized by third party from different locations of different places. Before storing the data in the cloud environment the user must and should give security from unauthorized access. So in this paper we proposed new algorithm to secure user data by using image Steganography and image Segmentation. In this the data is hidden different segments of image by using image segmentation. The performance of proposed algorithm is evaluated by considering various parameters like PSNR, MSE values and the results are compared with various existing algorithms for various sizes of images.

Keywords Steganography · Segmentation · PSNR · MSE

1 Introduction

Now a day's cloud computing is a popular technology for storing data. It provides so many facilities to the users. In this environment providing security for data is a major prominent issue because the data is available at different places and it comes from different vendors of various organizations so there was a chance to access users secured data by an unauthorized person. In this environment the security concerned at different places like data transmissions, data storage, data retrieving etc. So many techniques are available for providing security for's data in the cloud environment. In those techniques Steganography and Segmentation is also one of the secured technique for providing security for user's data from unauthorized users.

R. Kiran Kumar · D. Suneetha (✉)
Krishna University, Machilipatnam, Andhra Pradesh, India
e-mail: sunithadavuluri8@gmail.com

R. Kiran Kumar
e-mail: kirankreddi@gmail.com

© Springer Nature Singapore Pte Ltd. 2019   75
S. C. Satapathy et al. (eds.), *Information Systems Design and Intelligent Applications*,
Advances in Intelligent Systems and Computing 862,
https://doi.org/10.1007/978-981-13-3329-3_8

Steganography is a process of hiding data in a image so data is not visible by the human eye directy. There are so many ways of Steganography techniques are available like Image Steganography, Video Steganography and Audio Steganography. Due to the growing need of security we go for either one of these Steganography types because the data is not directly visible by typical human eye.

Segmentation is a one of the very important technique nowadays. In this the image is divided into number of segments or the various parts of the image are extracted. The goal of segmentation is to simplify or change the representation of an image into something that is more meaning full and easier to analyze.

We introduced an algorithm based on image steganography and image segmentation technique which can hide secret data only in the calculated or processed segments of the original cover image. The proposed algorithm has been analyzed and it provides a better security compared to various attacks from unauthorized persons. The performance of proposed algorithm is analyzed with various security algorithms by considering different parameters.

Further the paper is arranged as follows. Section 2 contains some existing algorithms for data storing in cloud environment; Sect. 3 discusses proposed algorithm i.e. data security for cloud data using image segmentation Sect. 4 shows Evaluation results and Sect. 5. Consists of conclusions and future work.

2   Literature Review

We have various security algorithms based on the image steganogarphic techniques in order to hide user's secret data securely.

In Sharma et al.'s [1] has proposed a new algorithm for storing user's data in cloud environment using image steganography and image cryptography. In which secret data is encrypted using DES algorithm and obtain a secured key again secured key is encrypted using S-DES algorithm. So final is generated from S-DES algorithm that is key hidden in selected pixels of the cover image. Results are good and it provides high security with acceptable PSNR values.

In Awwad, Yousef Bani et al.'s [2] has proposed an algorithm for hiding secret data in a image using genetic and blowfish algorithms. In which first the cover image is selected for those cover image the pixels are calculated using genetic population algorithm and the secret message is encrypted using blowfish algorithm. Finally the encrypted secret message is embed in selected pixels of the cover image. Results are good but it produces more noise ratio when compared to the various algorithms of steganography techniques.

In Christina et al.'s [3] proposed an algorithm for hiding data in a image using image steganographic algorithms i.e. optimized blowfish algorithm. In which data is encrypted using blowfish algorithm and secret key is obtained based on that secret key again data encryption techniques is applied finally that encrypted key is embed in a original cover image.

In Suneetha et al.'s [4] proposed a better steganography approach which improves the hiding capacity and the image quality by using optimal partition based LSB algorithm. In which proposed algorithm the original cover image is partition into different images for placing huge amount of data after that apply edge detection algorithm to select edge pixels of an image. After that apply LSB substitution for those edge pixels for hiding secret data into an image. The advantage of the proposed method is improving the secret message length by diving image into number of partition and it improves the quality of the stego image.

In Kiran et al.'s [5] proposed a new technique in spatial environment for a gray scale image. In this data is hidden different parts of image so unauthorized person cannot identity where the secret is actually hidden. The proposed algorithm produces better PSNR values when compared to the several existing algorithm.

In Sadeq AlHamouz et al.'s [6] proposed a new approach for hiding secret message in a image using back propagation neural network. In this papers two cover images are used one is secret image and the other one cover image, both the images are color images. The algorithm uses two different phases for embedding secret data one is data embedding process and other one is data extracting process. The selected pixels positions are calculated using Fibonacci linear feedback shift register. The results are compared with several exciting algorithms that high PSNR and low MSE value is obtained with more processing time and improve in quality of an image.

3 Proposed Technique

In proposed technique we use the concept of image segmentation for extracting different parts of an image. First, read the original cover color image and convert it into black and white or gray scale image with required threshold value. Then we identify the iris part of image by applying image segmentation techniques. Next we extract the iris part of the original image, extract pupil part of a iris image. Then we construct inner and outer circle of image. Then select edge pixels of the inner and outer circle of a iris of the image by using canny edge detection algorithm.

Store the secret data in a text file and encipher the original message using RSA encryption algorithm and get the secret key. Then embed the secret key into the selected pixels of the original gray scale image. Finally we have stego image that stego image is stored into the cloud environment.

In receiving process the stego is taken as input and the reverse operation is performed with respect all the steps server side then obtains the secret key from an image and decrypts the original secret data using secret key.

Algorithm for Sender Side:
Sender Side

Input: Original cover image, Text file (Which consists of secret data)
Output: Stego object

Step 1: Start.
Step 2: image = imread('coverimage.jpeg').
Step 3: a = rgb2gray(image)
Step 4: Convert the original cover image into black and white with sufficient threshold value for inner circle.
Step 5: Read original text and encrypt original secret data and obtain key using RSA encryption algorithm
Step 6: Obtain the pixels of iris part of the image by using vertical scan to get the tangent and hence radius of the circle
Step 7: Do the same process for getting outer circle with change in threshold value (For outer circle the threshold value is little bit high)
Step 8: reconstruct inner and outer circle using obtained values
Step 9: Apply the canny edge detection for inner and outer circle pixels
Step 8: Insert secret data into the selected pixels
Step 9: Finally image is obtained called it as stego image.
Step 10 Stop.

Algorithm for Receiver Side:

Input: Stego Image
Output: Original Secret key

Step1: Start
Step2: a = imread ('stegoimage.png')
Step3: Get the pixels of edge corners of a cover image.
Step4: Get the pixels of iris part of an stego image
Step5: Decrypt the hiding secret data from the obtained pixels of a image.
Step6: Decrypt original secret message using key based on decryption algorithm.
Step 7: Stop

4 Experimental Results and Analysis

The proposed secured segmentation technique is too used to hide the secret data in selected pixels of original cover image which provides a better flexibility to the users in all aspects like robustness and quality of an image. The images are taken from the data set http://sipi.usc.edu/database/. We have used different gray scale images with different sizes and for different length of the original message for justifying the process. Figure 1 shows the original image.

Fig. 1 Original cover image

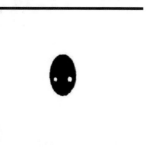

Fig. 2 Pupil part of the image

Fig. 3 Image after applying filtering process

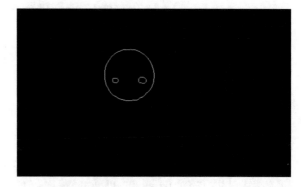

Read the color image of 512 * 512 sizes, then convert it into grayscale and stored in some file for further processing. In this process we identify two things one is inner circle for the iris and the other one is outer circle for the iris of a given image. For the inner circle we have to identify pupil region of an image. For this we select the image threshold value is 0.03 and apply canny edge detection algorithm for smoothen the image. Figure 2 shows the pupil part of the image after extraction process Fig. 3 shows the image after applying filtering process.

Fig. 4 Outer boundary of an
Iris

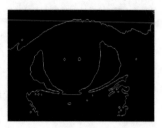

Fig. 5 Reconstructed outer
part of an image

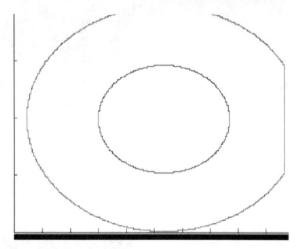

Determine the outer boundary of the iris from a given image. Applying the same process with change in threshold values from 0.03 to of 0.38. The next two equation are used for to construct the circles.

$$X = (\text{double})(\text{rad}) * \sin(\text{an}) + \text{double}(\text{cir}). \tag{1}$$

$$y = (\text{double})(\text{rad}) * \cos(\text{an}) + \text{double}(\text{cir}). \tag{2}$$

where 'rad' is a indication for radius of the corresponding image, 'cir' is a indication for center of the circle for an obtained image and 'an' is the indication for discrete angular value ranging from 0 to 2 * pi. Figure 4 shows outer boundary of iris part of an image. Figure 5 shows the reconstructed outer part of an image. Figure 6. Shows the stego image.

Evaluation of PSNR and MSE

For measurement of image quality for proposed algorithm two factors are calculated those are

$$\text{PSNR} = 10 \log_{10}(\text{MAX})_i^2 / \text{MSE}. \tag{3}$$

$$\text{MSE} = \sum M, N[I_1(m, n) - I_2(m, n)]^2 / M * N. \tag{4}$$

Fig. 6 Stego Image

Table 1 Comparison of various previous existing algorithms MSE values with proposed algorithm MSE values

Original image	Previous MSE	Proposed MSE
Mital	0.0048	0.0044
Minion	0.0041	0.0039
Bluefox	0.0021	0.0019
Blacing	0.0004	0.0001

5 Corresponding Comparison Table

In this region we present a comparison of various steganographic data hiding algorithms with proposed security algorithm using image steganographic and image segmentation. It is clearly calculated and show there was a significant change for proposed algorithm is better than several existing algorithms based on the image segmentation and image steganography. For evaluating various parameters and identifying difference between original cover image and stego image the MATLAB tools are used. Table 1 shows the comparison results of various existing algorithm MSE values with proposed algorithm MSE and Table 2 shows the comparison results of various existing algorithm PSNR values with proposed algorithm PSNR values.

Table 2 Comparison of various previous existing algorithms PSNR values with Proposed Algorithm PSNR values

Original image	Previous PSNR	Proposed PSNR
Mital	63.01	75.67
Minion	64.89	76.89
Bluefox	70.98	83.99
Blacing	56.01	84.02

6 Conclusion and Future Work

In this paper the data is hidden into the selected pixels of iris part of an image using image segmentation. This proposed algorithm has been produced better security for user's data and it produce significant results compared to existing algorithms. This proposed algorithm works for data with different formats like text, image etc. and for various sizes of secret data, it produces comparatively better results with existing algorithms. The proposed algorithm works on spatial domain with Gray scale image, in future if we implement the same process for RGB image it will get some better results.

References

1. V. Sharma, Madhusudhan, A two new steganography approaches for secure cloud data, in *Proceedings. Thrid International Conference on. Image Processing,* vol. 3. IEEE (2015)
2. Hammad, A new approach for data security using blowfish and symmetric algorithm. IJCSNS **17**(3):65 (2017)
3. L. Hristina, V.S. Irudayaraj, A New enhanement for data security using optimized blowfish technique. IJIRCCE **2**, 2320–9798 (2014)
4. D. Suneetha, Enhancement of data security for cloud using segmentation, in *International Conference on Recent Trends*
5. R. Kiran Kumar, A novel approach for data security in cloud environment using image segmentation and image steganography. IJCSI Issues, **9**(3, No. 1), 131–139 (2017)
6. Reyadh, Enhanced image steganography system based on discretewavelet transformation and resilient back-propagation. IJCSNS **15**(1) 2015

Suppression of Artifacts for Mobile ICG Using Nonlinear Adaptive Algorithms

Madhavi Mallam, A. GuruvaReddy, B. JanakiRamaiah, B. Ramesh Reddy and M. K. Lingamurthy

Abstract Impedance cardiography (ICG) is advantageous to identify various heart diseases. In recent times, mobile ICG has grabbed the attention of researchers and analysts for real-time supervision and automatic diagnosis. However, the attainment of ICG's survey system was degraded due to many interrupts created by subject movement, which paves to wrong diagnosis. Several attempts have been performed to abolish the noise from clinical ICG signal using distinct digital signal processing techniques. Those approaches are not directly synced to be used for the mobile ICG environment and regular noises. Basically, motion artifact still is an open problem in mobile ICG. In this paper, an advanced process of adaptive artifact elimination proposed through impedance cardiography (ICG) signals. This is a composite exemplary based on wavelets and adaptive filter. The prime aspect of this methodology is the realization of adaptive noise canceller (ANC) without any reference signal. In the real-time medical environment during critical conditions due to heartbeat disorders, the filter coefficients become negative. This convergence unbalance leads to low filtering capability. In order to solve this issue, one may incorporate NN adaptive algorithms in the suggested ANC. To enhance the attainment of ANC, error normalization is adapted to change filter coefficients automatically. Again, in order to minimize computational complexity and to avoid overlapping of data samples at the input stage of the filter, a hybrid version of nonnegative and sign sign-based

M. Mallam (✉) · B. Ramesh Reddy · M. K. Lingamurthy
Lakireddy Bali Reddy College of Engineering, Mylavaram, India
e-mail: gurumadhu432@gmail.com

B. Ramesh Reddy
e-mail: brrece73@gmail.com

M. K. Lingamurthy
e-mail: lingamurthy413@gmail.com

A. GuruvaReddy
DVR & Dr. HS MIC College of Technology, Kanchikacherla, India
e-mail: guruvareddy78@gmail.com

B. JanakiRamaiah
PVP Siddhartha Institute of Technology, Vijayawada, India
e-mail: bjanakiramaiah@gmail.com

© Springer Nature Singapore Pte Ltd. 2019
S. C. Satapathy et al. (eds.), *Information Systems Design and Intelligent Applications*,
Advances in Intelligent Systems and Computing 862,
https://doi.org/10.1007/978-981-13-3329-3_9

83

algorithms is considered for implementation. The resulting hybrid versions are exponential normalized nonnegative least mean square (eN^3LMS) algorithm, exponential normalized nonnegative sign regressor LMS (eN^3SRLMS) algorithm, exponential normalized nonnegative sign error LMS (eN^3SELMS) algorithm, and exponential normalized nonnegative sign sign LMS (eN^3SSLMS) algorithm. Finally, various ANCs are developed using these algorithms, and attainment measures are calculated and compared. Several implemented ANCs are verified on real impedance cardiogram signals.

Keywords Nonnegative algorithm · Remote healthcare artifact canceller Cardiovascular issues · Impedance cardiogram

1 Introduction

Cardiovascular diseases (CVDs) refer huge medical problems related to the functionality of the heart. In the 2015 annual report, WHO states that nearly 50% of all non-transmissible disease (NTD) deaths are due to CVDs [1, 2]. Among these, most of the deaths are on the outside of the hospital; the reason is that the patient is not treated timely. Hence, the investigation of cardiovascular healthcare technology becomes an intensive area. ICG is one of the promising techniques among various methods of cardiac functionality study. ICG noninvasive method measures the total electrical potential of the thorax and its variations in time to the number of cardiodynamic specification such as stroke intensity (SI), heart rate (HR), and stroke volume (SV) in medical scenario [3]. ICG provides the impedance variations that occur at thorax due to high-frequency, low-magnitude current flows through the thorax between two pairs of electrodes located outside the measured segment. However, during the extraction, the desired ICG signal meets with physiological and nonphysiological noises known as baseline Wander noise (BW), electro-muscle noises (EM), and impedance mismatch noises (IM). These artifacts affect the shape of signal and tiny characteristics, which are the main parameters for diagnosis of disease [4, 5]. For efficient clinical investigations, the ICG signals needed to be free from artifacts. As most of the biomedical artifacts are nonstationary in their nature, adaptive filtering techniques provide a better solution in this application. The adaptive filter weights are updated automatically in accordance with the noise level of the input signal. Several researchers contributed their work to the study of ICG using signal processing techniques. In these contributions, mainly the adaptive filtering part focused on the LMS and RLS algorithms [5–7]. But in the critical conditions, these algorithms suffer from some drawbacks. The ICG signal amplitude levels vary significantly due to abnormal heart rhythms; this leads to filter weights become negative. Due to the negative filter weights, the balanced convergence and effective filtering performance are not possible. This problem can be overcome by introducing a vector of input diagonally in the weight update equation of LMS algorithm. This algorithm is named as nonnegative LMS (eN^3LMS) algorithm; here, filter coefficients become nonnegative. We can normalize the step size with reference to error for improving the performance of the N^2LMS algorithm in terms of concurrence and excess mean square error (EMSE). We merge the nonnegative algorithms with three simplified algorithms. These simplified algorithms are based on reoccurrence of LMS known as sign error, sign regressor, and sign LMS algorithms [8]. To make the proposed algorithms less complex in computing, we combine the EN^3LMS algorithm with SRLMS, SELMS, and SSLMS. This results in EN^3SRLMS, EN^3SELMS, and EN^3SSLMS algorithms [9, 10]. In view of evaluating the attainment of the abovementioned algorithms which are applicable to real-time clinical scenario, DWT-based AAEs are developed based on these algorithms and tested on real ICG signals. The performance measures considered are convergence characteristics, signal-to-noise ratio (SNR), EMSE, and misadjustment (MSD). The assessment of the proposed algorithms and experimental outcome of the various applications are shown later.

2 Nonnegative Adaptive Algorithms for ICG Enhancement

In conventional ICG which is applicable to remote healthcare visualization system, few cognitive and noncognitive contaminations encounter with in the heart functionality graph, during signal acquisition, and lead to unambiguous diagnosis and assessments. Along with these contaminations, channel noise is also one major problem. The channel noise may mask the tiny characteristics of the ICG signal [7]. The main noises that encounter with the functioning of the heart are baseline wander noise (BW), electro-muscle noise (EM), and impedance mismatch noise (IM). BW is the small change of the ICG signal due to respiratory activity. EM is due to the muscle activity, and IM is due to impedance dissimilarity between electrodes and skin, also due to dissimilarity of the electrodes.

Figure 1 illustrates the schematic block diagram of an AAE based on a wavelet technique for remote health monitoring systems.

The following equation represents the desired ICG signals contaminated with noises:

$$ICG(k) = s_1(k) + n(k) \tag{1}$$

where $ICG(k)$ is ICG signal recorded; $s_1(k)$ is the desired ICG signal of the heart functionality; $n(k)$ is the noise parameter. In a remote healthcare monitoring system, the channel noise may also include in this parameter. The basic principle on which the recommended AAC operates is the raw ICG signal recorded $ICG(k)$ which is the input to DWT decomposition section. This generated signal is given as a reference to the algorithm in order to update the filter weights automatically. In this approach, the proposed AAC plays an important role in the realization of an intellectual remote healthcare supervision system.

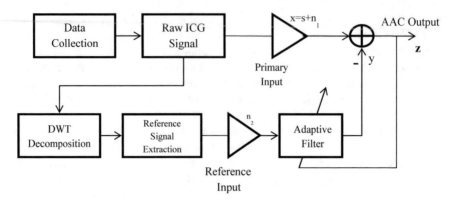

Fig. 1 Structure of adaptive artifact canceller for ICG signal enhancement using wavelet decomposition

The AAC comprises an FIR filter of length L taps using weight update principle; the coefficients are updated in filters used. The basic LMS algorithm weighted equation is written as follows:

$$M(p + 1) = M(p) + \eta\, a(p)e(p), \qquad (2)$$

$e(p)$ is the error occurred, which is given as a feedback.
$(p + 1)$ is the successive weight coefficient,
$a(p)$ is the input signal to the adaptive filter,
$M(p)$ is the previous weights, and
H is the step size.

Due to the malformation in the ICG signal and fast changes in the signal, the weighting coefficients may become negative. Because of this performance in terms of convergence, stability and filtering capability of the adaptive algorithm is poor. The exponential nonnegative LMS (N^2LMS) algorithm overcomes these drawbacks. The weight update mechanism of N^2LMS is given by

$$M(p + 1) = M(p) + \eta D(p)a(p)e(p), \qquad (3)$$

Here, $D(p)$ is the diagonal matrix of the weight.

First, normalization is performed with respect to the input IGG signal. This leads to exponential normalized nonnegative LMS (eN^3LMS) algorithm. In this algorithm, the step size "s" is variable rather than constant. The weight update equation for this algorithm is written as follows:

$$\mathbf{M}(n + 1) = \mathbf{M}(n) + s(n)\mathbf{x}(n)e(n)M^{\gamma}(n) \qquad (4)$$

whereas (n) is a variable step size with respect to the reference input as follows:

$$s(n) = \frac{s}{c + x^t(n)x(n)} \tag{5}$$

where *c is* a small constant used for avoiding numerical difficulties.

The amount of the change to weight vector $\mathbf{M}(n)$ is inversely proportional to pattern of input vector $x(n)$. The $x(n)$ with a large norm will normally tend to a very small change in $\mathbf{M}(n)$ than a vector with a minute norm. This normalization results in smaller step size outcome than conventional N^2LMS. This algorithm is usually faster than the *LMS* algorithm, since it uses a variable convergence factor directed for the reduction of the output error. Hence, eN^3LMS can be applied to very high-speed application such as biotelemetry. The step size specification μ, of this algorithm, is independent of input power. But this algorithm is in need of more computations to evaluate the normalization term $\|x(n)\|^2$. At the same time, eN^3LMS requires less priority information than *LMS*. Thus, the resulting mean square error of eN^3LMS is greater than that of N^2LMS.

Inspite of using data vector for normalization, norm of the error vector can be utilized. So this algorithm is known as exponential error normalized nonnegative LMS (e^2N^3LMS) algorithm. In this algorithm, the step size is inversely proportional to the norm of the error vector than the input data as in the eN^3LMS. This algorithm provides good developments in minimizing signal noise. The advantage of the e^2N^3LMS algorithm is that the step size can be chosen independent on input signal power and the number of weights. Hence, e^2N^3LMS has a steady-state error and convergence rate better than N^2LMS algorithm.

The weight update equation for this algorithm is written as follows:

$$\mathbf{M}(n + 1) = \mathbf{M}(n) + s_e(n)x(n)e(n)M^\gamma(n) \tag{6}$$

where $s(n)$ is a variable step size w.r.t input as follows:

$$s_e(n) = \frac{s}{c + e^t(n)e(n)} \tag{7}$$

To make them more suitable for healthcare applications and minimize the computational complexity of the above algorithms, we merge the eN^3LMS and e^2N^3LMS algorithms with the signum-based algorithms described by Boroujeny. These are three signum-based algorithms available. In these algorithms, the signum function (F) is applied to data vector or error or both data and error quantities.

His results sign regressor algorithm (SRA) when signum is applied to data only, sign error algorithm (SA) when signum is applied to error only, and sign sign algorithm (SSA) when the signum is applied to both data and error.

The kth coefficient in the sign of the data vector may be written as follows:

$$F\{x(n - k)\} = \frac{F(n - k)}{|F(n - k)|} \tag{7}$$

In normalized algorithms, the weight vector is normalized by instantaneous data or error. For example, in the case of SRA, the algorithm individually normalizes each coefficient of the weight vector, so this outcome is better than the other algorithms. Another feature of SRA is that its multiplications are independent of filter length.

2.1 Less Computational Complexity Algorithms for Remote Healthcare System

By combing eN^3LMS and signum algorithms, we can minimize the computational complexity of the eN^3LMS algorithm [11]. These variant algorithms are as follows.

(i) The hybrid version of eN^3LMS and SRA is exponential normalized nonnegative sign regressor LMS (eN^3SRLMS) algorithm. Its weight update equation is written as

$$\mathbf{M}(n+1) = \mathbf{M}(n) + s(n)F\{\mathbf{x}(n)\}e(n)M^{\gamma}(n) \qquad (8)$$

(ii) The combination of eN^3LMS and SA is exponential normalized nonnegative sign LMS (eN^3SLMS) algorithm. Its weight update equation is written as

$$\mathbf{M}(n+1) = \mathbf{M}(n) + s(n)\mathbf{x}(n)F\{e(n)\}M^{\gamma}(n) \qquad (9)$$

(iii) The combination of eN^3LMS and SSA is exponential normalized nonnegative sign LMS (eN^3SSLMS) algorithm. Its weight update equation is written as

$$\mathbf{M}(n+1) = \mathbf{M}(n) + s(n)F\{\mathbf{x}(n)\}F\{e(n)\}M^{\gamma}(n) \qquad (10)$$

In the three above equations, the variable step size $s(n)$ involves a data normalization.

3 Simulation Results

For the purpose of remote health monitoring systems, we have introduced a new model for cancelation of noises in the ICG signals. In our analysis, the filter length is chosen as 10, and the step size is considered as 0.1 (Fig. 2).

Removal of Baseline Wander Artifact from ICG Signals Using New AAEs
This experiment illustrates the cancelation of baseline artifact from ICG signal. For this, the raw ICG signal is given as input. A reference signal is constructed by DWT decomposition. This reference signal in effective is used as a reference to the AAE; as shown in diagram, this is considered as n_2. By taking feedback from the output $z(n)$, the algorithm trains n_2 such that it is closely correlated with the

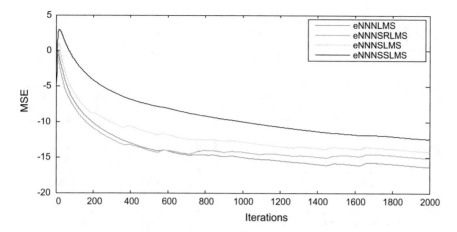

Fig. 2 Convergence characteristics of exponential normalized nonnegative LMS(eN^3LMS) and its variants in Gaussian environment

contaminated component n_1 in the input ICG signal $x(n)$. To validate the performance of the DWT-based AAE for BWA cancelation, we considered SNR, EMSE, and MSD as performance measures. Among the algorithms considered, EN^3SRLMS is better than EN^3LMS. This is due to the two normalization operations involved in the weight update recursion because of error normalization and data normalization in the signum function. The residual error component after filtering with various algorithms is shown in Fig. 3. From Fig. 3d, is it clear that the residual error in the case of EN^3SRLMS is less when compared to other algorithms. Finally, based on the experimental results, it can be concluded that EN^3SRLMS is better than the other algorithms, and hence it can be used as adaptive algorithm in a practical healthcare monitoring system for remote applications.

4 Conclusion

In this work for the purpose of remote health monitoring application, a new method was developed for enhancing the ICG signals. In this proposed method, the adaptive artifact canceller does not require a reference signal externally; it can construct the reference signal using DWT decomposition method. To update the filter weight coefficients in order to eliminate the noise, the constructed reference signal is used. Due to the negative weights in order to avoid the divergence at the period of abnormal heart conditions, we developed various nonnegative LMS algorithms. Due to the negative weights occurred during the period of abnormal heart conditions to avoid divergence, we have developed various nonnegative LMS algorithms: eN^3LMS, eN^3SRLMS, eN^3SELMS, and eN^3SSLMS algorithms. Out of all these algorithms, sign regressor-based algorithms need a less number of multiplications as well as achieving better

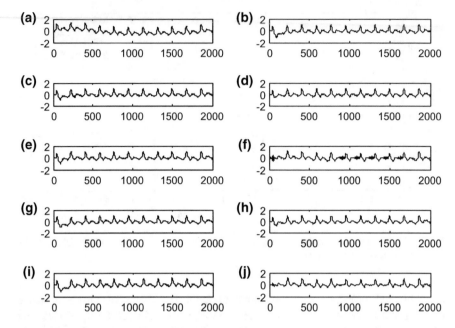

Fig. 3 Results of filtering BWA using various AACs

convergence. The sign regressor version is little bit inferior to its unsigned version regarding its feature of convergence; from above all the inferences and results, it is clear that eN^3SRLMS-based AAC exhibits well than the other algorithms.

Acknowledgements This research was supported by Lakireddy Bali Reddy College of Engineering, Mylavaram. So I would like to express my sincere thanks to our Management, Principal and Dean R&D, who provided insight and encouragement that greatly assisted this research work.

References

1. Heart Disease and Stroke Statistics, *American Heart Association* (2015)
2. World Health Organization Fact Sheets on Cardiovascular Diseases, *World Health Organization*, Fact sheet N317, January (2015)
3. H.H. Woltjer, H.J. Bogaard, P.M.J.M. de Vries, The technique of impedance cardiography. Eur. Heart J. **18**(9), 1396–1403 (1997)
4. J.H. Nagel, L.Y. Shyu, S.P. Reddy, B.E. Hurwitz, P.M. McCabe, N. Schneiderman, New signal processing techniques for improved precession of non invasive impedance cardiography. Ann. Biomed. Eng. **17**, 517–534 (1989)
5. X. Wang, H.H. Sun, J.M.V.D. Water, An advanced signal processing technique for impedance cardiography. IEEE Trans. Biomed. Eng. **42**(2), 224–230 (1995)
6. A. Scherhag, J.J. Kaden, E. Kentschke, T. Sueselbeck, M. Borggrefe, Comparison of impedance cardiography and thermodilution derived measurement of stroke volume and cardiac output at rest and during exercise testing. Cardiovasc. Drugs Ther. **19**, 141–147 (2005)

7. B.B. Sramek, Electrical bioimpedance. Med. Electron. **14**, 95–103 (1983)
8. M. Muzi, T.J. Ebert, F.E. Tristani, D.C. Jeutter, J.A. Barney, J.J. Smith, Determination of cardiac output using ensemble-averaged impedance cardiograms. J. Appl. Physiol. **58**, 200–205 (1985)
9. W.G. Kubicek, J.N. Karnegis, R.P. Patterson, D.A. Witsoe, R.H. Mattson, Development and evaluation of an impedance cardiography output system. Aerosp. Med. **37**, 1208–1212 (1966)
10. W.G. Kubicek, F.J. Kottke, M.U. Ramos, The Minnesota impedance cardiograph theory and applications. Biomed. Eng. **9**, 410–416 (1974)
11. Health in 2015: from millennium development goals to sustainable development goals, *World Health Organization* (2015)

Review on Big Data and Its Impact on Business Intelligence

C. S. Pavan Kumar and L. D. Dhinesh Babu

Abstract Over the last 10 years most of the organizations use Big Data to improve their standards with respect to quality and cost. Big Data is a broad and mosaic set of unstructured and structured data which sizes over exabytes ≈ 1016. A significant amount of digital data is created when the organizations convert their data from analog to digital. The data keeps on increasing and petabytes of information are generated every year, which leads to complexity in handling data. A major issue in Big Data is volume apart from the other six issues. There are many dynamic design challenges which lead to no comprehensive design strategy for Big Data. Many open sources and commercial data analysis tools are developed and are significant. Investments on Big Data have a steep hike year by year, which is a good sign in the perspective of business intelligence and decision-making capabilities of the organizations.

Keywords Big Data · Challenges · Design · Data processing engine
Big data market · BI · Revenue

1 Introduction to Big Data

Big Data as the name indicates deals with the humongous volume of data. Candidly talking, these chunks of data that are too thick to meet in an excel sheet. In vocational terms, handling large data using conventional database management systems is a bit tricky. Big Data needs special kind of tools to compute these massive quantities of evidence. We need even more computing ability to collect the full data and extra tools to study them. Here, the large blocks of documents are accumulated from many sources; therefore, the information is entirely bleak and in most instances it

C. S. Pavan Kumar (✉)
School of Computer Science and Engineering, Vellore Institute of Technology, Vellore, India
e-mail: pavan540.mic@gmail.com

L. D. Dhinesh Babu
School of Information and Technology, Vellore Institute of Technology, Vellore, India
e-mail: lddhineshbabu@gmail.com

© Springer Nature Singapore Pte Ltd. 2019
S. C. Satapathy et al. (eds.), *Information Systems Design and Intelligent Applications*,
Advances in Intelligent Systems and Computing 862,
https://doi.org/10.1007/978-981-13-3329-3_10

is amorphous. If we desire to use the unstructured data for an objective, it will not serve us. Data preprocessing has to be executed as an initial step and will require just about nonconventional methods to store the information equally well. Many of us have a small confession regarding data mining and Big Data. Data mining also deals with vast quantities of data and Big Data as well. In Big Data most of the data is unstructured. In general words, we can differentiate Big Data as an "Asset" and data mining as a "handler" of the property.

According to IBM Infographic view in 2010, the digital data in the world are around 2.7 zeta bytes. The roots of Big Data are categorized as enterprise data, sensor-generated (machine generated), and communicative data (social) according to Liu [1–3]. Enterprise data includes the data of customer relationships in companies and sensor generated that includes CCTV recordings, weblogs, and climate-related data from weather stations and online trading info. Communicative data includes data from social media such as Twitter, Facebook, Instagram, and many microblogging websites like FriendFeed, Tumblr.com, Flattr, and Meetme.

Thither are many definitions for Big Data. The most interesting one is too big, too fast, or too hard to process the data with already available resources. In general, "Too big" means an enormous amount of data to handle, i.e., systems must be able to manage the vast quantity of information which is being collected daily from many sources (sensors, CCTV cameras, Facebook, Twitter, mobile phones, and genetics).

"Too Fast" means the data (structured, unstructured, or semi-structured) has to be treated rapidly. Look at the position of reading an advertisement (ad) to a user on the web page, the processing of the data has to be very fast. "Too hard" refers to that the existing tools may not fit for the vast amount of data. There are many open-source technologies to solve the big data problem, but most of the service companies cannot afford much time to start learning and solve their problems using some tailored software. Some organizations are targeting those problems and making commercial enterprise by getting to some merchandise.

2   Characteristics of Big Data

From the conventional definition, we can hold just three features, namely, volume, variety, and velocity. As the research is moving forward, many new characteristics are summed to the Big Data Problem [4] which are addressed in Fig. 2.

2.1   Volume

Hundreds of petabytes information are uploaded to the Facebook every day. Close to 1.7 billion users are using Facebook monthly; Akamai is a content distribution network (CDN) which analyzes 75 million cases to target online ads. Walmart across the globe will handle over 2 million transactions per day. 90% of the entire data accumulated over the past 2 years. Volume is one of the major concerns of Big Data. As the data keeps on increasing, the tools for data storage cannot handle as large calculations are required.

2.2 Velocity

In 1999, Walmart data warehouse had over 1000 terabytes of information; by 2012, it was increased to two and a half petabytes (2,500,000 of gigabytes). Every day, thousands of gigabytes of information are uploaded to social media such as Facebook, Twitter, and YouTube. Every day, millions of people upload gigabytes of videos to YouTube, lakhs of tweets, and also send millions of emails. Here, velocity indicates the speed at which the data gets accumulated (store), processed, and analyzed by the relational databases. The processing of real-time data which facilitates the targeting of advertisements, according to the user will be possible. For a successful Big Data analytics, we have to get out of the box and use nontraditional techniques to achieve faster access.

2.3 Variety

Big Data involves many forms of data: structured data, unstructured, and semi-structured data. Aggregating everything and processing the data is one of the major challenges [5]. For example, a person uploaded pics during his visit to any one of the eight wonders they will have a strong correlation with tourism position and travel information. Processing of streaming data is required in such cases. Mark van Rijmenam proposed another four characteristics of Big Data [6].

2.4 Variability

Variability refers to the data where the meaning of the data continuously changes from time to time. The substance of the word will change per context wildly. An establishment is performing sentiment analysis on one of their products by considering tweets on that merchandise. An organization has to prepare programs to examine the real meaning of the context and change the actual substance of the words through it. It is challenging, but not inconceivable. IPL last season conducted fan battles through Twitter using a keyword to each team like "@IPL." They performed sentiment analysis on the tweets based on keyword and used to plot charts according to the region of the person from where he is tweeting. Figure 1 is taken from an official IPL website [4, 7, 8, 9].

Fig. 1 Fan battle made by IPL using sentiment analysis

2.5 Veracity

If the data is uncertain, considering the data and performing analysis on that data are worthless. This sort of data gives worse results in programs like computerized decision-making and training algorithms like unsupervised machine learning. For example, the position of a "data janitor" attempting to get some useful patterns in the customer purchase behavior to help market basket analysis has to invest exact dataset even before the starting time of the study.

2.6 Visualization

Data visualization means representing data in a pictorial or graphical format. It enables to take the decision by watching the analysis, pictorial representation so that they can find the hidden patterns or difficult concepts quickly. Organizations struggled to see issues like customer purchase pattern and fraud, where Big Data helped to master this challenge.

2.7 Value

The value of Big Data is enormous. It helps a lot for research in health care. Analyzing the data in motion helps the doctors to track their clients 24×7. Examining the streaming data and data in repositories (ancestors' data) helps doctors monitor the patients.

3 Importance of Big Data

The value of Big Data keeps on increasing, according to the context, either homogeneous volume of data or different type. The productivity of the firms is improved in the aspects of sales and getting better with the already manufactured products (Fig. 2).

Following are some areas where Big Data are used efficiently [10]:

- It helps data scientist to improve security and analyze patterns from existing logs for troubleshooting.
- Customer satisfaction organizations may explain their previous purchase behavior to provide them new customized services to make clients happy. Many telecom service providers use this scheme and offer customers' plans, agreeing to their function.
- Selling of products according to the user demand and satisfaction using social media so that the firm can modify its product to increase the customer base in a big plate.
- Identifying fraudulence data in online logs of any commercial.
- Assessment of hazard in financial management by analyzing the previous online transactions.

4 Challenges of Big Data

There are many emerging challenges on Big Data [5, 11, 12]; we are trying to discuss a few of them.

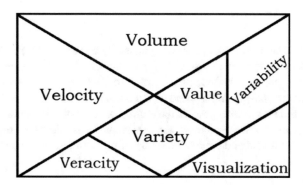

Fig. 2 7 V's of Big Data

4.1   Transport and Storage Issues

The data is being generated not just by professionals but also by many sources like security cameras, social media (Facebook, Twitter, and Instagram). Granting to the current technology one disk may carry 4 terabytes of data. If we want to store one exabyte of information, we need 25,000 of disks. To transfer the entire data from source to processing point it takes around 2800 h, which means moving of data takes much time than it takes for complete processing. To deflect this kind of issue we have to post the code near the information to process and send the result of the processing unit. Getting the data to the system—traditional approach of processing will not operate in the case of Big Data; we prefer to choose the alternative way of getting the code to data.

4.2   Management Issues

Data management is one of the major issues connected to Big Data. Presume that the large data is being scrambled and stored at various locations; to manage that data we need to observe lots of complex protocols. The data is maintained with accuracy and validity at the highest levels all the time. The major stumbling blocks are regarding the accessing of the data, maintaining metadata, and governance. They are different sorts of sources for this data which are varied by format, spacewise, and by the method of solicitation. Till today there is no perfect Big Data management tool/system.

4.3   Processing Issue

Let us speculate an example of processing an exabyte of raw data. For better under- standing, assume that the data distributed into chunks which consist of eight words each—which means one exabyte nearly equal to 1000 petabytes. Assume that the processor is about at 5-gigahertz speed which takes 20 ns for end-to-end processing, then the total time it needs to process the entire 1 k petabytes, it would take approx- imately 600 years. So, we need more active systems to handle an enormous amount of data to extract useful and actionable information from raw or unstructured data.

4.4   Design Challenges

When we are attempting to develop a system for Big Data, lot of subjects will come up in operation. Some of them are explained in the further sections.

4.4.1 Input and Output Processes

One of the major drawbacks in the process of Big Data design is output process [13]. Jacobs summarized that the process or tools which are used by traditional database systems are sufficient for Big Data storage also. Whereas tools used for processing transactions like add, update, and retrieve will work better for a small quantity of data. To extract huge volume of data and process the data within no time may not be satisfied with the existing traditional database management tools. In other words, we can say that storing the data is much easier than processing the data according to the request (query).

4.4.2 Quality over Quantity

Another future problem for Big Data users is quality versus quantity. Collecting data has been an addiction for some people. Some issues need better quality results than over amount of data. There are many concerns regarding this dilemma: what data to be decided irrelevant to the problem, is the data what we are considering to be consistent and error free.

4.4.3 Legal Issues

Many legal issues will arise while data scientists are trying to analyze the data or finding hidden patterns [14]. Legal issues for assigned problems are easily explained by Navetta [12]. Some sensitive information will be available in large chunks of data that are available for the data scientist, where personal feelings of an individual or an organization may get hurt where the data scientist may get penalized.

5 Data Processing Engines/Tools

Tools are used to collect and manipulate items related to the data to produce useful and essential information [15]. There are many data processing engines which are available in the market, some of them are open source, and some are commercial products by the respective organizations [15, 16]. In this section, we are going to discuss different data management tools [17] on HADOOP framework [18, 19]. Before going into the actualities of the processing tools, let us have a little glance on evaluation metrics of the tools: latency, throughput, fault tolerance, usability, and scalability. Latency is the period between the system initiation and first results. If the execution time is not a constraint, in that case, one can take a batch processing system. Throughput: the amount of task performed over a period, this is a parameter which can occupy its superior status regarding efficiency. Fault tolerance: stability of the system even after getting some error from any of its sub-components.

Table 1 Data processing engines

Engine	Current stable version (as of 19-04-2018)	Model—processing	Supported languages	Fault tolerance	Enterprise support
Spark	2.0.0	Batch and streaming	Java, Python, R, Scala	✓	✓
Flink	1.4.2	Batch and streaming	Java, Scala	✓	×
Storm	1.2.1	Streaming	Java, Clojure	✓	×
H_2O	3.18.0.7	Batch	Java, Python, R, Scala	✓	✓
MapReduce	3.1.0	Batch	Java	✓	×

Usability: the people who want to analyze the data will spend a bunch of time on tools, and the interface has to be comfortable to apply. Installation, configuration, which languages it support and how much complexity is needed to write the program will arrive into the delineation. Scalability: It can be accessed by stability of the system even if there is a sudden increase of load. All the engines that we are extending to demonstrate in this paper are scalable. Table 1 gives the brief introduction regarding the data processing engines or instruments which are trending like a shot.

5.1 Spark

The people at Algorithms, Machines, and People Lab developed Spark, which focusses on Big Data at University of California, Berkeley. Later on, the project was handed over to the Apache Software Foundation. Spark was developed by "Apache Foundation" from 2009. Spark provides programmers an interface which is a resilient distributed data set (RDD) which is a read-only data structure and distributed over a cluster of storage units. API is available in Java, Python, R, and Scala [19].

5.2 Flink

[20] Flink is an open-source framework for Big Data analytics just like Spark and HADOOP. Flink is mainly employed in data processing of streaming data. It is very easy to use in API of Scala and Java. Fault tolerance, consistency, high availability, high throughput, and low latency are the advantages of Flink. It allows both batch and streaming data processing. It does not bring in any storage mechanism on its own. The data has to be there in a distributed file system like HADOOP distributed file system (HDFS) or HBase.

5.3 Storm

BackType is a social media analytics company started developing Storm [21]. Later the project was taken over by Twitter and then it was declared as open-source software and handed over to the Apache Software Foundation. Strom was developed by the programmers using Clojure language. This application software or tool was developed using the design of directed acyclic graph (DAG). It got a chance in Apache's top projects list in September 2014.

5.4 H₂O

It is an open-source software (OSS) for Big Data analytics. It was initiated and trained by a start-up company called H_2O [22, 23] and written using Python, Java, and R. It uses divide and conquer technique which is almost suited for distributed and massively parallel tasks. There is an API for R, through which programmers can easily communicate with core software, which increases the usability of the tool which is a parameter for evaluation. By researchers of all the platforms (operating systems), H_2O is implemented. It does not handle streaming data, intended for only batch data processing.

5.5 MapReduce

It is an open-source framework which can utilize for processing parallelizable problems [19]. The concept is similar and can compare with message passing interface. The MapReduce program can be divided into map function and reduce function. The data can be processed only as a batch. These kinds of tools cannot help streaming data for processing. This framework can only be used with Java language. There are no APIs available except for Java.

6 Big Data in Market

When—"Niels Bohr," a Danish physicist reminds me that predicting the investment on Big Data is also difficult. In 2012, Barak Obama's Government of United States of America identified the potential of Big Data and announced a 200 M$ (million dollars) grant to different government bodies NSF, NIH, DOE, DOD, DARPA, and USGS [24] as an "Office of Science and Technology Policy."

Despite significant advantages of Big Data to the industry, there are some challenges as well which are as follows:

– Infrastructure does not support the Big Data problem,
– High capital investments,
– Great investments on nonstrategic actions,
– Restrictions due to vendor lock in, and
– Short of experts.

Equally, most of the data is semi-structured (Big Data). The traditional systems like customer relation management (CRM), enterprise resource planning (ERP), regular databases (SQL, NoSQL), and tools employed for data analytics may not be useful to handle Big Data analytics [25]. Collecting, refining, storing, and processing the Big Data may bring down the performance of the conventional computer networking mechanism, storage racks, and traditional database management systems. Even the efforts of making replication and scaling the existing conventional system also would not work for Big Data paradigm [26]. Suppose a team in an IT giant wants to work on Big Data, where they do not have any resources to work on Big Data, all the resources have to be arranged by the chief financial officer to start the new type of working style. Due to lack of wisdom, they cannot expect the need for resources (adding of resources at peak time) when there is fast growth in data. When more resources are added to the organization, they have to hire the candidates who have expertise in that area which leads to increase in the expenditure for nonstrategical aspects of the organization. Vendor lock in is another problem when the team wants to move from one strategy to another. The amount invested in resources would be in the soup if the resources will not fit for future purposes. Each and every organization will look over return over investment (ROI), and in that case purchasing some software or hardware to solve any one of the particular issues may not work. Another huge hurdle for technology organizations is to find the employees with the apt skill set for a particular problem. In a 2012 Gartner analysis [27] says that there were 4.4 million jobs available across the globe and only one-third employed, due to lack of skills required to carry out the responsibility of Big Data scientists. Even the companies have to be very dynamic in paying them, as there was a shortage of experts, they may leave the company due to huge offerings from the neighboring companies.

6.1 *Big Data Market Forecast 2012–2026 (Fig. 3)*

Many organizations spend lots of money in the area of Big Data; investment over Big Data started earlier that the government of the United States started funding for different organizations in 2012 itself. As the time keeps on going, the data is piling day by day. Companies do not have a choice except to invest massive amounts of money on Big Data. Major players in the technology are spending vast amounts of Big Data from the past 10 years. Statisticians are in the role to find out the impact of Big Data on stocks and economy of the organizations. Organizations like Statista,

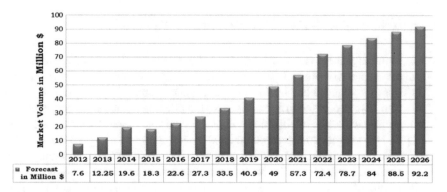

Fig. 3 Big Data market forecast, 2012–2026

Table 2 Impact of business intelligence	Social BI	Cloud BI	Mobile BI
2014	22.5	28	25
2015	23.8	30	25.5
2016	24.8	31	26
2017	25.8	32	26.8
2018	26.4	33	27.1

Mckinsey, Gartner, Brain, SAS, ids, and Forbes did many reviews on the economic aspects of Big Data (Table 2).

Statista collects opinions from business organizations, government institutions, and different individuals on various aspects and makes a report on over 1800 issues every year, which covers all the emerging and important aspects [28]. Investing on Big Data has become a major problem for every organization due to an increase in the volume of data in the process of converting from analog records to digital data. According to statistics, the investments on Big Data may touch 90 plus million dollars by 2016. Their forecast was shown in Graph 1 "Big Data Market Forecast." Even though the investments are increasing day by day, the return of investment on that achieved by only 2% of the organizations. Marr [29] explained why the investments on Big Data are not paying back the investors. Economist intelligence unit (EIU) ran a survey on 448 higher officials in the US agencies, where the majority of the analysts expressed their importance towards sales and business analytics [30]. The Big Data market is going to hit over 48.6 billion dollars in the year of 2019 [31]. Companies want to earn huge profits by investing more of Big Data even the ROI not achieved by the majority of the organizations [29, 32].

6.2 Influence of Big Data in Business Intelligence

Business intelligence is very crucial for kind of business model, analyzing the data is one of the biggest priorities. As the volume of the data keeps on increasing, the issue of business intelligence combined with the Big Data problem. Analyzing the Big Data is the first step in the firm intelligence; the effect of social intelligence, cloud intelligence, and mobile intelligence will play a vital role in the Big Data market as well [33, 34]. Gartner and Redwood capital investigated and analyzed the amount spent on business intelligence. Market analyzer's, investors, trade experts, and share brokers have a keen eye on the business intelligence of each and every company as the results of business intelligence show the real significance of that particular organization. As Big Data is one of the important aspects of BI, the amount of investment on BI, shown in Fig. 4.

Figure 3 indicates the amount spent on business intelligence on the critical aspects of technology and that indirectly implies on market of Big Data. To take any particular decision business intelligence, BI plays a vital role. To analyze, process, and predict Big Data takes a crucial role in BI. Big Data helps the business intelligence to make decision-driven decisions and to identify the unknown facts which indeed contributes to taking a step in the process of making a statement in profits aspects of the organization [35, 36].

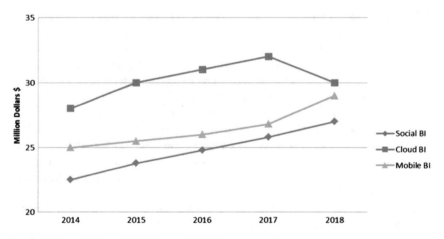

Fig. 4 Market of business intelligence, 2014–18

6.3 Alternative Aspects Which Play Vital Role in the Big Data Market

There are many alternative elements which correlate with Big Data market. Investments regarding networking, NoSQL, infrastructure software, SQL, cloud, storage, computation, apps and analytics, and professional services (salaries) will also come under the Big Data market. Analyzing with data centric may not work with Big Data; explaining has to be done using a query-centric manner with Big Data. To have a better query-centric data evaluation, the networking plays a vital role to maintain the reliability of network organizations focused on new contemporary networking techniques. Team to commence the work on Big Data, special hardware or software has to be purchased which even the infrastructure investments also may go high of all organizations which are signing for Big Data. Investments on NoSQL, cloud, apps and their analytics, and storage are escalating year after year. The expenditure on professional services is inevitable; the financial. Officers of the organizations are not able to pick the appropriate persons with the right skills to assign the data scientist role. Professionals who are said to be experts in the area of Big Data have no proof of their work. If an expert wants to leave the organization to full fill his accomplishments, no third party organizations are available to access the credibility of the candidate, which leaves the current organization in dilemma. A subtype forecast of alternative aspects which play a vital role in the Big Data market [37, 38] is shown in the figure forecast—aspects which imply Big Data. Figure 4 indicates the key factors which indicate the market of Big Data. Pictorial representation of Table 1 is shown in Fig. 5 (Table 3).

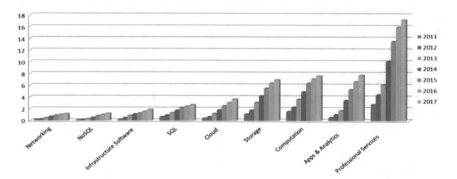

Fig. 5 Aspects which imply Big Data

Table 3 Alternative aspects which impact on Big Data market

Impact/Year	2011	2012	2013	2014	2015	2016	2017
Networking	0.15	0.23	0.42	0.65	0.85	1.01	1.15
NoSQL	0.07	0.13	0.29	0.5	0.8	1	1.2
Infrastructure Software	0.14	0.44	0.83	1.08	1.25	1.6	1.9
SQL	0.62	0.88	1.31	1.75	2.25	2.45	2.7
Cloud	0.36	0.62	1.19	1.82	2.52	3.05	3.65
Storage	1.1	1.75	3.09	4.2	5.5	6.4	6.95
Computation	1.53	2.29	3.65	4.92	6.4	7.1	7.6
Apps & Analytics	0.52	0.99	1.69	3.45	5.29	6.65	7.75
Professional Services	2.8	4.42	6.15	10.1	13.5	16	17.2

6.4   Vendor's Big Data Market and Their Revenue

Many organizations started investing in Big Data over the last decade, but the investments seen regarding ROI, i.e., revenue. Organizations spend a lot on Big Data (Fig. 6), but the financial officer alone knows the risk of investing the money without having the proper plan of action to have enough revenue. Giant organizations like Oracle have entered into Big Data and released a product which helps Big Data "One Fast, Secure SQL Query on All Your Data" [39] is the caption for Oracle Big Data. Keeping the revenue that is generated by Big Data products and the rankings were given by Wikibon [37]. SAP HANA [40] platform is one of the leading products to solve the business analytics problems to the customers. They made one of the biggest problems of data processing, bottleneck issues between the application, and the processor which made a huge change in the analytics world. IBM is offering many products on data collecting and warehousing. Palantir's products Gotham and Metropolis started to bring huge revenue to the organization. Many other teams gained significant income, and the data was collected and projected by Wikibon in one of its recent reports.

7   Conclusion

Big Data is an evergreen area in the field of research. Handling and maintaining the data is the biggest concern of many organizations. The name Big Data came into existence far back with three problems, as the research is escalating in the domain, many more characteristics came into the picture (7 V's). There is much scope for research in security, transport, management, and design of Big Data. Different data processing engines are available in the market in which we have to choose one which

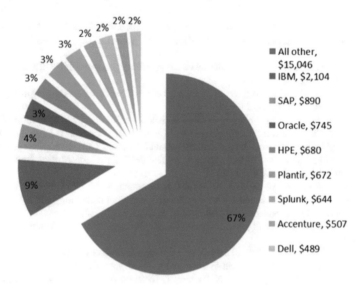

Fig. 6 Big Data vendor by revenue

suits to our need. Spark, Scala, and SAP's HANA are the most modern data processing engines which are attracting the entire world toward them. Financial stability is very crucial for each and every organization's existence. There is no suitable business model for investing in Big Data analytics, keeping the risk in mind the groups were spending lots of money in the name of unknown revenue generator—Big Data. Our future work is to design a business model which brings much income, with which the organizations can reach new heights in the market.

References

1. Z. Liu, P. Yang, L. Zhang, A sketch of big data technologies, in *Seventh International Conference on Internet Computing for Engineering and Science*, pp. 26–29 (2013)
2. A. Navint Partners White Paper, *Why is Big Data Important?* Online: May 2012, http://www.navint.com/images/Big.Data.pdf. Accessed 31 Aug 2016
3. J. Woodard, Big data, and ag-analytics: an open source, open data platform for agricultural & environmental finance, insurance, and risk. Agric. Fin. Rev. **76**(1), 15–26 (2016)
4. Rob Livingstone Advisory, 7 Vs of Big Data Online: http://mbitm.uts.edu.au/feed/7-vs-big-data. Accessed 31 Aug 2016
5. S. Kaiser, F. Armour, J.A. Espinosa, W. Money, Big data: issues and challenges moving forward, in *46th Hawaii International Conference on System Sciences*, pp. 995–1004 (2013)
6. M. van Rijmenam, *Think Bigger: Developing a Successful Big Data Strategy for Your Business*, 1st edn. (2013)
7. P. O'Donovan*, K. Leahy, K. Bruton, D.T.J. O'Sullivan, An industrial big data pipeline for data-driven analytics maintenance applications in large-scale smart manufacturing facilities. J. Big Data. pp. 2–25 (2015)

8. R. Wegener, V. Sinha, *The Value of Big Data: How Analytics Differentiates Winners*. Online: http://www.bain.com/Images/BAIN%20_BRIEF_The_value_of_Big_Data.pdf. Accessed 31 Aug 2016

9. http://bridgei2i.com/blog/fan-map-faux-pax-indian-premier-league-ipl/,IPLT20.com. Accessed 11 Aug 2016

10. E.G. Uluru, F.C. Puican, A. Apostu, M. Velicanu, Perspectives on big data and big data analytics. Database Syst. J. **III**(4), 3–13 (2012)

11. Avita Katal, Mohammad Wazid, R H Goudar, Big data: issues, challenges, tools and good practices, in *Sixth International Conference on Contemporary Computing*, pp. 404–409 (2013)

12. Legal Implications of Big Data A Primer—David Navetta, Online: https://c.ymcdn.com/sites/www.issa.org/resource/resmgr/journalpdfs/feature0313.pdf. Information Systems Security Association, Accessed 4 Aug 2016

13. A. Jacobs, Pathologies of big data. ACM Queues Commun. ACM **52**(8), 36–44 (2009)

14. C. Park, T. Wang, Big data and NSA surveillance—survey of technology and legal issues, in *IEEE International Symposium on Multimedia*, pp. 516–517 (2013)

15. C. French, *Data Processing and Information Technology*, 10th edn. Thomson. p. ISBN 1844801004

16. A. Marinheiro, J. Bernardino, Analysis of open source business intelligence suites, in *8th Iberian Conference on Information Systems and Technologies (CISTI)*, pp 1–8 (2013)

17. S. Landset, T.M. Khoshgoftaar, A.N. Richter, T. Hasanin, A survey of open source tools for machine learning with Big Data in the HADOOP ecosystem. J. Big Data, pp. 2–24 (2015)

18. T. White, *HADOOP The Definitive Guide*, 2nd edn. (O'Reilly Media Inc, United States, 2010)

19. A.B. Patel, M. Birla, U. Nair, Addressing big data problem using HADOOP and map reduce, in *2012 Nirma University International Conference on Engineering, NUiCONE-2012*, pp. 8–17 06–08 December

20. M.A. Alsheikh, D. Niyato, S. Lin, H.-P. Tan, Z. Han, *Mobile Big Data Analytics Using Deep Learning and Apache Spark*, IEEE Network, pp. 1–8 (2016)

21. A. Katsifodimos, S. Schelter, A. Flink, Stream analytics at scale, in *2016 IEEE International Conference on Cloud Engineering Workshop*, pp. 1–5

22. Z. Chena, N. Chena, J. Gong, Agro-Geoinformatics (Agro-geoinformatics), Design and implementation of the real-time GIS data model and sensor web service platform for environmental big data management with the apache storm, in *Fourth International Conference*, pp. 1–4 (2015)

23. http://www.h2o.ai/. Accessed 19 Mar 2018

24. Public announcement by Barak Obama, Online: https://www.whitehouse.gov/sites/default/files/microsites/op/big_data_press_release_final_2.pdf. Accessed: 22 Aug 2016

25. Anand Bhadouria Sr. *IT Strategist, Enterprise Cloud Solutions*. Online: https://support.rackspace.com/white-paper/turning-big-data-into-big-dollars/. Accessed 19 Mar 2018

26. T. Davenport, P. Barth, R. Bean, *How 'Big Data' is Different, MIT Sloan Management review*, 30 July 2012, Online: http://sloanreview.mit.edu/article/how-big-data-is-different/. Accessed 5 Aug 2016

27. Gartner Reveals Top Predictions for IT Organizations and Users for 2013 and beyond, press release, Gartner, 24 October 2012, Online: http://www.gartner.com/newsroom/id/2211115. Accessed 13 Mar 2018

28. Forecast of Big Data market size, based on revenue, from 2011 to 2026 (in billion U.S. dollars) Online: http://www.statista.com/statistics/254266/global-big-data-market-forecast/. Accessed: 19 Mar 2018

29. B. Marr, *Why Investments in Big Data and Analytics Are Not Yet Paying Off*, Online: http://www.forbes.com/sites/bernardmarr/2016/06/27/why-investments-in-big-data-and-analytics-are-not-yet-paying-off/#de9b60780a25. Accessed 19 Aug 2016

30. EIU- Economic Intelligence Unit- Online Http://www.zsassociates.com//media/files/publications/public/broken-links-why-analytics-investments-have-yet-to-pay-off.pdf?la=en. Accessed 19 Aug 2016

31. FRAMINGHAM, *New IDC Forecast Sees Worldwide Big Data Technology and Services Market Growing to $48.6 Billion in 2019, Driven by Wide Adoption Across Industries, Mass*, 9 Nov 2015, http://www.idc.com/getdoc.jsp?containerId=prUS40560115. Accessed 19 Mar 2018
32. S.F. Wamba, A. Gunasekaran, S. Akter, S.J.F. Ren, R. Dubey, S.J. Childe, Big data analytics and firm performance: effects of dynamic capabilities. J. Bus. Res. **70**, 356–365 (2017)
33. D. Opresnik, M. Taisch, The value of big data in servitization. **165**, 1 July 2015, Article number 5966, pp. 174–184
34. G.C. Nobre, E. Tavares, Scientific literature analysis on big data and internet of things applications on circular economy: a bibliometric study. Scientometrics **111**(1), 463–492 (2017)
35. A. Raj Purohit, Big data for business managers—bridging the gap between potential and value, in *2013 IEEE International Conference on Big Data*, pp. 1–3
36. U. Sivarajah, M.M. Kamal, Z. Irani, V. Weerakkody, Critical analysis of big data challenges and analytical methods. J. Bus. Res. **70**, 263–286 (2017)
37. J. Kelly, *Big Data Vendor Revenue and Market Forecast 2013–2017* (2014), http://wikibon.org/wiki/v/Big_Data_Vendor_Revenue_and_Market_Forecast_2013-2017. Accessed 31 Aug 2016
38. S. Ji-fan Ren, S. Fosso Wamba, S. Akter, R. Dubey, S.J. Childe, Modelling quality dynamics, business value and firm performance in a big data analytics environment. Int. J. Prod. Res. **55**(17), 5011–5026 (2017)
39. https://www.oracle.com/database/big-data-sql/index.html. Accessed 19 Mar 2018
40. https://hana.sap.com/abouthana.html. Accessed 19 Mar 2018

Big Data Analysis for Anomaly Detection in Telecommunication Using Clustering Techniques

C. Gunavathi, R. M. Swarna Priya and S. L. Aarthy

Abstract The recent development with respect to Information and Communication Technology (ICT) has a very high impact on the social well-being, economic-growth as well as national security. The ICT includes all the recent technologies like computers, mobile-devices and networks. This also includes few people who have the intent to attack maliciously and they are generally called as network intruders, cyber-criminals, etc. Confronting these detrimental cyber activities has become the highest priority internationally and hence the focused research area. For this kind of confront, anomaly detection plays a major role. This is an important task in data analysis which helps in detecting these kinds of intrusions. It helps in identifying the abnormal patterns in various domains like finance, computer networks, human behaviour, gene expression etc. This paper focuses on detecting the abnormalities in the telecommunication domain using the Call Detail Records (CDR). The abnormalities are identified using the clustering techniques namely k-means clustering, hierarchical clustering and PAM clustering. The results obtained are discussed and the clustering technique which is suited better in identifying the anomaly accurately is suggested.

Keywords Big data analytics · Anomaly detection · Clustering
User behaviour analysis

C. Gunavathi (✉) · R. M. Swarna Priya (✉) · S. L. Aarthy
School of Information Technology and Engineering,
Vellore Institute of Technology, Vellore, Tamil Nadu, India
e-mail: gunavathi.cm@vit.ac.in

R. M. Swarna Priya
e-mail: swarnapriya.rm@vit.ac.in

S. L. Aarthy
e-mail: aarthy.sl@vit.ac.in

© Springer Nature Singapore Pte Ltd. 2019
S. C. Satapathy et al. (eds.), *Information Systems Design and Intelligent Applications*,
Advances in Intelligent Systems and Computing 862,
https://doi.org/10.1007/978-981-13-3329-3_11

1 Introduction

Though there is high development in technology and computer usage in everyday life, this has further given rise to security threats like zero-day vulnerabilities, mobile threats, etc. Though so many Internet protocols were designed, they were not designed giving priority to security issues. Hence, it is the job of the network administrators to take care of the various intrusions and handle them with care [1].

Among the various tasks In Data Analysis, anomaly detection is the most important task which detects the unusual patterns or behaviour or anomalous data in a given data set. Anomalies are considered to the important task by most of the researchers because the rare events can be identified and critical actions can be prompted in various applications like unusual traffic pattern in a network could say that the computer has been hacked by unauthorized users, unusual or anomalous behaviour in transaction of credit card usage can indicate that the credit card is lost, etc. [2]. Thus anomaly detection could be used in varieties of applications including medical and public health, fraud detection, intrusion detection, industrial damage, image processing, sensor networks, robots behaviour, astronomy, etc.

Massive amount of data are being produced daily by the people as a result of their interaction all over the world. Due to the usage of smart devices, the interaction between people as well as machine-to-machine has increased to a large extent which has resulted in Big Data. The data which is generated in one month amounts to approximately 24.3 Exabyte [2]. For the sake of managing this unconventionally generated data, Big data Analytics is required.

Big data analytics is a term which has various methods and technologies for collection, managing and analyzing a very high amount of structured as well as unstructured data in real-time environment. Big data analytics analyses the entire data whereas sample data is taken in traditional data analytics. In traditional data analytics, a portion of sample data which forms the representative of the entire data is taken and analyzed. The results obtained from the analysis of the sample data gives inaccurate and incomplete information and hence the decisions taken are suboptimal. Especially telecommunication domain requires a real network analysis for providing prompt and quick solution. Hence this kind of solution could be provided only when the whole/Big data is analyzed.

In case of the mobile wireless network, a lot of data and parameters are exchanged continuously from/at the user end (UE) towards the core-network (CN) of long-term evolution advanced (LTE-A) network. For example, Call Detail Record (CDR), reference signal received power, radio link failure report, location information etc., are exchanged.

In this paper, we use the CDR information which is collected from CN of a real mobile cellular network. By analyzing CDR, we will get to know the specific activity of a user in a specific region at a particular date and time. When there is an unusual or abnormal behaviour in a particular user or when there is a change in the activity pattern of the user in the region, we consider it as anomaly. When the anomalous behaviour is detected and treated properly, various benefits like identification of chokes in bandwidth, identification of Region of Interest (ROI) for pro-active allotment of resources etc., could be achieved.

Our contributions towards anomaly detection in telecommunication domain are as follows:

(i)   We try to detect the anomalous behaviour of the users in a region using clustering techniques like k-means clustering, Hierarchical clustering and PAM clustering
(ii) We verify the results obtained with the ground truth values
(iii) We compare the results of all the three clustering techniques and give a suggestion on which clustering technique is well-suited.

The rest of the paper is organized as follows. Section 2 discusses the various related works that have been done by researchers in different timestamps. Section 3 highlights the preliminaries required for understanding the research focus. Section 4 discusses about the system model used and the data pre-processing strategy followed. Section 5 elaborates the clustering methodologies used in the work, and finally the paper is concluded.

2   Related Work

The previous research was based on detecting the anomalies using anyone of the supervised, semi-supervised and unsupervised learning methodologies. A framework was designed by Naboulsi et al. [3] in which the entire CDR data was categorized into call types and then they were analyzed for classifying the usage. This work also helped in identifying the unexpected traffic and that is considered to be anomaly. Another research Article [4] tried to identify the anomaly for the large sized data using the nearest neighbour, clustering and statistical methodologies. He concluded that the clustering methodology was best suited for real-time data compared to the nearest neighbour methodology.

Other authors [5–7] tried to use the k-means clustering methodology for the purpose of detecting anomaly in industrial parks, commercial and office areas, nightlife areas, etc. These authors were able to detect the anomaly using k-means clustering and they also achieved better results. Few authors [8, 9] also used k-means clustering methodology for detecting the anomaly with respect to volume using the traffic data.

Cici et al. [10] used the agglomerative hierarchical clustering methodology for performing segmentation process which helped in detecting the anomaly with the Pearson correlation as metric. In [11], the author used a rule-based approach and detected the anomaly in CDR data. Later few authors [12–15] improved the approach proposed by [11] with some security functions.

Though these above-discussed works helped in detecting the anomalies better, they were not used in real time because of the dynamicity of the network in the world and the behaviour of the user. To fulfil this need, the researchers tried to provide some solutions with less time complexity and more dynamicity. Having this as the requirement, few researchers [16–18] tried to work with the detection using the vehicular network data. Similarly, in recent years researchers focus on various data analytics schemes for improving the performance of available methodologies [19].

3 Preliminary Discussion

3.1 Types of Anomalies

Anomalies are usually referred to as the patterns of data which do not stand at par with the well-defined characteristics of normal patterns. Anomalies can be broadly classified as follows [2].

Point Anomaly

If the unusual pattern or behaviour is because of a particular instance of a data in the entire dataset, then this is considered to be point anomaly. For example, when a person's usual normal behaviour is that he fills 5 l of petrol daily and in a random day, he has filled 50 l of petrol, it is considered to be point anomaly.

Contextual Anomaly

If a data instance behaves unusually in a particular context, then it is considered to be contextual anomaly. For example, a user spending high during festive season using his credit card is usual pattern whereas if his credit card is being used for high expenditure during a non-festive season, might be considered to be anomaly.

Collective Anomaly

If the unusual behaviour is with respect to a group of similar data instances in a data set, then the group of data instances are considered to be under collective anomaly. For example, when ECG is taken for a particular person, only when there is continuous low values in a time period, it is considered to be anomalous behaviour of the heart otherwise not.

3.2 Clustering

Clustering is a kind of unsupervised learning algorithm. The major advantage of clustering is that it does not require a well-defined pre-labelled data for extracting rules to group similar data items or instances [20]. The clustering techniques are broadly classified as follows:

a. Regular clustering
   Regular clustering is one in which the clusters are formed only with the rows available in the data set. For example, k-means.
b. Co-clustering
   Co-clustering is a technique in which clusters are formed using both rows and columns simultaneously in the data set. For example, EM clustering.

4 System Model and Data Set Pre-processing

The model which is used here consists of the LTE-A cellular network. The architecture of the LTE-A cellular network is shown in Fig. 1.

The model consists of three layers namely User end (UE), access network (E-UTRAN) and CN. The CDR data information is usually collected from the CN layer of the network model. This data is utilized for understanding the activity pattern of the user and identify the anomalous behaviour.

4.1 Description of the Data Set

The CDR Dataset that is used in this paper is obtained from the real network Telecom Italia [21]. The dataset gives the information about the telecommunication activity that occurs in the city of Milan. The entire city Milan is subdivided into equal sized square grids of about 100×100. The length of each grid is about 0.235 km and covers an area of 0.055 km^2. The dataset has details about the various activities performed by the user in the region which includes call-in, call-out details from or to the grids at every 10 min time interval as shown in Table 1. Before pre-processing, the CDR Dataset had the following details in it:

(i) Square ID: This specifies the region ID in the city of Milan
(ii) Province: This specifies the name of province within the grid 100×100

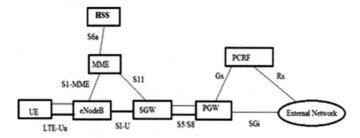

Fig. 1 LTE-network architecture

Table 1 Data set before pre-processing

Square ID	Province	Time stamp	Call-in	Call-out
1	Agrigento	1383321600000	0.052274849	
1	Agrigento	1383322200000	0.026137424	
1	Alessandria	1383291000000		0.00178731
1	Alessandria	1383306000000		0.00178731
1	Alessandria	1383320400000	0.00178731	

(iii) Time Stamp: The data of the activity logs were recorded in 10 min time interval
(iv) Call-in: This field specifies the duration for which the user had received calls in the time interval of 10 min.
 (v) Call-out: This field indicates the duration for which there were outgoing calls in the region within the 10 min time stamp.

4.2 Pre-processing of Dataset

The dataset obtained from the Telecom service provider was in raw form which has to be pre-processed for the purpose of analyzing. The pre-processing is a process by which the data is cleaned by filling in the missing values, merging few fields or removing certain unwanted information as per the requirements. The dataset had the activity logs for every 10 min interval separately for call-in and call-out. There were so many missing (or) blank values in the instances. We first filled in the blank fields with zeroes since the blank values specify that the user was inactive during that time stamp. We summed up the call-in and call-out details to form a new data field as "Activity". This gave the activity as a whole for the users in the region for a time stamp of 10 min intervals.

We further summed up the activities to calculate the log details for 1 h time interval. We retained the square ID, Province, Call-in, Call-out and Activity fields in the dataset. The processed dataset that is further used for anomaly detection is shown in Table 2. In the next section, we discuss about the various clustering schemes that are used to detect the anomalous behaviour of the user in a province within a grid.

Table 2 Data set after pre-processing

Square ID	Call-out	Call-in	Activity
1	0.080199583	0.030875085	0.111074668
1	0.05460093	0.130687121	0.185288051
1	0.055225199	0.079575313	0.134800513
1	0.05638824	0.054062159	0.110450398

5   Clustering Algorithms for Anomaly Detection

In the telecommunication network, the anomalies are those behaviours of the user in the network that are different or unusual from their usual or expected actions. In the forthcoming sections, we discuss about three clustering algorithms and how they are used to detect anomalies.

5.1   K-Means Clustering

K-means clustering is a kind of unsupervised methodology which is well-suited for grouping similar data in a large dataset. This technique comes under the partitioning method where the entire dataset is subdivided into "n" number of clusters. The number of clusters is always a fixed number which is usually decided before analyzing the dataset. Hence an optimal "k" value has to be decided which is calculated using the gap-statistics in our work. After the optimal "k" is decided the dataset is categorized into set of "k" clusters which are similar data instances. The k-means algorithm follows the following steps:

(1)   A random "k" values are decided. These values represent the centroid values initially.
(2)   Each and every data instance is compared to the centroid with respect to a distance metric either Euclidean or Manhattan. The instance which is very close to the centroid is considered to be a part of the cluster.
(3)   The new centroid is calculated for the new cluster formed as a result of step (2) by calculating the mean value.
(4)   The steps (2) and (4) are repeated till the centroids do not change or vary.

At the end of clustering, there will be some data instances which do not become a part of any cluster and these are considered to be anomalies.

5.2   Hierarchical Clustering

This is also a kind of partitioning method similar to that of k-means. Here, the clusters are formed by hierarchy either using bottom-up also called as Agglomerative approach or using top-down called Divisive approach. In case of Agglomerative approach, each and every data item is considered to be a cluster and when there is a nearby data item is similar, then that instance is grouped along with it, whereas in the case of Divisive approach the entire dataset is considered to be one cluster and the set is further divided into partitions to form smaller clusters. The agglomerative approach is widely used when compared to divisive approach. The algorithm works as per the following steps:

I   If there are "n" numbers of objects in the dataset then each object is considered to be one cluster. Hence "n" clusters will be formed.

II   The distance between the objects are calculated and those objects which are closest to each other are merged together to form one cluster.

III The process is repeated till there is no way to merge the clusters.

5.3   PAM Clustering

PAM stands for "Partition Around Medoids". The algorithm finds a sequence of objects which are called as "medoids". These medoids are located at the centre of the cluster. The data instances of the data set which are initially considered to be medoids are grouped in a set "S". If we consider that "b" is the set of objects which are selected as medoids, then $u = S - b$ is considered to be the set of objects that are not selected. There are two phases in the algorithm.

1. The first phase is called as BUILD where the set S is fixed.

2. The SWAP is the second phase, in which the selected objects are tried to be replaced with unselected objects to increase the quality of cluster formation.

The next section discusses about the experimental results obtained when the big data is clustered using these three techniques.

6   Experimental Results

The pre-processed entire data set is analyzed initially to find the optimal number of clusters using the gap-stat method for each and every clustering methodology. Using this statistic, the optimal clusters obtained for k-means clustering was 6, for hierarchical clustering was 3 and for PAM was also 3. This optimal value varies from one data set to the other data set.

Based on the optimal value obtained for k-means clustering, the entire data set was formed into six clusters as shown in Fig. 2. When the result obtained was compared with the ground truth value, it was found that these gas unusual patterns.

Similarly, clusters were formed using the hierarchical clustering and PAM clustering technique. The results obtained are shown in Fig. 3. When compared to the h-clustering, PAM clustering shows the anomalies which have unusual patterns.

To calculate and analyze the performance of the clustering schemes over the entire data set, the measure named as Average Silhouette Width is calculated. The results achieved by k-means clustering, h-clustering and PAM clustering are shown in Fig. 4.

As per the results obtained, the average silhouette width for every clustering scheme is tabulated in Table 3. As per the metric, the h-cluster clearly provides the outlier information. Hence, h-cluster is best suited for anomaly detection of the telecommunication CDR dataset.

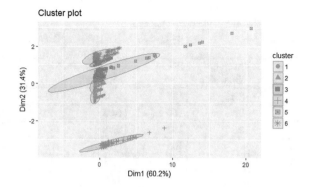

Fig. 2 Cluster formation using k-means clustering

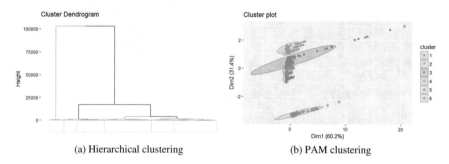

(a) Hierarchical clustering   (b) PAM clustering

Fig. 3 Cluster formation using hierarchical and PAM clustering

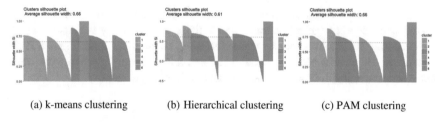

(a) k-means clustering   (b) Hierarchical clustering   (c) PAM clustering

Fig. 4 Average Silhouette width calculation

When the clusters are formed as per the optimal value obtained for h-cluster and PAM cluster, the results obtained still showed data instances which were not part of any cluster as shown in Fig. 5. But the average silhouette width obtained was 0.87 which is higher than the one obtained in six clusters. Though the width is higher, the outliers are not clearly visualized here.

Hence as per the results obtained, h-cluster is best suited for detecting anomaly in the telecommunication domain.

Table 3 Average Silhouette width for clustering

Cluster number	K-means		h-cluster		PAM	
	Size	Average Sil. width	Size	Average Sil. width	Size	Average Sil. width
1	173	0.62	127	0.67	127	0.67
2	173	0.52	60	0.85	60	0.85
3	60	0.85	195	0.48	173	0.62
4	68	1	105	0.72	173	0.51
5	173	0.62	219	0.45	173	0.62
6	127	0.67	68	1.00	68	1.00

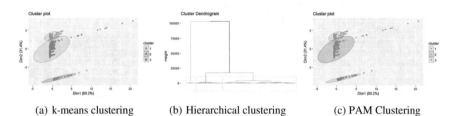

(a) k-means clustering   (b) Hierarchical clustering   (c) PAM Clustering

Fig. 5 Cluster formation with three clusters

7 Conclusion

The paper focused on identifying the anomalous behaviour of user in a particular region with respect to telecommunication activity. The anomaly is detected using the clustering schemes namely k-means clustering, hierarchical clustering and PAM clustering. The results obtained show that hierarchical clustering scheme performs better in identifying or detecting the anomaly in the CDR dataset.

References

1. E.E. Papalexakis, A. Beutel, P. Steenkiste, Network anomaly detection using co-clustering, in *IEEE International Conference on Advances in Social Networks Analysis and Mining*, pp. 403–410. Washington, DC, USA (2012)
2. M. Ahmed, A. Mahmood, Clustering based semantic data summarization technique: a new approach, in *9th IEEE Conference on Industrial Electronics and Applications*, pp. 1780–1785. Hangzhou, China (2014)
3. D. Naboulsi, R. Stanica, M. Fiore, Classifying call profiles in largescale mobile traffic datasets, in *IEEE Conference on Computer Communications*, pp. 1806–1814. Toronto, ON, Canada (2014)
4. M. Ahmed, A. Anwar, A.N. Mahmood, Z. Shah, M.J. Maher, An investigation of performance analysis of anomaly detection techniques for big data in SCADA systems. EAI Endorsed Trans. Ind. Netw. Intell. Syst. **15**(3), 1–16 (2015)

5. V. Soto, E. Frías-Martínez, Automated land use identification using cell-phone records, in *3rd ACM International Workshop on MobiArch*, pp. 17–22. (2011)
6. M. Amer, *Comparison of Unsupervised Anomaly Detection Techniques*. Bachelor Thesis (2011)
7. A. Zoha, A. Saeed, A. Imran, M.A. Imran, A. Abu-Dayya, A SON solution for sleeping cell detection using low-dimensional embedding of MDT measurements, in *IEEE 25th Annual International Symposium on Personal, Indoor, and Mobile Radio Communication*, pp. 1626–1630 (2014)
8. G. Münz, S. Li, G. Carle, Traffic anomaly detection using k-means clustering, in *GI/ITG Workshop MMBnet*, pp. 13–14 (2007)
9. M.F. Lima, B.B. Zarpelao, L.D. Sampaio, J.J. Rodrigues, T. Abrao, M.L. Proença, Anomaly detection using baseline and k-means clustering, in *IEEE International Conference on Software, Telecommunications and Computer Networks*, pp. 305–309 (2010)
10. B. Cici, M. Gjoka, A. Markopoulou, C.T. Butts, On the decomposition of cell phone activity patterns and their connection with urban ecology, in *16th ACM International Symposium on Mobile Ad Hoc Networking and Computing*, pp. 317–326 (2015)
11. I.A. Karatepe, E. Zeydan, Anomaly detection in cellular network data using big data analytics, in *20th European Wireless Conference*, pp. 1–5 (2014)
12. Y. Sun, H. Song, A.J. Jara, R. Bie, Internet of things and big data analytics for smart and connected communities. IEEE Access **4**, 766–773 (2016)
13. D.B. Rawat, S.R. Reddy, Software defined networking architecture, security and energy efficiency: a survey. Environment **3**(5), 325–346 (2017)
14. R.K. Sharma, D.B. Rawat, Advances on security threats and countermeasures for cognitive radio networks: a survey. IEEE Commun. Surv. Tutorials **17**(2), 1023–1043 (2015)
15. X. Xiong, D. Jiang, Y. Wu, L. He, H. Song, Z. Lv, Empirical analysis and modeling of the activity dilemmas in big social networks. IEEE Access **5**, 967–974 (2017)
16. N. Cordeschi, D. Amendola, M. Shojafar, E. Baccarelli, Distributed and adaptive resource management in cloud-assisted cognitive radio vehicular networks with hard reliability guarantees. Veh. Commun. **2**(1), 1–12 (2015)
17. M. Shojafar, N. Cordeschi, E. Baccarelli, Energy-efficient adaptive resource management for real-time vehicular cloud services. IEEE Trans. Cloud Comput. **99**, 1–14 (2016)
18. W. Li, H. Song, ART: an attack-resistant trust management scheme for securing vehicular ad-hoc networks. IEEE Trans. Intell. Transp. Syst. **17**(4), 960–969 (2016)
19. H. Song, M. Brandt-Pearce, Range of influence and impact of physical impairments in long-haul DWDM systems. J. Lightwave Technol. **31**(6), 846–854 (2013)
20. A.K. Jain, M.N. Murty, P.J. Flynn, Data clustering: a review. ACM Comput. Surv. **31**(3), 264–323 (1999)
21. CDR dataset information, https://dandelion.eu

Accurate Facial Ethnicity Classification Using Artificial Neural Networks Trained with Galactic Swarm Optimization Algorithm

Chhandak Bagchi, D. Geraldine Bessie Amali and M. Dinakaran

Abstract Facial images convey important demographic information such as ethnicity and gender. In this paper, machine learning approach is taken to solve the ethnicity classification problem. Artificial neural networks trained by state of the art optimization algorithms are used to classify faces as Caucasian or non-Caucasian based on the color of the skin. A feedforward neural network is trained using Galactic Swarm Optimization (GSO) algorithm which gives superior performance to other training algorithms such as backpropagation and Particle Swarm Optimization (PSO) which have been used earlier. In this paper, the RGB values of the skin are taken as inputs to the neural network. Each pixel of the image will be classified according to their RGB values and the class having the maximum number of pixels will be the output. Simulation results indicate that the neural network trained with GSO gives a more accurate classification and converges faster than the other state of the art optimization algorithms.

Keywords Artificial neural network · Galactic swarm optimization
Backpropagation · Image segmentation · Particle swarm optimization

1 Introduction

Humans can easily judge a person's ethnicity from their face. The face of a human being is a very important feature to understand a person's ethnicity, age, and gender [1]. In this paper, we will be exploring how a computer can identify a person's ethnicity from their face. According to psychology, whenever a human being encounters

---

C. Bagchi · D. Geraldine Bessie Amali (✉)
School of Computer Science and Engineering, Vellore Institute of Technology, Vellore, India
e-mail: geraldine.amali@vit.ac.in

M. Dinakaran
School of Information Technology, Vellore Institute of Technology, Vellore, India

© Springer Nature Singapore Pte Ltd. 2019
S. C. Satapathy et al. (eds.), *Information Systems Design and Intelligent Applications*,
Advances in Intelligent Systems and Computing 862,
https://doi.org/10.1007/978-981-13-3329-3_12

123

a stranger, the first things they try to evaluate is their race, age, and gender. These have consequential effects on the perceiver and the perceived [1–4].

Analysis of facial images has been an important problem in the field of image processing and computer vision [5]. But one should not only be concerned with recognizing a face but also try to understand the different type of demographic information they provide such as race, gender, and age. Among these attributes, ethnicity is one which remains constant throughout a person's lifetime and it also helps facial recognition systems. Hence, automatic classification of ethnicities has been a research problem of long-standing interest. The machine learning approach used in this paper will be greatly useful in face detection algorithms. Different databases can be maintained for each ethnicity and then perform a search on the database which has the same ethnicity as the query image. This will greatly increase the search speed. Previously, this task has been done using convolutional neural nets [6, 7], image filters [8, 9], and other ensemble machine learning algorithms [10–12]. Oriented Gradient Maps (OGMs) were used to analyze texture variations in the face and also help highlight local geometry in [13]. A Linear Discriminant Analysis (LDA) [14] based algorithm has also been developed for classification between Asians and non-Asians. This method was effective but slow.

In this paper, a single-layered neural network is used for the classification task where we classify each face as Caucasian or non-Caucasian based on the chromatic representation of the face image. This greatly increases the speed in both the training and inference phases over the previously used methods. State of the art optimization algorithms like Galactic Swarm Optimization (GSO), backpropogation, and PSO are used to optimize the mean-square error cost function. The algorithms are compared for the accuracy of the results and the rate of convergence.

2   Proposed Model for Ethnicity Classification Using Artificial Neural Networks

Our proposed approach goes through three phases. First, the face is extracted from the image. Then the skin is segmented out from the face image. This is given as input to the ethnicity classifier which then classifies and outputs the ethnicity. The three stages are depicted in Fig. 1.

2.1   Face Detection and Extraction

Accuracy of the prediction depends largely on the face images. So, it is very important to properly detect and extract faces. Figure 2 depicts the process of extracting the faces from the original image using the Cascade Object Detector feature in MATLAB.

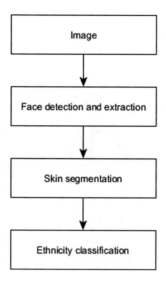

Fig. 1 Proposed architecture for ethnicity classification

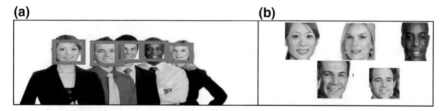

Fig. 2 a Detected faces marked in red. **b** Faces extracted from the image (*Source*: https://www.appcoda.com/photo-editing-extensions/)

2.2 Skin Segmentation

The next step in the proposed model is extracting the skin of each person from the facial images that have been generated in the previous step. For this process, a single-layered feedforward neural network is trained that takes as input RGB values of the pixels in the image and classifies them as skin or non-skin. A fully connected two layered feedforward neural network with six hidden layer neurons and one output neuron is used. Each connection has a weight associated with it which determines the effect of a neuron in one layer to the output of a neuron in the next layer. The activation function of choice is hyperbolic tangent function. The output neuron classifies and outputs whether the pixel is skin or non-skin. The neural network architecture is shown in Fig. 3. The output of this step are the RGB values of the pixels that are classified as skin. This will be used in the next step to predict the ethnicity of the facial image (Fig. 4).

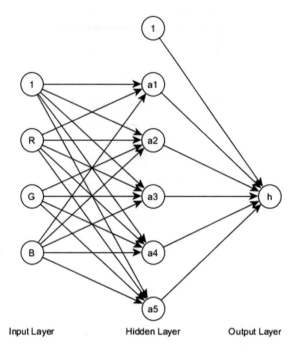

Input Layer   Hidden Layer   Output Layer

Fig. 3 Neural network architecture for skin segmentation

(a)   **(b)**

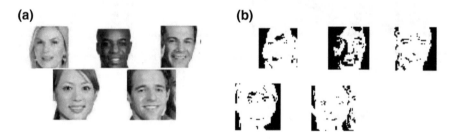

Fig. 4 Skin segmentation. **a** Extracted faces. **b** Skin segmented from the faces

2.3 Ethnicity Classification

This step takes as input the RGB values of the pixels that are classified as skin in the previous step. Here, a single-layered feedforward neural network classifies whether the skin pixel is Caucasian or non-Caucasian. After all the skin pixels have been classified, the number of pixels that are Caucasian and non-Caucasian is calculated. The class with the maximum number of pixels is given as the final output given in Eq. 1.

$$\text{Output} = \begin{cases} \text{Caucasian,} & \text{C} > \text{NC} \\ \text{Non-Caucasian,} & \text{NC} > \text{C} \end{cases} \qquad (1)$$

where C is the total number of pixels classified as Caucasian and NC is the total number of pixels classified as non-Caucasian. The neural network architecture for this step is the same as that for skin segmentation with six hidden layer neurons and one output layer neuron. The goal is to minimize the mean squared error (MSE) run through all the training samples given in Eq. 2. It can be formulated as follows

$$\text{MSE} = \frac{\sum (h_j - y_j)^2}{2m} \qquad (2)$$

where h_j is the output of our neural network, y_j is the target output and m is the number of samples. The optimization algorithms essentially tweak the weights through consecutive iterations in the neural network to give the best output. In this paper, the optimization of the cost function is done using three optimization algorithms: backpropagation, Particle Swarm Optimization (PSO) and GSO. The algorithms are compared for the accuracy of results and rate of convergence.

2.3.1 Backpropagation Algorithm

The backpropagation algorithm by Rumelhart et al. [15] solved problems that were previously unsolvable. This algorithm came to the limelight with the advent of neural networks and is vastly used to solve a lot of machine learning problems. The algorithm's objective is to find the minima of the cost function C given in Eq. 3. The backpropagation algorithm is used to perform gradient descent on the cost function.

$$C = \frac{\sum (a_j^{(L)} - y_j)^2}{2m} \qquad (3)$$

where $a_i(L)$ represents the result obtained from the network's final layer, m is the total number of samples present in the dataset, and y_j is the expected output. The neural network is calculated in the forward pass and the error δ is calculated in the backward pass. The weights are then subsequently updated as in Eq. 4. α denotes the learning rate.

$$w_{ij}^l := w_{ij}^l - \alpha a_i^{l-1} \delta_j^l \qquad (4)$$

The forward and backward passes are performed continuously till we find the optimum value. The drawback of backpropagation algorithm is that it gets stuck in local minima very often. Stochastic optimization algorithms like PSO and GSO, on the other hand, explore the search space more and overcome this problem.

2.3.2 Particle Swarm Optimization

PSO involves interactions between populations of agents with their environment as well as each other to find the global minimum [16, 17]. Initialization of the particles are done randomly and then the local minima of the cost function is searched by updating generations. Each particle is associated with the personal best which is the best fitness value it has achieved thus far and the other is the global best which is the best solution achieved by any particle. After finding personal best and global best, we update the velocity and the present position of the particle using the following formulae given in Eqs. 5 and 6

$$v_i(k+1) = \omega \cdot v_i(k) + \varphi_{1i} * \text{rand}() * \big(\text{pbest}_i - x_i(k)\big)$$
$$+ \varphi_{2i} * \text{rand}() * (\text{gbest} - x_i(k)) \tag{5}$$
$$x_i(k+1) = x_i(k) + v_i(k+1) \tag{6}$$

where $v_i(k)$ is the velocity of particle i at time k, $x_i(k)$ is the position of particle i at time k, pbest_i is the personal best of particle i, gbest is the global best, ω is the inertial factor, φ_{1i} and φ_{2i} learning factors. This process is repeated until we reach a minimum error value or maximum number of iterations. Although PSO has performed well in a variety of optimization problems, since it does a random walk of the search space it converges slowly.

2.3.3 Galactic Swarm Optimization

GSO draws inspiration from movement of stars and galaxies under the influence of gravity [18]. It works in two phases: exploration and exploitation. In the exploration phase, particles form subpopulations search the vector space for an optimal solution. In the exploitation phase, the best solution of each subpopulation moves toward the global best solution. The galaxies, in turn, can be treated as point masses on a large enough scale. These galaxies are again attracted toward larger masses to form superclusters of galaxies. The approach begins by dividing the particles into M subswarms. PSO is run on each of the subswarms. The velocity and positions are updated using the following formulae given in Eqs. 7 and 8

$$v_i(k+1) = \omega_1 \cdot v_i(k) + \varphi_{1i} * \text{rand}() * \big(\text{pbest}_i - x_i(k)\big)$$
$$+ \varphi_{2i} * \text{rand}() * (\text{gbest} - x_i(k)) \tag{7}$$
$$x_i(k+1) = x_i(k) + v_i(k+1) \tag{8}$$

where $v_i(k)$ is the velocity of particle i at time k, $x_i(k)$ is the position of particle i at time k, pbest_i is the personal best of particle i, gbest is the global best, $\omega 1$ is the inertial factor, φ_{1i} and φ_{2i} learning factors, k is the iteration number which is in the range 0 to L_1.

The M subswarms have M global bests. These form a superswarm Y containing the global bests of each subswarm. Again PSO is applied on Y. The update formulae are given by

$$v_i(k+1) = \omega_1 \cdot v_i(k) + \varphi_{1i} * \text{rand}() * \left(\text{pbest}_i - y_i(k)\right)$$
$$+ \varphi_{2i} * \text{rand}() * (\text{gbest} - y_i(k)) \tag{9}$$
$$y_i(k+1) = y_i(k) + v_i(k+1) \tag{10}$$

The subswarms X_i explores the search space. The superswarm Y exploits the information from the subswarms to find the global best. In this paper, GSO algorithm is used to train the neural network for ethnicity classification.

3 Results and Discussion

The dataset contains 8000 samples. It is divided into training and test datasets containing 70 and 30% of the samples, respectively. Backpropagation was used with learning rate of $3 * 10-5$. For PSO, the population size was set to 20 particles. Both PSO and GSO are run on a 26-dimension vector space. Figure 5a depicts the process of ethnicity classification for a black face. The faces are identified, then the skin is segmented from the facial image, the RGB values of the skin pixels are then classified as Caucasian or non-Caucasian. In the final image, the black pixels represent non-Caucasian ethnicity, white pixels represent Caucasian ethnicity and the grey pixels are the background. Figure 5a, b clearly shows that the number of black pixels for a non-Caucasian face is higher in the final image whereas white pixels are higher for a Caucasian face, respectively.

Three optimization algorithms have been used, namely backpropagation, PSO, and GSO, on our dataset to train the neural network. The accuracy of the neural network for any of the optimization algorithms is compared. The difference among them is in the training time. Backpropagation is very slow compared to PSO and GSO. PSO is faster than backpropagation. But using GSO we can train the neural network the fastest as it uses the concept of PSO but forms superswarms using the best fitness values of the subswarms that are created. This boosts the speed of the algorithm

(a)   **(b)**

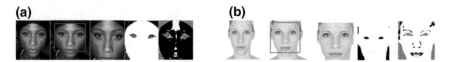

Fig. 5 a Non-Caucasian classification (*Source*: https://www.videoblocks.com/video/beautiful-woman-aging-process-portrait-female-model-with-pink-lipstick-with-aging-skin-and-growing-face-wrinkles-1920x1080-1080p-full-hd-footage-tscftk1). **b**. Caucasian classification (*Source*: http://www.taaz.com/makeover/e7qqu84q2udylhap_fnib57epa2tkn4o.html)

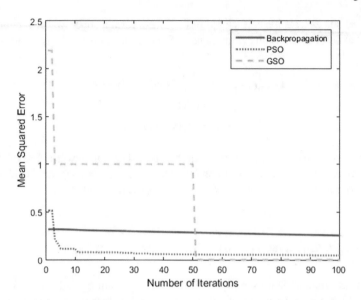

Fig. 6 Comparison between number of iterations required by backpropagation, PSO, and GSO to converge to the best solution

Table 1 Comparison of the cost of the best solutions obtained by backpropagation, PSO and GSO

Algorithm	MSE	Accuracy (%)
Backpropagation	0.1684	95.6
PSO	0.0468	98.25
GSO	0	98.42

making it converge faster to the lowest value. Figure 6 depicts very clearly the rate with which GSO converges to the optimum solution is much faster compared to the other two algorithms. Table 1 clearly shows that neural network trained with GSO has superior performance compared to the other two algorithms. Backpropogation gives an accuracy of 95.6%. PSO, on the other hand, gives an accuracy of 98.25%, whereas GSO starts with a higher cost but the cost reduces quickly and converges to a value less than machine precision after the 50th iteration. The classifier accuracy is seen to be 98.42% with GSO.

4   Conclusion

This paper proposes an effective way to classify facial images according to their ethnicities using feedforward artificial neural networks. GSO is used to optimize the cost function which shows superior performance over previously used algorithms like PSO and BP algorithm. The neural network for skin segmentation is trained on the "Skin Segmentation Dataset" from UCI Machine Learning Repository. The accuracy achieved for classification between Caucasian and non-Caucasian using backpropagation, PSO, and GSO are 95.6, 98.25 and 98.42%, respectively. GSO converges faster and provides a more accurate result than the other two algorithms. Parallelizing the GSO algorithm and taking advantage of the multi-core architecture and GPUs will be considered for future work.

References

1. R. Malpass, J. Kravitz, Recognition for faces of own and other race. J. Pers. Soc. Psychol. **13**(4), 330 (1969)
2. A.J. O'toole et al., Structural aspects of face recognition and the other-race effect. Mem. Cogn. **22**(2), 208–224 (1994)
3. A.J. Calder, A.W. Young, Understanding the recognition of facial identity and facial expression. Nat. Rev. Neurosci. **6**(8), 641–651 (2005)
4. S. Fu, H. He, Learning race from face: a survey. IEEE Trans. Pattern Anal. Mach. Intell. **36**(12) (2014)
5. P.J. Phillips et al., Overview of the face recognition grand challenge, in *Proceeding of IEEE Computer Society Conference on Computer Vision and Pattern Recognition*, vol. 1 (2005)
6. S. Lawrence, C.L. Giles, A.C. Tsoi, A.D. Back, Face recognition: a convolutional neural-network approach. IEEE Trans. Neural Netw. **8**(1), 98–113 (1997)
7. W. Wang, H. Feixiang, Q. Zhao, Facial ethnicity classification with deep convolutional neural networks, in *Chinese Conference on Biometric Recognition*. Springer International Publishing (2016)
8. H. Ding, D. Huang, Facial ethnicity classification based on boosted local texture and shape descriptions, in *IEEE International Conference and Workshops on Automatic Face and Gesture Recognition (FG)*, Shanghai, China (2013)
9. Z. Yang, H. Ai, Demographic classification with local binary patterns. Adv. Biometr., 464–473 (2007)
10. X. Lu, A.K. Jain, Ethnicity identification from face images. Proc. SPIE—Int. Soc. Opt. Eng. **5404**, 114–123 (2004)
11. S. Hosoi, E. Takikawa, M. Kawade, Ethnicity estimation with facial images, in *Proceedings of Sixth IEEE International Conference on Automatic Face and Gesture Recognition*, vol. 6 (2004)
12. G. Toderici, S.M. O'Malley, G. Passalis, T. Theoharis, I.A. Kakadiaris, Ethnicity-and gender-based subject retrieval using 3-D face-recognition techniques. Int. J. Comput. Vis. **89**(2), 382–391 (2010)
13. D. Huang, M. Ardabilian, Y.L. Wang, L. Chen, Oriented gradient maps based automatic asymmetric 3D-2D face recognition, in *IAPR International Conference on Biometrics (ICB)*, vol. 5 (2012)
14. S.K. Bhattacharyya, K. Rahul, Face recognition by linear discriminant analysis. Int. J. Commun. Netw. Secur. **2**(2), 31–35 (2013)
15. D.E. Rumelhart, R. Durbin, R. Golden, Y. Chauvin, *Backpropagation: The Basic Theory. Backpropagation: Theory, Architectures and Applications*, pp. 1–34 (1995)
16. J. Kennedy, *Particle Swarm Optimization. Encyclopedia of Machine Learning* (Springer, US, 2011), pp. 760–766

17. R. Poli, J. Kennedy, T. Blackwell, Particle swarm optimization. Swarm Intell. **1**, 33–57 (2007)
18. V. Muthiah-Nakarajan, M.M. Noel, Galactic swarm optimization: a new global optimization metaheuristic inspired by galactic motion. Appl. Soft Comput. **38**, 771–787 (2016)

Machine Learning on Medical Dataset

M. P. Gopinath, S. L. Aarthy, Aditya Manchanda and Rishabh

Abstract Machine learning for medical decision-making is helpful when it is possible to achieve two conditions: a well-defined way to reach the diagnosis and high prediction accuracy. With the introduction of sensors in every aspect of daily live, there is a huge amount of data compiled (the present study concentrates on data from health), such quantity without any processing/information extraction technique means nothing. However, not every processing makes it possible to obtain something helpful from the raw data. In the present, the document evaluated three methodologies about machine learning (ANN + PSO, SVM, and RVM) in order to make a valuation on the performance over a set of data, and after the tests, the RVM methodology behaves the best.

Keywords Machine learning · Metrics · Accuracy · ANN · PSO · SVM · RVM

1 Introduction

Nowadays with the inclusion of sensors in every human activity, there is a huge amount of data without any value, until the use of data mining and pattern recognition techniques on them in order to obtain some meaningful information for a particular activity ([2, 5, 6, 9, 10, 12, 16]; Leao dos santos 2013). As every technological advance, it is rooted on military applications migrating to industry, health,

The original version of this chapter was revised: The author name "Rishadh" has been changed to "Rishabh". The correction to this chapter is available at
https://doi.org/10.1007/978-981-13-3329-3_53

M. P. Gopinath (✉) · A. Manchanda · Rishabh
School of Computer Science and Engineering,
Vellore Institute of Technology, Vellore, India
e-mail: mpgopinath@vit.ac.in

S. L. Aarthy
School of Information Technology,
Vellore Institute of Technology, Vellore, Tamil Nadu, India

© Springer Nature Singapore Pte Ltd. 2019
S. C. Satapathy et al. (eds.), *Information Systems Design and Intelligent Applications*,
Advances in Intelligent Systems and Computing 862,
https://doi.org/10.1007/978-981-13-3329-3_13

and academy; the inclusion of techniques related pretends to provide quality data for posterior analysis [6, 12, 16].

Many related papers including the one made for the author are tested using academic public databases as the machine learning repository at University da California Irvine (UCI) (UCI NA) and Kraggle (NA NA). This functional prototype based on Relevance Machine Vector (RVM) is applied on breast cancer, liver, diabetes, and Parkinson's UCI dataset, further information about the dataset discussed on experimental section. Also as a comparative study ANN with PSO weight optimization and Support Vector Machine (SVM) methods is implemented.

This paper is structured as follows. Section 2 contains the description of the procedure and the basis from ML applied. Section 3 presents the benchmark and dataset information. In Sect. 4, the results of the application of the procedure and conclusions are presented.

2 Background

The automatic extraction of information from data corresponds to an area of knowledge called Machine Learning [1, 4, 7, 12–14]. Although there are a lot of philosophical questions about learning, in computer area, the most are related with "how to learn from data", "what is the appropriate test to evaluate the knowledge acquired?", and how to improve the performance? Keeping this in mind, there are several applications for Machine Learning (ML) [1, 2, 5, 6, 8]. Every instance in a dataset to be used as input for a ML system is represented using features (continuous, categorical, or binary) [4, 7, 8]. If the inputs have known labels (and corresponding correct outputs), the learning process is called supervised, and in the other case, it is called unsupervised [1, 4, 7, 14].

2.1 Knowledge Representation

The use of an "appropriate" knowledge representation is fundamental to all algorithms. It dictates how efficiently and effectively an algorithm tackles a task [4]. Some possible representations are propositional and predicate logic, decision list and trees, neural nets, genetic algorithms, semantic nets, and others. Most of the systems use propositional logic [1, 4, 7].

2.2 Levels of Abstraction for Learning

The learning process can be modeled using different levels of abstractions and subsymbolic "Dietterich (1986) concludes that there are two types of learning, knowledge level learning, which equates to knowledge acquisition and symbol-level learning which equates to speed-up learning" [4].

- Neural nets,
- Genetic algorithms
- Statistical approaches
- Reinforcement learning
- Hybrid approaches
- Linear separability.

2.2.1 ANN

Inspired by biological neural networks, the concept consists of an extremely large number of simple processors with many interconnections ([3, 11]; Gupta 2013). The model is composed of inputs which are multiplied by weights, associated to each flow of information. The weights are defined through a mathematical function which is defined to activate the component (neuron). Another function computes the output of the neuron and sums the inputs in order to generate an output (Castrounis NA) ([8]; Gupta 2013) (Fig. 1).

Depending on the output of those weights, which could be adjusted manually or automatically (the most used), the process of adjusting is called training or learning. The most used function for adjusting is called backpropagation [8], however, it is not the best one. In order to improve the behavior of the network, some hybrid approaches are designed [4, 8], and one of them is Particle Swarm optimization (PSO), described on the next section.

PSO

The initial ideas on particle aimed at producing computational intelligence by exploiting simple analogues of social interaction, rather than purely individual cognitive

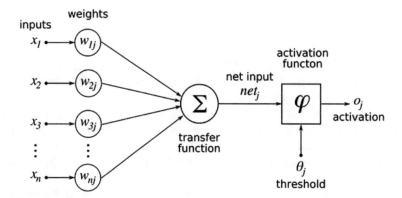

Fig. 1 Artificial neural network (Castrounis NA)

abilities [15, 17]. In PSO, a number of members (particles) are located on a search space and it evaluated the objective function in the current location. Then, each particle updates its movement and position based on the previous one and the best (best fitness) locations with those of one or more members of the swarm, with some random perturbations [8, 9, 13, 15]. Basic algorithmPSO ([15, 17]; Khare and Rangnekar 2013).

1. Initialize a population array of particles with random positions and velocities on N dimensions in the search space.
2. Loop.
3. For each particle, it calculates the desired optimization fitness function in N variables.
4. Compare particle fitness evaluation with the pbest$_i$. If the current value is better than pbest$_i$ then reset with current value, and $\overrightarrow{p_i}$ equals to the current location $\overrightarrow{x_i}$ in N dimensional space.
5. Identify the particle in the neighborhood with the best success so far, and assign its index to the variable g.
6. Change the velocity and position of the particle according to

$$\begin{cases} \overrightarrow{v}_i \leftarrow \overrightarrow{v}_i + \overrightarrow{U}(0, \phi_1) \otimes \left(\overrightarrow{p}_i - \overrightarrow{x}_i\right) + \overrightarrow{U}(0, \phi_2) \otimes \left(\overrightarrow{p}_g - \overrightarrow{x}_i\right) \\ \overrightarrow{x}_i \leftarrow \overrightarrow{x}_i + \overrightarrow{v}_i. \end{cases}$$

7. If the criterion is met, exit loop.
8. End loop.

where
$\overrightarrow{U}(0, \phi_1)$ represents a vector of random numbers uniformly distributed in $[0, \phi_2]$, which is randomly generated at each iteration and for each particle [9, 15]

$- \otimes$ is component $-$ wise multiplication

2.2.2 SVM

The Support Vector Machine (SVM) is a discriminative classifier defined by a separating hyperplane, so that the given labeled data for training the algorithm maps the points to one optimal hyperplane (Fig. 2).

The concept is rooted in 1960s [10]. The classifiers have a margin which is the distance from the classification boundary to the nearest data point. And, the points included on the closest classes to the classification boundary are known as support vectors. The process of training can be explained as (keeping in mind a 2D situation) [6, 7].

Given a dataset $(x_1, y_1), \dots (x_m, y_m)$, where $y_{i=} \pm 1$ (-1 for members of class 0 and $+1$ for members of class 1). Based on the vector line equation, the classification boundary is defined as $\vec{w}.\vec{x}+b = 0$, where $\vec{w}$ and $\vec{x}$ are 2D vectors With $\vec{w}.\overrightarrow{x_n}+b = -1$

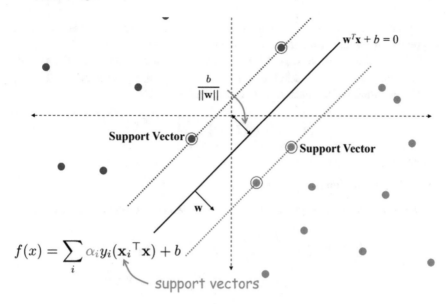

Fig. 2 SVM hyperplane

for the negative classification boundary and $\vec{w}.\vec{x}_p + b = 1$ for the positive one. The distance between the boundaries is $\frac{2}{w}$ and the size of margin M is $M = \frac{1}{w}$.

In order to maximize M, is necessary to minimize $\vec{w}$ considering the following constraints $\vec{w}.\vec{x}_i + b \leq -1$ for $\vec{x}_i$ inside class 0 and $\vec{w}.\vec{x}_i + b \geq 1$ for $\vec{x}_i$ inside class 1. Obtaining an optimization problem described as Minimize $\vec{w}$ subject to $(w.x_i - b) \geq 1$ for $I = 1, \ldots, n$. This approach is very common in medicine and engineering.

2.2.3 RVM

RVM is a Bayesian framework with the same functional form as SVM is used to obtain sparse solutions in regression and classification jobs employing linear models as parameters. The implementation is based on the algorithm described by tipping (2001). The advantage is a possible good generalization performance with sparse predictors called Relevant Vectors (RV). Compared with SVM approach, the formulation of RVM eliminates the necessity of free parameters.

RVM Learning Process

Given a set of data for training D with N inputs, the calculations can be made through the following steps.

Hyper parametrization for estimating $\bar{w}$ and σ^2: The identification of RVs is supported by hyper parametrization of the weight vector ($\bar{w}$).

Based on the vector of inspections $(\overline{T})$ and degradation states $(\bar{f})$, with a conditional probability of observations of $\bar{f}$ $p(\bar{f}\overline{T})$ as a Gaussian process $N(\bar{f}y(\overline{T}), \sigma^2)$.

The likelihood of the dataset can be described as

$$p(\bar{f}\bar{w}, \sigma^2) = (2\pi\sigma^2)^{-1/2} e^{\left\{\frac{-1}{2\sigma^2}\bar{f} - \overline{\Phi}\bar{w}^2\right\}}$$

where

$\overline{\Phi}$ is the $j \times j$ design basis function matrix

Using Bayes' rule, the posterior overall unknowns can be calculated using

$$p(\bar{w}\bar{f}, \bar{\alpha}\sigma^2) = \frac{p(\bar{f}\bar{w}, \sigma^2)p(\bar{w}, \bar{\alpha})}{p(\bar{f}\bar{\alpha}, \sigma^2)}$$

$$= (2\pi)^{\frac{-(j+1)}{2}}\left|\sum\right|^{-1/2} e^{\left\{\frac{-1}{2}(\bar{w} - =\bar{\mu})^T \sum^{-1}(\bar{w} - \bar{\mu})\right\}}$$

With $\sum = \left(\overline{\Phi}^T \overline{\overline{\beta}}\overline{\overline{\Phi}} + \bar{\alpha}\right)^{-1}$ and $\bar{\mu} = \overline{\sum \Phi}^T \bar{\beta}\bar{f}$

where $\bar{\alpha} = \text{diag}(\alpha_0, \alpha_1, \ldots, \alpha_N)$ and $\bar{\beta} = \sigma^{-2}I_N$

By integration of the weights, the marginal likelihood or evidence is obtained

$$p(\bar{f}\bar{\alpha}\sigma^2) = (2\pi)^{-N/2}\left|\bar{\beta}^{-1} + \overline{\Phi}^T \bar{\alpha}^{-1}\overline{\Phi}\right|^{-1/2} e^{\left\{-\frac{1}{2}\bar{f}^T\left(\bar{\beta}^{-1} + \overline{\Phi}^T \bar{\alpha}^{-1}\overline{\Phi}\right)^{-1}\bar{f}\right\}}$$

Optimization of Hyper Parameters

The values for optimization cannot be obtained in a closed form, so the iterative way is usually employed.

Relevance Vector Classification

In order to use RVM for classification, the linear model is generalized by applying the logistic sigmoid function as

$$P(t|w) = \prod_{n=1}^{N} \sigma\{y(X_n)\}^{t_n}[1 - \{y(X_n)\}]^{I-t_n}$$

As was commented on previous paragraphs, the weights cannot be determined analytically.

3 Results and Discussion

The inclusion of many features can affect the classification in a negative way.

With the present system evaluates whole set of data leads to extract unwanted features.

3.1 Datasets

Data set obtained from UCI repository and split on two subsets (66% for training and 34% for test).

3.1.1 Information About Datasets

They considered four datasets: breast cancer, liver disease, diabetes disease, and Parkinson's disease. Considering a twofold system and basic configurations, the following results are obtained (Fig. 3).

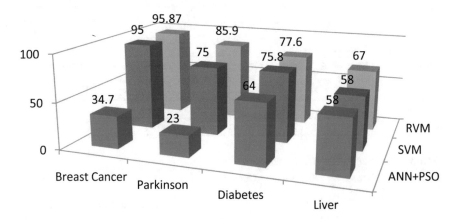

Fig. 3 Classification accuracy on medical dataset

3.1.2 Analysis on Reduced Set

Considering twofold system, the basic configuration and the inclusion of the following features are compared with [16].

3.1.3 Results

Result Comparison (Tables 1, 2, 3, 4 and 5)

Table 1 Accuracy on whole set

Database	Number of classes	Samples	Number of features	Accuracy		
				ANN+PSO (%)	SVM (%)	RVM
Breast cancer	2	155	19	34.7	95	95.87
Parkinson's	2	195	23	23	75	85.9
Diabetes	2	768	8	64	75.8	77.6
Liver	2	345	6	58	58	67

Table 2 Selected features

Dataset	Number of classes	Samples	Number of features	Number of selected features	Selected features
Diabetes	2	768	8	6	2, 4, 5, 6, 7, 8
Liver	2	345	6	5	1, 2, 3, 4, 5

Table 3 Results on dataset after feature extraction

Database	Number of classes	Samples	Number of features	Accuracy		
				ANN+PSO (%)	SVM (%)	RVM
Diabetes	2	768	8	63.3	74.9	75.8
Liver	2	345	6	57.6	58.1	67.8

Table 4 Classification accuracy over liver dataset

Author	Method	Accuracy (%)	Ranking
Lee and Mangsarian	SSVM (10-fold)	70.33	6
Van gestel et al.	SVM with GP	69.7	8
Goncalves et al.	HNBF	73.3	4
	Awais (10-fold)	70.17	7
	AIS with hybrid	60.57	11
Ozsen and gunes	AIS with Manhattan distance	60.21	12
	AIS with Euclidean distance	60.00	13
Li et al.	Fuzzy + SVM	70.85	5
Chen et al.	PSO + 1 NN	68.99	9
Chang et al.	CBR + PSO	76.81	2
	ABCFS + SVM	82.55	1
Serter et al.	ABCFS + SVM	74.81	3
Present job	ANN + PSO	57.6	15
	SVM (twofold)	58.1	14
	RVM	67.8	10

Table 5 Classification accuracy over diabetes dataset

Author	Method	Accuracy	
Sahan et al.	AWAIS	75.87	9
Polat and gunes	PCA and ANFIS	89.47	2
	LS-SVM	78.21	8
Polat et al.	GDA-LS-SVM	82.05	5
Kahramanli and Allahverdi	ANN and FNN	84.2	4
Patil et al.	HPM with reduced dataset	92.38	1
Isa and Mamat	Clustered	80.59	7
Aibinu et al.	AR1 + NN	81.28	6
	ABCSFS + SVM	86.97	3
Serter et al.	ABCFS + SVM	74.81	12
Present job	ANN + PSO	63.3	13
	SVM (twofold)	74.9	11
	RVM	75.8	10

4 Conclusion

This study was designed to evaluate the performance of RVM over other methodologies applied on medical datasets. Comparison between the three methodologies ANN, PSO, SVM, and RVM shows the superior accuracy of RVM on the four cases. In the general background related to Machine Learning, it is common to consider the presence of redundant and low impact features inside datasets available on the academic repositories. Also, the comparison with other approaches was made (in case of liver and diabetes datasets) shows improvement needed in feature extraction part.

References

1. R. Agarwal, V. Dhar, Big data, data science, and analytics: the opportunity and challenge for IS research, 443–448(2014)
2. A.C. Azubogu et al., Wireless sensor networks for long distance pipeline monitoring. World Acad. Sci. Eng. Technol. Int. J. Electr. Comput. Energ. Electron. Commun. Eng. 7(3), 285–289 (2013)
3. W.G. Baxt, Use of an artificial neural network for the diagnosis of myocardial infarction. Ann. Intern. Med. 115(11), 843–848 (1991)
4. D.M. Dutton, G.V. Conroy, A review of machine learning. Knowl. Eng. Rev. 12(4), 341–367 (1997)
5. J.D. Halamka, Early experiences with big data at an academic medical center. Health Aff. 33(7), 1132–1138 (2014)
6. L. Han, S. Luo, J. Yu, L. Pan, S. Chen, Rule extraction from support vector machines using ensemble learning approach: an application for diagnosis of diabetes. IEEE J. Biomed. Health Inform. 19(2), 728–734 (2015)
7. M.I. Jordan, T.M. Mitchell, Machine learning: trends, perspectives, and prospects. Science 349(6245), 255–260 (2015)
8. S.B. Kotsiantis, I.D. Zaharakis, P.E. Pintelas, Machine learning: a review of classification and combining techniques. Artif. Intell. Rev. 26(3), 159–190 (2006)
9. R.V. Kulkarni, G.K. Venayagamoorthy, Particle swarm optimization in wireless-sensor networks: a brief survey. IEEE Trans. Syst. Man Cybern. Part C (Appl. Rev.) 41(2), 262–267 (2011)
10. V.A. Kumari, R. Chitra, Classification of diabetes disease using support vector machine. Int. J. Eng. Res. Appl. 3(2), 1797–1801 (2013)
11. N. Murata, S. Yoshizawa, S.I. Amari, Network information criterion-determining the number of hidden units for an artificial neural network model. IEEE Trans. Neural Netw. 5(6), 865–872 (1994)
12. E. Naidus, L.A. Celi, Big data in healthcare: are we close to it? Rev. Bras. de Terapia Intensiva 28(1), 8–10 (2016)
13. P. Neirotti, A. De Marco, A.C. Cagliano, G. Mangano, F. Scorrano, Current trends in smart city initiatives: some stylised facts. Cities 38, 25–36 (2014)

14. L.G. Nongxa, Mathematical and statistical foundations and challenges of (big) data sciences. S. Afr. J. Sci. **113**(3–4), 1–4 (2017)
15. R. Poli, J. Kennedy, T. Blackwell, Particle swarm optimization. Swarm Intell. **1**(1), 33–57 (2007)
16. M.S. Uzer, N. Yilmaz, O. Inan, Feature selection method based on artificial bee colony algorithm and support vector machines for medical datasets classification. Sci. World J. (2013)
17. B. Xue, M. Zhang, W.N. Browne, Particle swarm optimization for feature selection in classification: a multi-objective approach. IEEE Trans. Cybern. **43**(6), 1656–1671 (2013)

Analysis of Diverse Open Source Digital Tools and Learning Management System Users in Academics

J. Karthikeyan, W. Christopher Rajasekaran and Panadda Unyapho

Abstract Digital footprints in the teaching–learning process will enrich learning which in turn will nurture education in a fast-growing economy like India. Technology in the form of digital tools, Learning Management System (LMS) and Learning Object Repository (LOR) will make education more informative and interesting that will realize aptitude among the young generation. This flipped classroom model being very successful among the G8 countries is emerging in countries that are enthusiastic for a revolution in education. The study revealed how technology is integrated effectively into everyday classroom teaching and how it is well-received by the learners. It also exposed various open sources that help teachers and students.

Keywords LMS · Digital tools · Evaluation · Documentation · Analysis

1 Introduction

Teachers dealing with Gen 'Z' learners have to adopt new techniques that maintain the tempo of their classroom teaching and learning process. Failing to update and maintain the rhythm, they are branded as inefficient teachers in spite of comprehensive subject knowledge and other subject skills. This problem is so common in Asian Universities because of the high student population in these countries. Academic designers and Institutional heads always scrutinize to identify a pertinent solution to resolve these challenges within their economic limitations. The western world and many countries in the European Union have experimented [1] new methodolo-

J. Karthikeyan (✉) · W. Christopher Rajasekaran
SSL, Vellore Institute of Technology, Vellore 632014, India
e-mail: jkarthikeyan@vit.ac.in

W. Christopher Rajasekaran
e-mail: cristo_wilson13@vit.ac.in

P. Unyapho
IAO, Bangkok University International, Bangkok, Thailand
e-mail: panadda.u@bu.ac.th

© Springer Nature Singapore Pte Ltd. 2019
S. C. Satapathy et al. (eds.), *Information Systems Design and Intelligent Applications*,
Advances in Intelligent Systems and Computing 862,
https://doi.org/10.1007/978-981-13-3329-3_14

145

gies in teaching and learning supported by technology for a very long period. Their investments were huge and their policies have accepted failures that occurred during testing and implementing stages. Their investments in software and digital tools that help teaching and learning were enormous in recent days which is quite impossible in developing and upcoming countries with reference to the economy. So the better solution for these economically downtrodden countries to bridge the gap between students and teachers in improving the standard of education is to track the availability of open source tool and software that will be resourceful and support the teaching and learning. This paper is an outcome of an empirical study conducted among the teachers from different disciplines of Engineering, Language and Social Sciences to understand the effectiveness of various digital tools they started to use in recent years to match with the interest of their learners. It is revealed that Open Source Digital Tools help the teachers for conducting quizzes, collecting assignments, off class discussions, maintaining records and to manage the Classroom online. It is identified from this study that the learners are very comfortable in this kind of interaction between the teachers and them not only inside the class but also outside the class hours. This method is found unique that matches with the interest of the Gen 'Z' learners who are skillful enough using digital contraptions and online resources. The outcome of this study will be beneficial for other institutions across the globe who thinks of implementing digital tools in teaching and learning process. The study also revealed specific tools that are widely used by teachers and learners.

2 Literary Survey

A case study conducted by Study North Carolina State University [2] professors revealed how the use of google forms helped the administrators to access the quality of teaching in regular intervals and how it helped teachers to give feedback to their students continuously after their weekly discussions and assessments. It was reported that after initiating the use of Google forms time and money spend on paper works were considerably reduced.

Empirical study conducted in King Mongkut's University of Technology Thonburi [3], Thailand reveals how Edmodo offers a modest way for the faculty invariable of their departments to generate, manage and connect students with their classmates and teachers from anywhere and any anytime just with the access to the internet either in their mobile phones, tablets, laptop or the Desktop. Moreover, it also maintains the records of discussions and all other activities between them in a secured manner. The students enter the online class with the code generated and shared by the teacher. The teacher has the complete control over the class preventing trespassers. From this study, it was understood that their students were fond of using such tools that match with their interest.

Centro Escolar University (CEU), Makati City, Philippines conducted [4] a pilot study among their students to identify the efficacy of Schoology (Learning Management System—LMS) in improving the skill of the university students in business writing. During this study, the Pre, Progressive and Post-tests were conducted through Schoology and analysis were also made using the inbuilt facility in it. The outcome of the study suggested their teachers utilize the Schoology as an additional resource to the traditional method of teaching.

GoToMeeting [5] is a feasible option for anyone tangled in meetings that comprise travel by the contributors. School of Social Work, University of Michigan (UM) uses GoToMeeting as a standard tool for its research activities that involve meetings and discussions with national and international research partners, interviews and as a teaching tool that involve students in small group projects. It also provides prospects to organize online workshops and discussions. The Office of Student Services also makes use of it because it is a cost-effective and efficient way to outreach and orientate students to the School.

Action research conducted by Ph.D. scholars in the Curtin University of Technology, Australia exposed [6] how Moodle was significant amongst students for sharing their exclusive curricular knowledge. It was also noted that the sense of accountability amongst the students for ensuring that they engaged one another in mutually productive dialogue. This study also revealed how the positive relationship between the teacher and their students flourished.

3　Problem Statement

The biggest challenge for the existing teachers in higher education institutions across the globe is to satisfy the needs of every single learner and assure there is a considerable augmentation in their subject skill. Though the curriculum and the syllabus are designed to meet the expectation of the teaching–learning process, most of the teachers fail to achieve the objective due to various reasons such as massive classroom strengths, varied proficiency level students in one class, stipulated class hours, teaching and evaluation method and lack of continuous monitoring of students' progress. The uttermost of all is the generation gap. The physical age difference between a student in higher education institute and an experienced teacher is not less than 10–20 years hence the thought process and approach towards any action is different. The teacher sometimes is stubborn with their traditional way of teaching whereas the learner expects the teacher to be in their pace. There is a tug of war between the conventional way of teaching and technological way of learning and communicating. In this war, the learners win and the teacher is expected to update and stay on par with the expectation of the student as the education itself is students centric. It is a known fact that the modern generation so-called 'Techno Generation' feel comfortable in using technical facets rather than sitting in a classroom and listening to lectures for hours together. The right mix of technology in teaching will definitely make the learning effective. Hence the teachers are expected to add technology in some form to make teaching–learning more enjoyable and eloquent that suites the contemporary learners.

| Questionnaire → Sampling→ Data Collection→ Data Analysis→ Findings |

Fig. 1 Methodology of the study

4 Limitations of the Study

This study was conducted in Vellore Institute of Technology, Vellore, India which is one among the premier technical institutes in India which has over thirty thousand versatile students and over one thousand four hundred Teachers from 29 states and three Union territories of the country. The foremost reason for selecting this University for the study is since the year 2015 it is mandatory for the teachers to use Digital tools of their choice either to conduct quiz, collect assignment, off class discussion, maintain records, conduct the virtual class or to manage the Classroom online. Moreover, teachers and students are exposed to use Digital Pads for writing exams and evaluation, virtual classes connecting students and experts from around the globe, effortless access to digital resources and top of all, Gen ' Z' students who are fair enough to use technology at its best.

5 The Methodology of the Study

See Fig. 1.

5.1 Questionnaire

The questionnaire was designed very specifically to identify various open source tools and Learning Management Systems (LMS) used by teachers for teaching–learning process (Virtual Class), evaluation (Quizzes and Assignments) and documentation (maintain data). The questionnaire also aimed at getting information on how comfortable the teachers and students feel using the digital tools and Learning Management System in the day to day classes. It also will reveal how often the teachers use digital tools, various tools they use in each semester, teachers eagerness to use new tools and their comfort level while using these open sources. The questionnaire was not derived from any other source rather it is customized as per the need of the study.

5.2   Sampling

This study adopted the most common approach of quantitative sampling [7] that is to use random samples method in which the sample population is demarcated and all affiliates have an equal chance of selection and the results of studying the sample can then be widespread back to the population. The extent of the test may be decided eventually by perusing those ideal amount fundamental to empower substantial inferences to make settled on something like those number. The bigger the sample size, the more diminutive the opportunity of an arbitrary examining lapse [8]. The designed questionnaire was sent randomly to the teachers of the University in the following Schools

- School of Social Sciences and Languages (SSL)
- School of Information Technology & Engineering (SITE)
- School of Advanced Sciences (SAS)
- School of Computer Science and Engineering (SCOPE)
- School of Electronics Engineering (SENSE)
- School of Electrical Engineering (SELECT)
- School of Bio Sciences and Technology (SBST)
- School of Mechanical Engineering (SMEC)
- School of Civil and Chemical Engineering (SCALE)
- VIT Business School (VITBS).

5.3   Data Collection

Several online free tools are available to conduct surveys. Some of them are Survey Planet, Typeform, Zoho Survey, Google Forms, Survey Gizmo and Survey Monkey. The researcher should identify and choose an apt tool that suits their type of survey. In this study, the questionnaire was shared via Google Forms with the faculty of ten schools which has an average of one hundred faculties and above. Google [1] forms eventually saved cost involved in purchase and printing of questionnaire, time involved in data collection like meeting people explaining them the intention of the questionnaire and organizing the collected data and transferring them from hard copies to soft copy for analysis and further study.

Nine major schools were selected as shown in Fig. 2 and the questionnaire link was shared with an average of 25% the teachers in these schools. The teachers for whom the questionnaire was sent had a teaching experience ranges between 1 and 20 years and above. The percentage of teachers who has 5 years and less experience did not exceed 5%. The teachers furnished their names and School they belong to. This information helped to classify the data based on School.

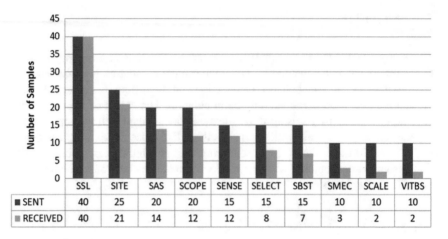

Fig. 2 Detail of questionnaire sent and received

5.4 Detailed Analysis

It is a clear evidence from Fig. 3 that well-experienced teachers are involved in this survey who represent the earlier generations bit away from using technology for teaching and learning process and who believes in traditional teaching method. The intention of this study is to find how these teachers are ready to switch to new methodologies by understanding the needs of the current generation students who ate technosavy. The majority of samples have teaching experience exceeding 10 years and very few who has less teaching experience.

It is revealed from Fig. 4 that a large sum of teachers is comfortable in using GoToMeeting which is an open source tool generally preferred by corporates to conduct meetings. But it is one among the better tool for the teachers to conduct virtual classes. The advantages include details of students attending the discussions can be viewed. The whole lecture can be scheduled well in advance and the access code of the scheduled class can be shared with the students. The teacher can interact with the students individually during the course of presentation and vice-versa. The audio and video quality is good compared to other open source tools. Some of the teachers also use YouTube live which is a free service provided by Google. But the general issue with this is streaming. During lecture, if the signal is slightly weak at any end the progress is put on hold. It takes a lot of time to resume. There is no facility to record the details of the students who are attending the presentation. The same problem is with Skype, Zoom, WhatsApp live Chat, and Big Blue Button. There are other sources available to host virtual classes but with payments. The focus here is how teachers are using free resources effectively in their class.

Figure 5 uncover that a maximum number of teachers prefers to use Learning Management System (LMS) like Schoology, Moodle, EasyClass, Edmodo, and Classmaker for conducting Quiz because the questions used for assessment is made

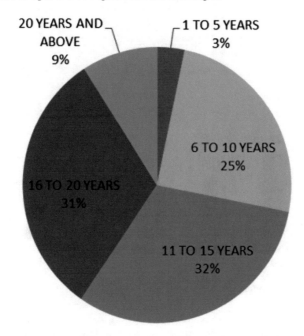

Fig. 3 Teaching experience of the samples

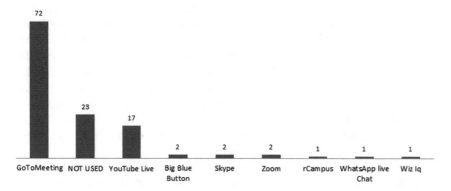

Fig. 4 Tool to conduct virtual class

available in the question bank and teachers can use the same questions for other classes following the same subject. They do not have to retype the same questions and the key again and again. This saves a lot of time and in a long run, the teacher will have large numbers of questions in the bank. Schoology is widely preferred as it is user friendly. Google Forms are also preferred by some teachers because it can effectively be used for alternative purpose like conducting the survey, meeting invitation and data collection in addition to conducting a quiz.

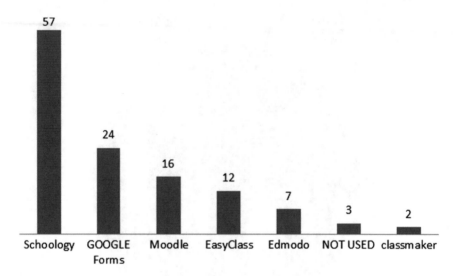

Fig. 5 Tool to conduct quiz

Figure 6 shows that collecting assignments and grading them in a colossal class environment is a Herculean task for the teachers. LMS is widely used by the faculty as it substitutes varied functions related to teaching, learning and evaluation process. The survey revealed over 80% of the teachers use LMS in which Schoology is frequently used by large numbers. Though there are other simple options like Google Drive, Dropbox or institutional customized medium to collect assignments, LMS tops the list as it is user-friendly and can store date up to 5 GB I free versions. Notification to the teacher on every submission is triggered with complete details of the submission. In addition to this organizing data is simplified automatically in all the LMS. As far as students, they will have the details of all their submissions, chats they had with the teacher as well as the other group members and all these information can be retrieved at any point of time.

The greatest challenge for the teachers especially in continents like Asia is evaluation. Invariable of countries, the class size is huge and the work load for each teacher is also more. Though the teachers are accustomed to teach passionately for these bigger classes they find it difficult during evaluation. The higher education scheme of evaluation in India has continuous assessment test that needs to be conducted before a student finally appears in the semester final exam. So if a teacher handles three courses in a semester with a class capacity of 60 each, they should evaluate 180 scripts in the form of hard copies or in digital version for every assessment. If there are three internal assessments before the final exams then it comes to 540 scripts. This is the main reason why faculty prefers to use LMS as it saves lot of time and energy being wasted in evaluation. It is a single time investment where the teachers upload the questions and answer keys for different forms of assessments. All the questions along with the key are stored in the database and the teacher can use it for the current

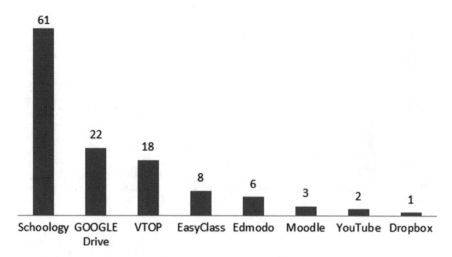

Fig. 6 Tool to collect assignments

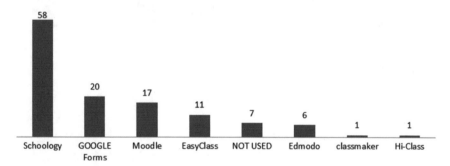

Fig. 7 Tool for automated evaluation

batch of students and the same can be used in near future where the teacher need not retype the questions rather they retrieve it from the database. Though there are many LMS that helps the teachers in the automated evaluation Schoology is extensively preferred which is revealed from Fig. 7. A Google Form is also equally significant for automated evaluation and it also can store the data which can be used for future assessment.

For the sake of analysis, documentation and further references the data related to teaching, learning and evaluation has to be stored by every faculty in educational institutions. Storing it in online has advantage as it can be viewed at anytime from anywhere. Figure 8 discloses that Google Drive is the top propriety for the teachers when it comes to storage as it can store varied forms of data that can easily be shared with students at any time. LMS stores data that is only related to the assessments, content related to the subject and information shared to the learners and teachers.

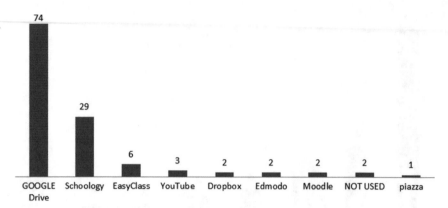

Fig. 8 Tool for storing data

| Since1 Year | Since 2 Years | Since 3 Years | Over 4 Years |

Fig. 9 Exposure to digital tools & LMS

Some of the other mode of storage used by teachers includes YouTube for storing and sharing video files, Dropbox, Moodle and piazza (Fig. 9).

Faculty in higher education institutions across India adopted to use of technology in teaching and learning in recent years to bridge the generation gap and to make the process more interesting and meaningful. This survey exposed that 28% of them are using digital tools since 4 years and over 20% of them have just started to use. The percentage of the faculty using Digital tools will be considerably increasing in all forms of educational institution in the years to come as there will be an urge from the learners for technological interwoven learning or blended learning.

It is very clear evidence from Fig. 10a, b that both the faculty and students are comfortable using the digital tools in the teaching and learning process. The main reason for the success rate is the whole communication or the interaction is made

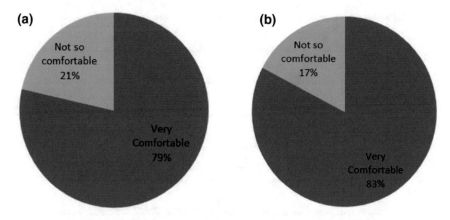

Fig. 10 A faculty comfort using digital tools B students comfort using digital tools

available online. The student is free enough to take the assessment or post their assignment from any place on this globe connected to the Internet and for the teacher posting and sharing of the information to their students is just a click away when they are connected to the Internet. The same can be accessed using their mobile phones as the app developer made it convenient to access the tools in varied formats.

Use of digital tools and LMS is habituated among teachers is a well-received outcome from this survey analysis depicted in Fig. 11. 98% of the faculty use one or more of these tools to make learning more effective and to simplify the teacher–student interactive process. There is also an urge from the current generation students to inculcate technology in education to meet the needs of the future. So the institutions across the globe encourage blended learning or the flipped classroom model of teaching to make the educational environment so inspiring.

5.5 Findings

The study made it very clear that Learning Management Systems (LMS) plays a vital role in assisting teachers to conduct Quiz, Collect Assignments, to have an online discussion of the classes, online discussions among students of the same classes and to maintain data of all the above said. It also helps to evaluate standardized test that has pre-defined answers and save a lot of time for physical evaluation for the teachers. Schoology is extensively used by the teachers as they felt that it is easy to administer various activities. Easy class, Google Classroom, Moodle, Edmodo, Classmaker, and Hi-Class are also used by few of them. Except for Hi-Class all other LMS are available free with basic functions which satisfy the needs of the teachers. From the part of students, as all these free LMS can be operated through mobile phone they

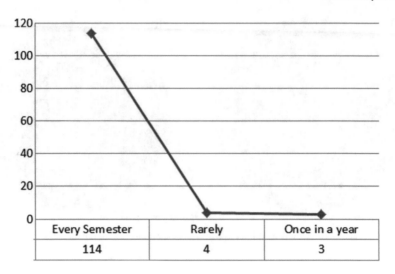

Every Semester	Rarely	Once in a year
114	4	3

Fig. 11 Frequency of using digital tools and LMS

are very comfortable using it. Students can also retrieve data they submitted in the form of quiz and assignments anytime.

Though LMS can save data such as supporting and additional reference materials related to the subject to be shared among students it has limitations in free versions. Hence the top priority of the teachers is to use Google Drive for storing and sharing the data. Though Dropbox and piazza stores data, G Drive tops the list of preference. Teachers also prefer to use Google components such as YouTube for sharing their video Google forms for conducting surveys and quiz, Google Docs for online inter-action and Blogger to update the subject information and other related info that add value to their teaching.

Among various online conferencing facilities which are used for virtual class, GoToMeeting is used and recommended by most of the teachers. Though YouTube live, Big Blue Button, Skype Zoom, rCampus, Wiz Iq and WhatsApp Live are used for virtual class GoToMeeting can collect details of the participants and make it available during the presentation. Moreover, it is easy for the speaker and the listener to interact individually or with the group during the presentation. The complete presentation and discussion can be recorded and saved for further references. This feature is made available in free version subjected to a limited time. This makes it unique and that is the reason many prefer this app.

This study discovered that experienced teachers are ready for the change that is to adopt flipped classroom assisted by technology for better reach and prospect.

6 Conclusion

This study revealed that teachers cannot restrict themselves to follow the traditional method of teaching where learning happens inside their class rather they are expected to follow flipped classroom technique which is a strategy of inculcating knowledge off the classes either through online resources or resources that will nurture their ability to understand the concepts. To be very successful, teachers are expected to go hand in hand with the change in the mindset of the student's behaviour, adoption to the technical gadget, social media and exposing to latest trends the students follow, for better reach and enhance productivity among their students. By adapting to such environments the teacher reduces the generation gap and teaching–learning process. Teachers embracing themselves to these changes are successful in saving time and energy which in turn used for other productive research and career development.

References

1. M. Van Selm, N.W. Jankowski, Conducting online surveys. Qual. Quant. **40**(3), 435–456 (2006)
2. E.F. Gehringer, Daily course evaluation with Google forms, in *ASEE, American Society for Engineering Education Annual Conference & Exposition* (2010)
3. C. Kongchan, How Edmodo and Google Docs can change traditional classrooms, in *The European Conference on Language Learning 2013* (2013)
4. A.S. Sicat, Enhancing college students' proficiency in business writing via schoology. Int. J. Educ. Res. **3**(1), 159–178 (2015)
5. B.E. Perron, M.C. Ruffolo, *A Review of a Web conferencing Technology: GoToMeeting. Research on Social Work Practice* (2010)
6. M. Dougiamas, P. Taylor, Moodle: using learning communities to create an open source course management system (2003)
7. M.N. Marshall, Sampling for qualitative research. Fam. Pract. **13**(6), 522–526 (1996)
8. K.B. Wright, Researching internet-based populations: advantages and disadvantages of online survey research, online questionnaire authoring software packages, and web survey services. J. Comput.-Mediated Commun. **10**(3), JCMC1034 (2005)

Secured Key Management with Trusted Certificate Revocation in MANET

Manikandan Narayanan, Anusha Kannan, Senthilkumaran Ulaganathan and Monesh Reddiar

Abstract MANET is an infrastructure-less network as the topology can change instantly. The nodes are connected wirelessly and are dependent on each other. Due to the existence of many devices that can receive and transmit signals, the use of MANET has increased and the possibility of an attack is high. In MANET, addition of node is done irrespective of the situation; for the node to communicate with another node there has to be a secure and trusted way. Trust is an opinion of a node on another node in a network, it is represented in numerical form. Trust is calculated depending on the previous interaction between two nodes. Trust is used to identify malicious node and revoke its certificate. This paper presents various ways to calculate trust and the methods that can take place in a MANET to revoke a certificate.

Keywords MANET · Certificate revocation · Trust management
Trust calculation · MANET security

1 Introduction

Mobile Ad Hoc Network does not need any preexisting infrastructure, so the demand for this type of network is increasing. MANET is used in military warfare and emer-

M. Narayanan (✉) · S. Ulaganathan
School of Information Technology and Engineering,
VIT University, Vellore, Tamil Nadu, India
e-mail: mkyadhav@yahoo.com

S. Ulaganathan
e-mail: usenthilkumaran@vit.ac.in

A. Kannan · M. Reddiar
School of Computer Science and Engineering,
VIT University, Chennai, Tamil Nadu, India
e-mail: Anusha.k@vit.ac.in

M. Reddiar
e-mail: moneshm.reddiar2016@vitstudent.ac.in

© Springer Nature Singapore Pte Ltd. 2019
S. C. Satapathy et al. (eds.), *Information Systems Design and Intelligent Applications*,
Advances in Intelligent Systems and Computing 862,
https://doi.org/10.1007/978-981-13-3329-3_15

gency/rescue operations as a medium of communication in a rapidly changing environment, locally it can be used in conferences to share a file to multiple devices, and it is also used in extending the access of the Internet from one device to another device without any infrastructure. Nodes in a MANET use certificate to authenticate and communicate with other node in a network. A device can enter and exit the network at any instance of time so securing the network from a malicious node is very important task. Communication between nodes in MANET takes place in multi-hop form, it is effective as each node can receive and transmit data. Mobile Ad Hoc Network is vulnerable to various attacks. Some attacks in MANET are Bad Mouthing Attack, Black Hole, Ballot Stuffing Attack, Time reliant attack, and location reliant attack.

- BMA—In this attack, the malicious node gives false information about other node in a network.
- Black Hole Attack—In this type of intrusion, the malicious node drops packet that passes through it.
- Ballot Stuffing Attack—This attack makes node give positive rating to a node that is performing bad.
- Time Reliant Attack—In this type of attack, the node attacks in a specified time that reduces the probability of the identification of malicious node
- Location Reliant Attack—This attack makes the node do malicious task from different location.

Trust is the amount of belief held by one node on other node. It is the variable by which a node can maintain the opinion on other node in a numerical form. Trust is used to identify malicious node. Each node in a network maintains the trust values of all other node in the network. There are two types of trust, viz., (DT) Direct Trust and (IDT) Indirect Trust. Direct trust means the personal experience of the node interaction with neighboring node. Direct Trust is updated whenever an interaction between two nodes takes place. Indirect trust is the recommendation of the node by other node, indirect trust is used to identify malicious node prior to interaction.

A certificate is assigned to a node by CA (Certificate Authority) whenever a new node enters the network. Certificate contains the authentication and keying details of the node for secure communication. To prevent the nodes in MANET to communicate with malicious node, the certificate of the malicious node is invalidated. The certificate is said to be revoked when the certificate authority updates its certificate revocation list (CRL) or makes the certificate invalid.

There are two approaches to revoke a certificate of malicious node—election-based and non-election based.

Election-Based Approach—If a node is reported malicious, an election is conducted, i.e., all nodes in network are asked to vote and according to the vote the malicious node is revoked, it is time-consuming but effective.

Non-Election Based Approach—If a node is reported malicious the malicious node is immediately revoked without an election it is fast but is more vulnerable compared to election based approach. In this paper, we have proposed an approach to calculate trust and revoke certificate of malicious node in an efficient manner.

2 Literature Survey

It is always a difficult task to reduce vulnerability of MANET due to the infrastructure-less network. Several kinds of (CR) certificate revocation procedures and trust calculation methods have been referred to improve security of the network in this literature. Shabut et al. [1] proposed a recommendation-based trust model. The objective of this paper was to include various properties like trust, confidence value, and deviation value while creating a cluster for secure communication of node in MANET. It also provided an efficient way to calculate trust. Liu et al. [2] presented a certificate revocation method to prevent intruders from contributing in network activities. The method included two type of list WL (Warning List) and BL (Black List) these lists are handled by CA. This method helped in reducing false acquisition. Dahshan et al. [3] proposed a trust-based threshold cryptography revocation scheme for MANETs. It describes ways to share private key of CA after applying hash chain function. Zhao et al. [4] proposed a way to calculate a trust graph that states when the nodes will communicate with each other by taking LCM of the first communication time and exchanging trust values during next interaction with the neighboring node. It reduced the overhead communication cost when a node moves in cyclic track. Harn and Ren [5] presented the concept of (GDC) generalized digital certificate the main purpose of GDC was to provide user authentication/identification and key agreement. They use (DL) Discrete logarithm based and integer factoring based protocols that can achieve authentication of user and secret key establishment. Mahmoud et al. [6] suggested an E-STAR concept for establishing reliable routes in assorted multi-hop wireless networks. E-STAR combines trust systems and payment with a trust-based and energy-aware routing protocol. Li and Liu [7] proposed a fully distributed IMKM. It is implemented by combining threshold cryptography and ID-based multiple secrets. It eradicates the necessity of certificates for authentication and gives more emphasis on effective key management. Haas et al. [8] present a series of steps that accomplish the goal to minimize CRL size, an effective strategy for finding out if certificate is present in CRL, and a way for CRL updates. Jiang et al. [9] proposed Efficient Distributed Trust Model (EDTM) that can assess the trustworthiness of sensor nodes more accurately and prevent the security violation more effectively. Chae et al. [10] proposed a trust calculation method that extensively deals with a harsh on–off attack scenario. Chang and Kuo [11] present a way to calculate trust using Markov Chain Trust model and way to keep a secondary CA on hold in case of failure of primary CA. Abbas et al. [12] proposed a lightweight IDS that defends against nodes using identity switch to cause attack. Venkataraman et al. [13] proposed a trust model based on regression to provide a secured routing. In Yu et al. [6], a method to identify internal attack using a routing algorithm is proposed. Forne et al. [14, 15] point the disadvantages of using CRL due to increasing memory space for every addition of node to the list and has provided an extended CRL concept as a solution. Marchang and Datta [16] provide a TBRP (Trust-Based Routing Protocol) that has minimal computing cost while securing MANET. Nguyen et al.

[17, 18] proved the delay caused in MANET is mostly due to the security blocks and computation added in the network.

3 Proposed Work

In this section, we propose a way to calculate trust and method to revoke certificate.

The proposed method consists of three stages—Creating Cluster, Calculating Trust and Certificate revocation of malicious node.

3.1 Creating Cluster

Cluster is grouping of devices in the network to subcategories. Cluster is formed by using the distance between two nodes. This stage will decrease the number of hops while passing message because the cluster will be closely coupled. The distance between two nodes is used while forming cluster because they are likely to experience the same conditions or environment so grouping them together will reduce the difference of trust values between them.

3.2 Calculating Trust

This paper proposes to calculate direct trust using number of positive and negative communication among the nodes.

$$DT = \alpha_{ij}/\left(\alpha_{ij} + b_{ij}\right) \tag{1}$$

For two nodes i and j, α is the number of successful interaction and β is the number of unsuccessful interaction. The trust value will always be $0 \leq DT \leq 1$.

Whenever a new node enters a network α_{ij} and b_{ij} is assigned 1, therefore, DT of the new node is set to 0.5 by default to give the node chance to communicate with other node at initial stage. In this model, the time is not taken under consideration and no depreciation is applied to trust in respect to time, change in trust occurs whenever there is an interaction between nodes. The node stores trust values in a vector, this allows easy access of the trust value of other nodes in a MANET by neighboring nodes.

CH (Cluster Head) is chosen among the node in cluster with the highest trust average. For 'n' cluster in a network, there will be '$n - 1$' CH because on CH will act as CA. CH is appointed by CA. The CA is replaced with the CH with the highest energy level.

Fig. 1 MANET with five nodes

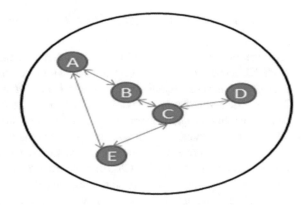

Classification of the interaction to successful or unsuccessful interactions depends on the purpose of MANET, all nodes participating in MANET must follow the specified rules for classification.

For nodes which do not have communication history with each other it uses indirect trust. Indirect trust is calculated using the recommended trust and the confidence value of the neighboring node. Confidence value is the number of times the node has interacted with the recommending node. Confidence value can be used to identify if a node is giving false information or not, if confidence value is high it is more reliable.

$$\text{IDT Check} = \text{Recommended Direct trust} * \text{confidence level} \qquad (2)$$

The recommended direct trust of node with highest IDT check is taken as indirect trust. This method also solves the problem of data sparsity. If the trust level is below the specified threshold in a cluster, it is found to be malicious node (Fig. 1).

Algorithm for cluster head	Algorithm for calculating trust for a node
Step 1-If a malicious node 'm' is reported to CH	Step 1-For each interaction of 'i' with node 'j'
Step 2-Check if node 'm' is in its cluster	Step 2-Calculate direct trust of the node interacted
Step 2.1-If true, report 'm' to CA	Step 3-Update trust of 'j' in 'i'
Step 2.2-else wait until CA call for vote	Step 4-Check IDT of the non-neighbor nodes 'k'
Step 3-If CA calls for vote respect to 'm'	Step 5-If IDT check of the node k is higher
Step 4-Get trust of 'm' from all nodes in cluster	Step 6-Update trust of 'k' in 'i'
Step 5-Compute average of received trust	
Step 6-Send it to CA.	

3.3 Certificate Revocation of Malicious Node

Certification Revocation List (CRL) is a list that is updated only by CA and all CH in MANET has a copy of CRL. The CRL with CA and CH is always same. The instance a node is found malicious, it is reported to CH and CH reports it to CA. The CA calls for a vote from other CH in a MANET. The CH calculates average trust toward the malicious node from all nodes in a cluster and sends it to CA. Restricting only CH to communicate with CA during revocation process leads to the decrease of communication overhead. The CA checks the response of CH and if its less than threshold, the node is added to CRL else the trust value of node is reinitiated to 0.5. It allows falsely accused node to start communicating again in the network (Fig. 2).

A node pretending to be trustworthy can be identified in this model as all nodes in the cluster have to communicate with each other because MANET uses hop by hop communication, so whenever node causes any problem the trust would decrease and would be identified as malicious node. The copy of CRL after any update by CA is sent to all CH in the network. CH checks the list and revokes certificate of the nodes present in the list (Fig. 3).

3.4 Selection of Threshold

The threshold selection is an important aspect in the above-proposed model as the security level depends on the threshold, if threshold is higher the security will be high because all the nodes less than threshold will be revoked. Threshold from 0.7 to 1.0 will provide less scope for malicious activity. Selecting high threshold value comes with some disadvantage, i.e., if a communication is being affected due to some environmental or external component it reduces the trust value of the corresponding node and will result in revocation of the node. Therefore selection of threshold should be analysed including various factors.

4 Implementation and Results

From the above comparison of overall throughput of existing system and the proposed system, we can see that the proposed system is effective (Fig. 5).

Direct Trust

Node	Successful Interaction	Unsuccessful Interaction	DT Value	IDT Value
A				
B	6	4	0.6	
C				
D				
E	4	4	0.5	

Node	Successful Interaction	Unsuccessful Interaction	DT Value	IDT Value
A	3	7	0.3	
B				
C	4	5	0.4444	
D				
E				

Node	Successful Interaction	Unsuccessful Interaction	DT Value	IDT Value
A				
B	9	0	1	
C				
D	1	1	0.5	
E	4	3	0.5714	

Node	Successful Interaction	Unsuccessful Interaction	DT Value	IDT Value
A				
B				
C	0	2	0	
D				
E				

Node	Successful Interaction	Unsuccessful Interaction	DT Value	IDT Value
A	3	5	0.375	
B				
C	6	1	0.8571	
D				
E				

Fig. 2 Table displays direct trust of nodes in Fig. 4 using random interaction classification

Indirect Trust

Node	Successful Interaction	Unsuccessful Interaction	DT Value	IDT Value
A				
B	6	4	0.6	
C				0.8571
D			0.5	
E	4	4	0.5	

Node	Successful Interaction	Unsuccessful Interaction	DT Value	IDT Value
A	3	7	0.3	
B				
C	4	5	0.4444	
D				0.5
E				0.5

Node	Successful Interaction	Unsuccessful Interaction	DT Value	IDT Value
A				0.3
B	9	0	1	
C				
D	1	1	0.5	
E	4	3	0.5714	

Node	Successful Interaction	Unsuccessful Interaction	DT Value	IDT Value
A				0.3
B				1
C	0	2	0	
D				
E				0.5714

Node	Successful Interaction	Unsuccessful Interaction	DT Value	IDT Value
A	3	5	0.375	
B				1
C	6	1	0.8571	
D				0.5
E				

Fig. 3 Table displays indirect trust of nodes in Fig. 4 using random interaction classification

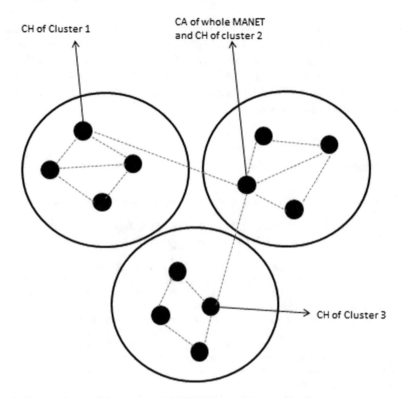

Fig. 4 Communication links of nodes in MANET for certificate revocation

Fig. 5 Comparative result

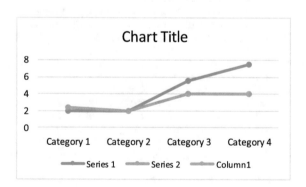

5 Conclusion

In this paper, we have proposed a way to calculate trust, revoke certificate and keeping the key secure after revocation of malicious node. In our design the communication overhead while revocation is less compared to recommendation based revocation or voting based revocation and automatically deals with false rating. The proposed model can be extended by implementing a cryptographic algorithm while assigning a key that won't allow the malicious node to decrypt private key of other nodes. The proposed model whenever simulated in terms of reliability and communication overhead. The simulated model will show less communication overhead and throughput while revocation and will be reliable against attack like bad mouthing by revoking certificate of the malicious node.

References

1. S. Abbas, M. Merabti, D. Llewellyn-Jones, K. Kifayat, Lightweight sybil attack detection in MANETs. IEEE Syst. J. **7**(2), 236–248 (2013)
2. W. Liu, H. Nishiyama, N. Ansari, J. Yang, N. Kato, Cluster-based certificate revocation with vindication capability for mobile ad hoc networks. IEEE Trans. Parallel Distrib. Syst. **24**(2), 239–249 (2013)
3. H. Dahshan, F. Elsayed, A. Rohiem, A. Elgmoghazy, J. Irvine, A trust based threshold revocation scheme for MANETs, in *IEEE 78th Vehicular Technology Conference (VTC Fall)* (2013)
4. H. Zhao, X. Yang, X. Li, CTrust: trust management in cyclic mobile ad hoc networks. IEEE Trans. Veh. Technol. **62**(6), 2792–2806 (2013)
5. L. Harn, J. Ren, Generalized digital certificate for user authentication and key establishment for secure communications. IEEE Trans. Wirel. Commun. **10**(7), 2372–2379 (2011)
6. M. Yu, M. Zhou, W. Su, A secure routing protocol against byzantine attacks for MANET in adversarial environments. IEEE Trans. Veh. Technol. **58**(1), 449–460 (2009)
7. L. Li, R. Liu, Securing cluster-based ad hoc networks with distributed authorities. IEEE Trans. Wirel. Commun. **9**(10), 3072–3081 (2010)
8. J.J. Haas, Y. Hu, K.P. Laberteaux, Efficient certificate revocation list organization and distribution. IEEE J. Sel. Areas Commun. **29**(3), 595–604 (2011)
9. J. Jiang, G. Han, F. Wang, L. Shu, M. Guizani, An efficient distributed trust model for wireless sensor networks. IEEE Trans. Parallel Distrib. Syst. **26**(5), 1228–1237 (2015)
10. Y. Chae, L.C. Dipippo, Y.L. Sun, Trust management for defending on-off attacks. IEEE Trans. Parallel Distrib. Syst. **26**(4), 1178–1191 (2015)
11. B.-J. Chang, S.-L. Kuo, Markov Chain trust model for trust-value analysis and key management in distributed multicast MANETs. IEEE Trans. Veh. Technol. **58**(4), 1846–1863 (2009)
12. A.M. Shabut, K.P. Dahal, S.K. Bista, I.U. Awan, Recommendation based trust model with an effective defence scheme for MANETs. IEEE Trans. Mob. Comput. **14**(10), 2101–2115 (2015)
13. R. Venkataraman, T. Rama Rao, M. Pushpalatha, Regression-based trust model for mobile ad hoc networks. IET Inf. Secur. **6**(3), 131–140 (2012)
14. K. Akkaya, K. Rabieh, M. Mahmoud, S. Tonyali, Customized certificate revocation lists for IEEE 802.11 s-based smart grid AMI networks. IEEE Trans. Smart Grid **6**(5), 2366–2374 (2015)
15. J. Forne, O. Esparza, F. Hinarejos, Certificate status validation in mobile ad hoc networks. IEEE Wirel. Commun. **16**(1), 55–62 (2009)
16. N. Marchang, R. Datta, Light-weight trust-based routing protocol for mobile ad hoc networks. IET Inf. Secur. **6**(2), 77 (2012)
17. M.M. Mahmoud, X. Lin, X. Shen, Secure and reliable routing protocols for heterogeneous multihop wireless networks. IEEE Trans. Parallel Distrib. Syst. **26**(4), 1140–1153 (2015)
18. D. Nguyen, M. Toulgoat, L. Lamont, Impact of trust-based security association and mobility on the delay metric in MANET. J. Commun. Netw. **18**(1), 105–111 (2016)

A New Technique for Accurate Segmentation, and Detection of Outfit Using Convolution Neural Networks

Priyal Jain, Abhishek Kankani and D. Geraldine Bessie Amali

Abstract Wearable Detection is a societally and economically critical yet a very challenging issue because of the number of layers and clothing someone could be wearing. Also layering, pose, body style, and shape become an issue. In this paper, we handle the wearable detection issue using recovery approaches. For model picture, we use the comparable styles from substantial database—labeled pictures and utilize cases to perceive dress things in the inquiry. Our tests come about moreover show that the general posture estimation issue can profit by apparel detection. In addition, for the correct detection and classification of what a person is wearing, we use the process of image segmentation and pose estimation to segment the image into superpixels and then analyze accordingly. In addition, we use a large novel dataset and tools for labeling garment items, to retrieve similar style to help with clothing classification.

Keywords Conditional Random Field model · Convoluted neural networks
Nearest neighbor · K—nearest neighbor · Segmentation

1 Introduction

Apparel decisions fluctuate broadly over the worldwide populace. For instance, one individual's style may define them as preppy while someone else's opinion might differ. Be that as it may, there are shared traits. For example, strolling through a school ground, you may see a great many students reliably wearing mixes of pants, shirts, sweatshirts, also, shoes. Indeed, even trendy people who indicate being autonomous in reasoning and dress, tend towards wearing comparable clothing comprising of minor departure from tight pants, catch shirts, and thick glasses. Now and again, clothing decisions are a solid prompt of visual acknowledgment [1]. As a solid case, shirts have an inconceivably extensive variety of appearances in view of cut,

P. Jain · A. Kankani · D. Geraldine Bessie Amali (✉)
School of Computer Science and Engineering, Vellore Institute
of Technology, Vellore, India
e-mail: geraldine.amali@vit.ac.in

© Springer Nature Singapore Pte Ltd. 2019
S. C. Satapathy et al. (eds.), *Information Systems Design and Intelligent Applications*,
Advances in Intelligent Systems and Computing 862,
https://doi.org/10.1007/978-981-13-3329-3_16

shading, sample, and material. This distinguishes some portion of the outfit as top exceptionally difficult. Fortunately, a specific decision of the parameters, e.g., green checked catch down, numerous tops with comparative looks. The visual compatibility and the presence of some consistency in the style decisions mentioned above are of use in the framework. Here, we adopt an information-driven strategy for attire detection [2]. We start by gathering an extensive, mind-boggling, genuine gathering of outfit pictures from an informal community centered on meld. Utilizing a little arrangement of hand parsed pictures in mix with the content labels related with each picture in the gathering, we can parse my huge database precisely. Presently, given an inquiry picture without any related content, we can anticipate a precise parse by recovering comparative clothing from detected gathering, making nearby models using recovered apparel things, and exchanging construed apparel things from the recovered examples to the inquiry picture. Last iterative smoothing produces my final product. In every one of these means, we exploit the connection amongst dress and body stance to compel forecast and create more precise detection.

1.1   Theory

Successful clothes detection and classification emphatically relies upon precise human body posture limitation. In this paper, we demonstrate exact proof that semantic garments division is advantageous for enhancing posture estimation [3]. Our objective is anticipating semantic name (e.g., pants, shirt, and shoes) by pixel on the individual. Apparel detection is a moderately new PC vision assignment, however, it is critical to empower applications, to develop helpful garments classifiers. Early works on attire classification demonstrated attire as a sentence structure of portray layouts. Other work adopted a subspace strategy to portray attire mistakes or deformable spatial priors. These methodologies for the most part center on how to show shape disfigurements for apparel acknowledgment. Attire detection is detailed as MAP estimation—superpixel names utilizing Conditional Random Field Model (CRF). The primary knowledge of this strategy is the utilization of people's posture estimation [3] for apparel detection [4].

1.2   Similar Works

The gap in the studies was the lack of a system for clothing. Various image recognition software programs have been created, but nothing has yet been made for various wearable for people. The previous approaches have been working on majorly focusing on either pose estimation or just detecting one type of clothing. None of them are doing it for such a huge range of wearable. Our paper approaches the same challenge of detecting several wearables with a good accuracy. This would be really useful for surveillance and to keep tabs on people by the government for safety. To pin people in crowds, fashion analysis, trend pattern analysis, and fashion show analysis. All of these are various other applications of this program. And most existing systems are on openCV [5], a heavier system and therefore slower. The best system would be using CRF (Conditional Random Field) [6].

This paper is organized as follows. Methodology is presented in Sect. 2. Implementation and the results observed in Sect. 3. Conclusion is presented in Sect. 4.

2 Methodology

This paper utilizes Fashionista dataset given in the extension called Paper Doll [7] dataset. Fashionista dataset gives 685 completely detected pictures which are used to regulate preparing and execution assessment, 456 for preparing and 229 to test. Preparation tests are utilized to learn highlight changes, to build worldwide attire models and modifying parameter. Testing tests are saved to assess. The Paper Doll dataset is an expansive gathering—labeled design pictures. We found a collection of gathered—more than 1 million—pictures from chictopia.com with related metadata labels signifying qualities, for example, shading, dress thing, or event. These outcomes in 339,797 pictures pitifully commented on with attire things and assessed posture. Despite the fact that the explanations are not generally total—clients frequently do not name every single delineated thing, particularly little things or extras—it is uncommon to discover pictures, where a commented on tag, is absent. We utilize the Paper Doll dataset for style recovery (Fig. 1).

2.1 Process

We calculated the pose distance first by introducing 27 joints assessed by posture estimator acquiring 14 focuses on the body. After acquiring focus on the 14 points, we calculate the log-remove change for each and every point. In this way, for an N-dimensional element vector, we generally learn $2N + 1$ parameters [8]. We discover

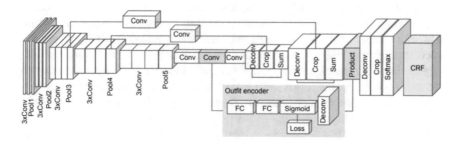

Fig. 1 Our segmentation model based on fully convoluted network architecture

Fig. 2 Position estimation

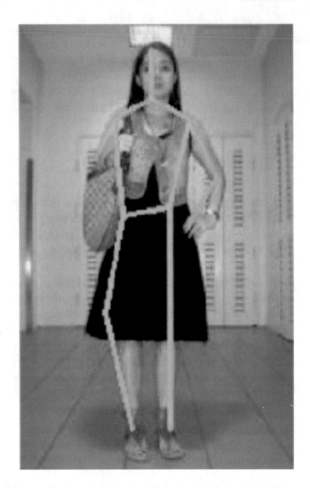

parameters of calculated relapses by 3-criss-cross approval inside preparing data of thick component writes (each detection technique utilizes some subset) (Fig. 2).

In view of enormous number of the pixels in dataset, we do subsample the pixels to prepare every one of all the strategic relapse models. Amid the subsampling, we attempt at testing the pixels with the goal that the subsequent name circulation is near uniform in each picture, keeping the taken subpixels in models from just anticipating substantial things. We use the recovered nearest neighbours (NN) [9] pictures to calculate a relapse, which is available as a second term in our model. Here we take a neighborhood and check for the presence of each wearable in view of cases that are like the inquiry, e.g., coats that seem to be like the question jacket since they were recovered by means of style likeness. These nearby models are vastly improved models for the inquiry picture than those prepared all inclusive.

We gained the model parameter locally using recovered examples, utilizing Boundary Distance, Gradient, RGB, Lab, MR8, and Pose Distance. In the progres-

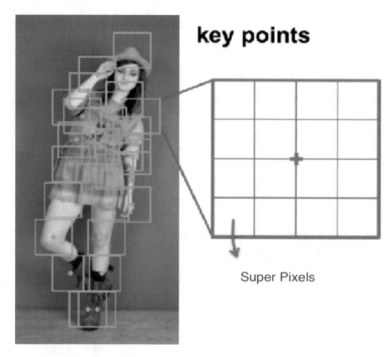

key points

Super Pixels

Fig. 3 Super pixel sample

sion, we learn neighborhood look models utilizing anticipated pixel recovered from examples figured amid pre-preparing. We then prepare the Nearest Neighbor (NN) models utilizing pixel (including subsampling) in recovered examples in one versus all form [10].

Our approach initially figures an over division of both inquiry and recovered pictures utilizing a quick and straightforward division calculation, at that point finds comparing sets of super pixels between the question and each recovered picture in light of posture and appearance: For every super pixel in the question, discover the five closest super pixels in each recovered picture utilizing Pose Distance [11] (Fig. 3).

Now we use the Conditional Random Field (CRF) [1] approach. In this paper, we define the issue is one of the deduction in an available Conditional Random Field (CRF) [12] which considers conditions amongst garments and human posture. We build a rich arrangement of possibilities in which to encode the individual's appearance to the shape and to perform figure/background division. This includes the shape and area probabilities for each of the and every article of clothing, which we call clothelets. This also includes the similitude between fragments establishing—shirt pixels—on the whole body to concur with the T-shirts' available pixels on each person's hand and arm [13]. Utilize a wide range of nearby highlights encoding material appearance, additionally in the neighborhood state of the individual's parts

[14]. We exhibit viability of the approach the available Fashionista dataset appearing that the approach essentially beats the current best.

2.2 Mathematical Formulation

Limb Segment: Use the per class of bias for every limb segment available for capturing location-specific bias, like, shoes on feet:

$$\emptyset_{p,j}^{\text{bias}}(l_p) = \begin{cases} 1, \text{if } l_p = j \\ 0, \text{otherwise} \end{cases} \tag{1}$$

Compatibility Segmentation: We find potentials limb that connects as limb segments with the superpixels nearby to encouraging agreement in the labels. With this goal in mind, we bring out a Gaussian mask situated inside the two joints. For 2 joints that are consecutive with the coordinates $J_a = (u_a, v_a)$ and $J_b = (u_b, v_b)$, we define a mask that is based on following Normal distribution:

$$M(J_a, J_b) = N\left(\frac{J_a + J_b}{2}, R\begin{pmatrix} q_1||J_a - J_b||0 \\ 0q_2 \end{pmatrix} R^T\right) \tag{2}$$

This is R, being 2D rotation matrix and q_1 and q_2 are two hyper parameters controlling spread of the mask in a longitudinally and the transversely available, respectively. The strength of the connection, it is always like an overlap between the super pixels and Gaussian mask:

$$\emptyset_{i,p}^{\text{comp}}(y_i, l_p) = \begin{cases} M(J_a, J_b).S_i \text{ if } y_i \neq 1 \text{ and } y_i = k_p \\ 0, \text{otherwise} \end{cases} \tag{3}$$

Full: We interpret energy of full model as sum of three types of energies, unary and pairwise potential depending on superpixel labeling, and term linking—limb segments and super pixels:

$$E(y, 1) = E_{\text{unary}}(y) + E_{\text{similarity}}(y) + E_{\text{limbs}}(y, 1) \tag{4}$$

3 Results and Discussion

The proposed system accurately detects all the wearables in a given image and color codes them and gives the final output image in this format. Our model beats the earlier work on numerous things, particularly significant frontal area things, for

Input	Truth	[1]	Ours

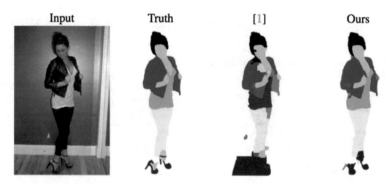

Fig. 4 Output sample

example, dress, pants, coat, shorts, or skirt. This outcome in a critical lift in closer view exactness and perceptually better detection outcomes (Fig. 4).

We assess detection execution on the testing tests from the Fashionista dataset. The undertaking is to foresee a name for each pixel where names speak to an arrangement of 56 unique classes—a vast and testing assortment of attire things. Execution is estimated as far as standard measurements: exactness, normal accuracy, normal review, and normal pixels [15]. What is more, we likewise incorporate closer view precision as measure of precisely every technique is at detection forefront districts (the pixels on top of the body, not on foundation). Note that normal measures—over non-purge names subsequent to computing pixel-based execution for each since a few names are absent in the test set. The underneath table outlines prescient execution of my detection strategy, including a breakdown of how well the middle of the road detection steps performs. We watch that my technique creates a parse that regards the genuine thing limit, regardless of whether a few things are inaccurately marked; e.g., foreseeing pants as pants, or coat as overcoat. In any case, frequently these disarrays are because of high similitude in appearance amongst things and now and again because of non-selectiveness in thing writes, i.e., pants are a kind of jeans. Our model beats the earlier work on numerous things, particularly significant frontal area things, for example, dress, pants, coat, shorts, or skirt. These outcomes in a critical lift, and in closer view, exactness and perceptually better detection outcomes [16].

Our system achieved an accuracy of 85.92% on the original Fashionista v0.2 Dataset [2] is presented in Table 1.

Table 1 Comparison of the accuracy obtained with the different methods on the Fashionista v0.2 dataset

Dataset	Method	Accuracy (%)
Fashionista v0.2	PaperDoll	82.68
	Clothelets CRF	83.71
	FCN-8	84.51
	+CRF	85.92

4 Conclusion

This paper proposes an effective method to produce an intricate and accurate parse of a person's outfit. It also discusses a methodology to parse, segment, detect, and classify all the clothing items being worn by the person in the given image. The segmentation works in combination with pose estimation to be able to function properly and accurately. This type of functionality could help with the classification of groups based on clothing to further be able to study the effect of the socioeconomic and the cultural background, on each individual's style of dressing. The future work would include adding estimation for partial body pictures and images where the pose cannot be estimated. Also, adding more accessories as well, such as bindi, earrings, etc. The model will be further optimized to get even more accurate. Also, adding a prevention mechanism to detect clothing that could possibly be a conflict, such as boots and shoes, or a dress and a skirt that tend not to be worn together.

References

1. X. He, R.S. Zemel, M.A. Carreira-Perpinan, Multiscale conditional random fields for image labeling, in *Proceedings of the 2004 IEEE Computer Society Conference on Computer Vision and Pattern Recognition, CVPR* (2004). https://doi.org/10.1109/cvpr.2004.1315232
2. G. Mori, X. Ren, A.A. Efros, J. Malik, *Recovering Human Body Configurations: Combining Segmentation and Recognition* (2004). https://doi.org/10.1109/cvpr.2004.1315182
3. Large-scale Fashion (DeepFashion) Database: http://mmlab.ie.cuhk.edu.hk/projects/DeepFashion.html
4. Y. Kalantidis, L. Kennedy, L.-J. Li, Getting the look-clothing recognition and segmentation for automatic product suggestions in everyday photos, in *International Conference on Multimedia Retrieval (ICMR), ACM*, Dallas, TX, (2013)
5. Open CV Tutorial Docs. https://docs.opencv.org/master/d9/df8/tutorial_root.html
6. L.-X. Chang, W.-D. Gao, X. Zhang, Discussion on fashion color forecasting researches for textile and fashion industries. J. Fiber Bioeng. Inform. **2**. https://doi.org/10.3993/jfbi06200902
7. J. Long, E. Shelhamer, T. Darrell, Convolutional networks for semantic segmentation, in *2015 IEEE Conference on Computer Vision and Pattern Recognition (CVPR)* (2015). https://doi.org/10.1109/cvpr.2015.7298965
8. P. Guan, A. Weiss, A.O. Balan, M.J. Black, Estimating human shape and pose from a single image. Published in *2009 IEEE 12th International Conference on Computer Vision*. https://doi.org/10.1109/iccv.2009.545930017

9. J. Shotton, A. Fitzgibbon, M. Cook, T. Sharp, M. Finocchio, Real-time human pose recognition in parts from single depth images, in CVPR 2011 (2011). https://doi.org/10.1109/cvpr.2011. 5995316
10. Conditional Random Fields, https://people.cs.umass.edu/~mccallum/papers/crf-tutorial.pdf
11. J. Lafferty, A. McCallum, A. McCallum, Conditional random fields: probabilistic models for segmenting and labeling sequence data, in *ICML'01 Proceedings of the Eighteenth International Conference on Machine Learning*, pp. 282–289 (2001)
12. S. Zheng, S. Jayasumana, B. Romera-Paredes, V. Vin, Conditional random fields as recurrent neural networks, in *2015 IEEE International Conference on Computer Vision (ICCV)* (2015). https://doi.org/10.1109/iccv.2015.179
13. PennLibraries, https://repository.upenn.edu/cgi/viewcontent.cgi?article=1162&context=cis_ papers
14. L. Wang, J. Shi, G. Song, Object detection combining recognition and segmentation, in *ACCV 2007: Computer Vision*, pp. 189–199 (2007)
15. B. Lao, K. Jagadeesh, Convolutional neural networks for fashion classification and object detection, in *CCCV 2015: Computer Vision*, pp. 120–129
16. Z. Liu, P. Luo, S. Qiu, X. Wang, X. Tang, DeepFashion: powering robust clothes recognition and retrieval with rich annotations, in *2016 IEEE Conference on Computer Vision and Pattern Recognition (CVPR)* (2016). https://doi.org/10.1109/cvpr.2016.124

An Investigation on Educational Data Mining to Analyze and Predict the Student's Academic Performance Using Visualization

J. Dheeraj Kumar, K. R. Shankar and R. A. K. Saravanaguru

Abstract Presently, educational institutions compile and store huge volumes of data such as student's enrollment details, academic history, attendance records, and as well as their examination results. Traditional data mining approaches cannot be directly applied for visualization so we are using Pandas software library framework for pre-processing of the academic's data and visualization of the data using matplotlib and seaborn libraries are used in this approach to get better results and easily understand and predict the outcomes from the data.

Keywords EDM · Academic performance · MatplotLib · Visualization

1 Introduction

Data visualization to analyze the educational datasets and to get the exact and appropriate insights is aimed in this paper. The EDM [1] is a very vast and huge data which are stored in various warehouses in various databases through the world in the different universities, government colleges, private colleges, etc. [2, 3] which stores the information of the student's educational background details, academic profile details, exam results, performance metrics, and so on. Which synthesis them for understanding the performance of any student in the upcoming examinations held in their colleges or any competitive exams or any technical exams held on the colleges and universities which helps them to either estimate or predict the performance of the student's with their available previous record's in the databases by performing

J. Dheeraj Kumar (✉) · K. R. Shankar · R. A. K. Saravanaguru
School of Computer Science & Engineering, Vellore Institute
of Technology, Vellore 632014, Tamil Nadu, India
e-mail: dheerajk994@gmail.com

K. R. Shankar
e-mail: shankar.kr@vit.ac.in

R. A. K. Saravanaguru
e-mail: saravanank@vit.ac.in

© Springer Nature Singapore Pte Ltd. 2019   179
S. C. Satapathy et al. (eds.), *Information Systems Design and Intelligent Applications*,
Advances in Intelligent Systems and Computing 862,
https://doi.org/10.1007/978-981-13-3329-3_17

the data mining algorithms such as clustering, prediction which helps the researchers to predict the outcomes of any student and make any sense-making decisions in the academic improvements in the universities and further produce any new algorithms for getting more insights in the data and make any good assumptions by preparing proper visualizations for the data insights outcomes by using appropriate tools such as Tableau for the better visualizations [4] and easy understandability [5]. And which further could also help to know whether how many students are interested to proceed to their higher education and carry their performance the same way [6, 7]. As their being a lot of advancement in the technology and software which also reflects in the Education of Learning analytics [8] and growth of the knowledge towards the future of technology into education which could be either Scientific, Information, Knowledge, and visual analytics.

1.1   Literature Survey

A.   Summary of the existing work in EDM

The outline of forecast models in the field of educational data mining there has been a dynamic research region in the past decades all the work related to educational data [9], the learning analytics and knowledge the linked data for the evaluation and analysis the massive datasets and small datasets have been worked out a lot (Table 1).

As the author Dietze [2, 10] in the paper talks about the survey research carried out about the use of the linked data for analysis of scientometrics and LAK dataset used here is all about the information they have carried from the journals, conferences, and even publications have been included for the analysis of the learning analytics in education. And the same way the educational institutions and the governing practices are being increased the growth of digital learning in education has also risen up a lot in the decade which helps the researchers to trace the student's performance with the help of digital education governance says author [7, 5] Williamson et al., and such as Pearson and knewton have told about visualizing [11] about the data and the emerging digital policy's.

B.   Visualization Types

Scientific: Scientific [12] visualization majorly focuses on the insights of the data that it particularly deals with data and about its content, and it is also referred to as [13] visual data analyzation.

Education: Education visualization is to be taught about to implement a picture of something with the help of simulation, for example, topic related to microorganisms which are very minute in size [14] which cannot be easily studied without the help of scientific equipment.

Information: Information visualization [15] helps in the visual modeling of the unreal data to strengthen up human for acquiring knowledge and exploring the [16] large amounts of data. For example, analyzing any data such as static or charts or maps with needed information (Table 2).

Table 1 Summary work in EDM

Year/Author	Concept/Technique	Outcome/Analysis	Open issues
2018 Villegas-Ch, Luján-Mora	Big Data, Smart Data and Data Lake	Data Lake is advantageous for educational environment, and the data collected from access card	If all the sensors and systems surround by the student converge into one
2018 Khare, K., Lam, H.	Classification, Cluster analysis, Knowledge tracing, Association rule mining (ART)	Online business learning education setting, Feedback for instructors, predicting student performance	Data privacy has been a big issue and the use of information
2017 Dietze S, Taibi	Learning analytics and knowledge	Machine processing, Scholarly resources	Five-star classification and vocabulary
2016 Williamson B	Machine readability and visualization	Digitization of Educational datasets	Statistical, Computational model
2015 Xing, Petakovic	Geogebra, Subject, Rules, Tools, Community, Division of labor, object	Activity based measures, Performance predictions, Sample model	Increase of prediction rate
2014 Srecko Natek, Moti	Building student success rate using MS Excel Tool, and MS SQL Server	Key Influencer for student, data interpretation, REPTree model	International student datasets
2013 Romero C, Sebastian	Sentiment analysis, Sense-making, Classification & clustering	Knowledge tracing, nonnegative Matrix factorization	EDM tools, Course designer tools
2012 Siemens, D. Baker	Success prediction, Social Network analysis, Bayesian modeling, Clustering	Model generalizability, multi-level cross-validation	Opportunities to influence nonacademic
2011 Khan M, Khan S	Scientific, Software, Information, Data, Concept, Strategic	Various categories of visual analysis such as interactive, visual forms	Should be able to represent good interactive
2010 Romero, Ventura	Analysis and visualization of data, providing feedback for supporting instructors	Difficultness of applying data mining algorithms with the initial parameters	several areas of e-learning, adaptive hypermedia, intelligent tutoring
2009 Chen M, Ebert	Perceptual and cognitive space, Computational space	Reduce the search space for optimal control parameters	Significant amount of ongoing development

(continued)

Table 1 (continued)

Year/Author	Concept/Technique	Outcome/Analysis	Open issues
2005 Ahrens, Geveci	Parallel and distributed visualization toolkit, Paraview, Level of detail and parallel rendering	Data partitioning and to demonstrate the multi-block datasets	Unable to process the datasets for this specific magnitude
2004 Peng, Ward	Analysis of parallel coordinates using clutter, Algorithm for computing clutters	With the help of heuristic algorithms, we can reorder the higher dimensions for good results	Dimension-Clutter reduction and hierarchical visualization and for high volume datasets
2001 Borner, Zhou	Implementing a software repositories and collecting the sites and the information for visualization's and doing projects	The repository was developed in java and the packages were provided with the references for	Implementation of new commercial applications
1996 Fayyad Piatetsky Shapiro	Preference criterion, model, the search algorithm, data mining and knowledge data discovery of the hidden	Model functions and representations and optimizing the algorithms	Handling the massive datasets with high dimensionality, user interaction
1990 Rew, Davis	The abstraction and interface of (SV) Scientific visualization and the dimensions, variables and attributes	Conventional access and appending to the NeTCDF, direct access and hyper slab access, unlimited dimension and the data format	There should be more comprehensive variable names
1988 Myers, Chandhok, Sharoon	Object–oriented, recursive display routines for visualization of objects, and Integration	Special and custom displays helped in the visualization of various objects with correct measure and proposed fit of the objects	The manipulation of actual data algorithms was quite difficult for the programmers to implement

Table 2 Various tools for visualization

Open Source	Proprietary
GoogleCharts	Tableau
Kibana	Qlik
RawGraphs	Sas
Tableau Public	DataWrappers
Lumify	Infogram

2 Implementation

Here, we are using Python Pandas framework for the prediction of students' academic performance. The data set characteristic is multivariate and number of instances is 22,242 records. The area under which it is included is e-learning, education, predictive models [17], and educational data mining. The attribute characteristics is integer/categorical and the number of attributes are 14, and the file format used here is the Experience API (X-API) Educational Data.csv, the analysis of the data is done by removing the duplicate records from the dataset and visualizing the data with matplotlib and seaborn libraries is used for better interactive visualizations for the end users. The tools required for the processing the data are anaconda continuum software with packages and libraries required for Python, Jupyter notebook, and Anaconda Navigator command prompt. The steps in preprocessing of data include are information, scientific and education data related, cleaning data with Pandas, exploring data with Pandas, conveying your data, changing data into interactive visualizations with seaborn and matplotlib, information integration and transformation of data, and data reduction is done (Fig. 1).

3 Results and Discussion

For the analysis of the dataset, here, we have visualized more than 20 plus different visual diagrams in different ways for interactive's such as bar charts, box plots, swarm plots, pair plots, etc. Students registered to all the 12 courses and the visualization of the courses is shown below (Fig. 2).

The different course topics are IT, Math, Arabic, Science, English, Quran, Spanish, French, History, Biology, Chemistry, and Zoology. Thus, the grades are classified into three different numerical intervals and the grades/marks are allotted on basis of these intervals and their courses registered. Student's performance grades in three numeric intervals with bar plot analysis. The box plot, swarm plot, and KDE plot analysis for students who raised hands in all course topics registered and attended the courses both the gender wise and from different nationality backgrounds is analyzed here also based on the students have listened to the discussions and announcement views and have they visited resources or not. The kernel density plots show the clear picture of students with 1 and 0 indications of the pass and fail candidates (Fig. 3).

Students who have listened to the discussions conducted while the class intervals the performance variation is shown below. It is clear that the lowest performers rarely visited the course resources. Now that, we see the student's academic behavior by marks and this pattern is consistent in the Geology class (Fig. 4).

Based on these, all the final analyzation is whether they are listening to the announcements view in the active hours of the classes or they are not listening to them (Fig. 5).

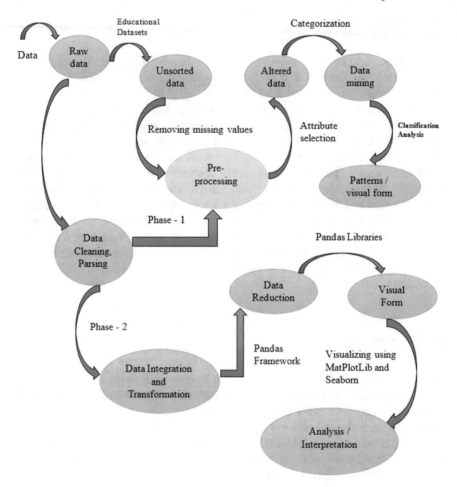

Fig. 1 Methodology framework

As there different stage id and assigned grade ids for different intervals for different courses, the students have obtained is visualized here. The students absence reports have also helped here for the better analyzation of the results and outcomes of the grades in all course topics as they are present continuously in a week or absence continuously in any of the courses registered (Fig. 6).

The parent school satisfaction survey report has also been conducted for the feedback and the analysis of the student's parents opinion and suggestion are taken into considerations here. The heat map analysis gives us the overview description of the whole dataset here in a clear picture.

	raisedhands	VisITedResources	AnnouncementsView	Discussion	Failed
Topic					
Arabic	32.0	65.0	41.0	38.0	0
Biology	78.5	88.5	54.0	47.0	0
Chemistry	79.0	84.5	47.0	30.5	0
English	55.0	50.0	33.0	36.0	0
French	35.0	80.0	23.0	21.0	0
Geology	80.0	82.0	68.5	60.5	0
History	69.0	84.0	72.0	65.0	0
IT	20.0	25.0	10.0	40.0	0
Math	28.0	15.0	19.0	40.0	0
Quran	65.0	75.0	50.0	45.0	0
Science	62.0	64.0	58.0	66.0	0
Spanish	27.0	51.0	40.0	20.0	0

Fig. 2 Subject-wise marks scored by students in all course topics

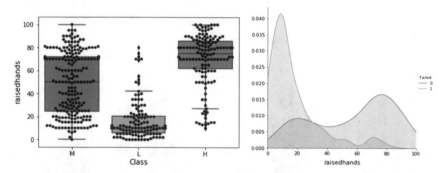

Fig. 3 Performers who are active and least active

4   Conclusions and Future Enhancements

C.   *Conclusion*

The data here used is collected with the tracker activity tool which is actually called as (X-API). The X-API is a component used for the architecture of learning and training which helps to auditor the progress of learning and the actions of learner's such as watching any video or reading any article. We could predict that from the dataset used here students X-API data where the insights are that the females in their academics were performing well when compared over with the males and the parent survey

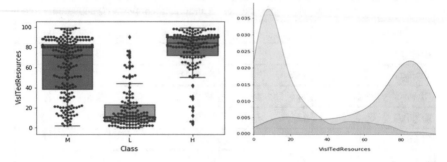

Fig. 4 Students who have visited the resources

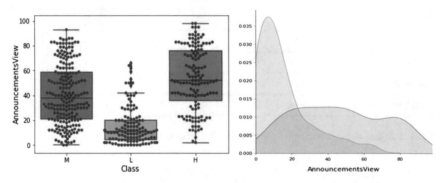

Fig. 5 Students who are active at listening announcements view

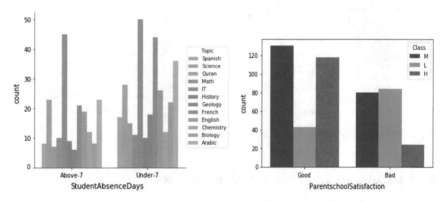

Fig. 6 Topic-wise absence report and degree of parent satisfaction from school

satisfaction gave the results about the mother who was more precautious in taking care of the female child while over when compared with father towards the male child. In all topics when compared over the students' performance and participation was good in geology subject with less failures, and as their attendance records have also helped to know which students could score good. Based upon the students using the resources or not like they have visited resources, listening to the announcements view, has raised hands in class or not and were they hearing to the discussions. Based on these activities, we have analyzed and predicted the results well. Also, the parent answering survey and parent school satisfaction has helped in knowing the students grade id and their performance in different semesters.

D. *Future Enhancements*

The further enhancement could be carried out here is using different model for analysis which may give even more precise results, and we could also use XGB xtreme gradient boosting algorithm for improving the performance, speed and getting a tree-based visualization.

Acknowledgements We undertake that we have the required permission to use images/dataset in our work from suitable authority and we shall be solely responsible if any conflicts arise in the future.

References

1. C. Romero, S. Ventura, Educational data mining: a review of the state of the art. IEEE Trans. Syst. Man Cybern. Part C (Appl. Rev.) **40**(6), 601–618 (2010)
2. W. Villegas-Ch, S. Luján-Mora, D. Buenaño-Fernandez, Palacios-Pacheco X, Big Data, The next step in the evolution of educational data analysis, in *International Conference on Information Theoretic Security*, (Springer, Cham, 2018), pp. 138–147
3. A. Peña-Ayala, Educational data mining: a survey and a data mining-based analysis of recent works. Expert Syst. Appl. **41**(4), 1432–1462 (2014)
4. B.A. Myers, R. Chandhok, A. Sareen, *Automatic Data Visualization for Novice Pascal Programmers*, pp. 192–198 (1988)
5. B. Williamson, Digital education governance: data visualization, predictive analytics, and 'real-time' policy instruments. J. Educ. Policy **31**(2), 123–141 (2016)
6. C. Romero, S. Ventura, Data mining in education. Wiley Interdis. Rev.: Data Min. Knowl. Discovery **3**(1), 12–27 (2013)
7. K. Khare, H. Lam, A. Khare, *Educational Data Mining (EDM): Researching Impact on Online Business Education* (Springer, Cham, 2018, In On the Line), pp. 37–53
8. G. Siemens, R.S. d Baker, Learning analytics and educational data mining: towards communication and collaboration, in *Proceedings of the 2nd International Conference on Learning Analytics and Knowledge* (ACM, USA, 2012), pp. 252–254
9. U. Fayyad, G. Piatetsky-Shapiro, P. Smyth, The KDD process for extracting useful knowledge from volumes of data. Commun. ACM **39**(11), 27–34 (1996)
10. S. Dietze, D. Taibi, M. d'Aquin, Facilitating scientometrics in learning analytics and educational data mining–the LAK dataset. Semant. Web **8**(3), 395–403 (2017)
11. J. Ahrens, B. Geveci, C. Law, Paraview: an end-user tool for large data visualization. Vis. Handb. **717** (2005)
12. R. Rew, G. Davis, NetCDF: an interface for scientific data access. IEEE Comput. Graphics Appl **10**(4), 76–82 (1990)
13. M. Chen, D. Ebert, H. Hagen, R.S. Laramee, R. Van Liere, K.L. Ma, ... D. Silver, Data, information, and knowledge in visualization. IEEE Comput. Graphics Appl. **29**(1) (2009)
14. W. Peng, M.O. Ward, E.A. Rundensteiner, Clutter reduction in multi-dimensional data visualization using dimension reordering, in *IEEE Symposium on Information Visualization, INFOVIS*, pp. 89–96 (2004)

15. M. Khan, S.S. Khan, Data and information visualization methods, and interactive mechanisms: a survey. Int. J. Comput. Appl. **34**(1), 1–14 (2011)
16. K. Borner, Y. Zhou, A software repository for education and research in information visualization, in *Fifth International Conference on IEEE Information Visualization Proceedings*, pp. 257–262 (2001)
17. W. Xing, R. Guo, E. Petakovic, S. Goggins, Participation-based student final performance prediction model through interpretable genetic programming: integrating learning analytics, educational data mining and theory. Comput. Hum. Behav. **47**, 168–181 (2015)

Real-Time Hierarchical Sensitivity Measure-Based Access Restriction for Efficient Data Retrieval in Cloud

A. Antonidoss

Abstract The problem of data retrieval from cloud environment has been discussed in many articles. However, restricting the cloud users from illegal access is identified as a challenging issue. Toward the problem of access restriction, different attribute-based approaches have been discussed earlier but suffer to achieve higher performance in data retrieval also. To solve this issue, an efficient real-time hierarchical sensitivity measure-based access restriction and data retrieval algorithm is presented in this paper. The method classifies the cloud data into different classes according to their importance, and for each user taxonomy of access has been maintained. At the query time, the method estimates hierarchical sensitivity support (HSS) according to the data items being accessed. Based on the HSS on each level, a cumulative data retrieval support (CDRS) is estimated. Estimated CDRS value has been used to perform access restriction and data retrieval. The proposed method improves the performance of data retrieval and access restriction.

Keywords Cloud data · Access restriction · Data retrieval · Hierarchical data HSM · HSS · CDRS

1 Introduction

Maintaining huge organization data in a distributed environment have been a routine in the last decade, where the organizations maintain different data servers in various geographic locations. Toward this, the organizations have spent huge amount in maintaining data servers. As the number of micro-, small-, and medium-scale organizations has increased, they suffer to spend such huge amount for the data maintenance. This problem of cost has been a challenging issue for their development and not all the organizations could invest in that. In the meantime, the development of

A. Antonidoss (✉)
Department of Computer Science & Engineering, Hindustan Institute
of Technology and Science, Chennai, India
e-mail: a.antoniphd@gmail.com

© Springer Nature Singapore Pte Ltd. 2019   189
S. C. Satapathy et al. (eds.), *Information Systems Design and Intelligent Applications*,
Advances in Intelligent Systems and Computing 862,
https://doi.org/10.1007/978-981-13-3329-3_18

information technology has introduced the cloud, which is a distributed computing paradigm where the organizations need not afford such huge amount in maintaining their data.

The cloud is the medium where there are service providers who maintain dedicated data servers which can access the data servers to store and retrieve their data. The organizations would store their data in the cloud which is provided by the cloud providers and through a set of services. The users of the organizations would access the data whenever they required. In real context, the service provider does not know the identity of the user who accesses the service. Also, the organization would maintain various forms of data which can be accessed by various types of users. However, it is necessary to restrict the users from malicious access because not all the user can access all organization data stored in the cloud. For example, in a hospital organization, there will be various types of users like doctors, pharmaceutical person, management, and nurses. The management person would access the overall data of management, accounts, and so on. Similarly, the doctor would access all the data related to the patients where the nurses would access basic data of patients and the medicines recommended. So all the types of users would have certain access limitation. The profile-based approaches in access restriction are enforced in different articles. The profile-based approach is not efficient in various situations where for the other user it would be necessary to access some important information related to a person. So, the access restriction issue needs a strategical approach to produce efficient data retrieval.

In the profile-based approach, the user from other profile would not obtain some basic information related to a person. This increases the necessity of organizing the data in different forms. The organizational data would contain various information related to their employee, customers, and other business information. In employee data, there will be many information like personal, general information like address, phone, salary, etc. Such data can be classified into different hierarchical orders. The first level would contain the person name, position, and contact number. The second level would contain address, and the third level would contain the family details. Similarly, any data can be arranged in a hierarchical manner where each level of data has different importance. To provide efficient access restriction, it is necessary to organize the data into different levels according to their importance. This paper focuses on such approach which organizes the data in a hierarchical way and explains how the restriction with data retrieval is performed.

The hierarchical sensitivity strength represents the support measure for the input query according to the user. The user would have access to different attributes and some of them would present in the query and some of them not. So based on the size/number of attributes, the user either has access or does not, and the HSS measure can be measured. Again, at each level, the HSS measure can be estimated. Once the HSS measure for all the levels of the hierarchy has been measured, the cumulative data retrieval support measure can be estimated. The detailed approach is discussed in the next section.

2 Related Works

The problem of access restriction has been handled with various approaches. A set of methods has been reviewed toward efficient access restriction.

In [1], the author presented an access control approach based on attributes. The method uses the size of ciphertext in a constant manner. This reduces the time complexity of both encryption and decryption. Also, the method restricts the user in a hierarchical manner where a single user has been allowed to access a certain level of attributes and not to the further levels.

In [2], the role-based access control has been enforced based on the availability and security of the attributes. The method estimates the relationship on the trust based on the network availability. Similarly, in [3], the trust relationship has been measured to restrict the user access in the cloud. To measure the trust degree, a trust weight has been estimated based on the feedback obtained.

A session-based ABAC is presented in [4] toward the video sharing. To access the video contents of the cloud, the user should have enough keys which are applicable for the specific time window. On the other side, the revocation of user access has been enforced for untrusted cloud in [5]. In this approach, the access of invalid user at any level has been revoked.

In [6], a survey of ABAC has been presented which compares the functionality and characteristics of various attribute-based access control mechanisms. The method maintains taxonomy of attributes and their mode of access control enforced. This improves the performance of access control and presents a view of how the ABAC should be enforced.

In [7], the OpenStack has been used to enforce a predicate-based access control in service-oriented architectures. This system achieves higher performance in fine-grained access control toward IaaS clouds.

To provide guarantee to the security of user data in cloud, an efficient approach is presented in [8], which maintains security of data in domain level. In this approach, the data have been stored in different domains which can be accessed in a trusted manner. The system provides virtual machine to access the data from remote site based on the trust evaluation.

The problem of cloud service access restriction has been approached using reasoning techniques in [9] which present two techniques for the assessment of trust, namely, QHP and QTP. Similarly, a trust-based flexible access control method has been presented in [10]. The method estimates the user trust based on the trust given by various reputation centers according to encryption techniques.

To improve the performance in privacy preservation and access control approach is presented in [11]. The method uses various privileges for different users which have been used to perform access control. The PS-ACS approach separates users based on their domain as public and private. According to that, different aggregation schemes have been adopted. An efficient temporal access control scheme is presented in [12], using an integer comparison and proxy re-encryption schemes. Based on the integer range selected, different encryption schemes have been selected. Similarly, a keyword search approach with access control has been presented in [13]. The method retrieves the file based on the keyword search and the data have been encrypted, and uses cryptographic techniques to generate the original file.

A multi-authority storage approach for efficient ABAC is presented in [14], which overrides the need of central authority for key maintenance. The authorities can issue security keys independently to ensure security of the data and to enforce access control. Also, the authority can revoke the permission for the users independently. Similarly, the user accountability-based fine-grained access control is presented in [15]. The method overrides the illegal access key sharing with other users based on different policies of keys sharing.

All the above-discussed methods have the problem of restricting the user in an efficient way and suffer to produce higher data retrieval efficiency.

3   Hierarchical Sensitivity-Based Access Restriction and Data Retrieval

The proposed real-time hierarchical sensitivity-based access restriction algorithm receives the user request. From the user request, the list of data being impacted has been identified. For each data identified, the method identifies the hierarchy level and sensitivity rank. Using these information, for each hierarchy, a sensitivity support has been measured. Finally, based on the HSS measures, a CDRS measure is estimated to perform data retrieval.

Figure 1 shows the architecture of HSARD-based data retrieval system and shows various components of the proposed system.

Preprocessing

The request generated by the user has been received, and the method identifies the list of data attributes being accessed. The method identifies the list of attributes that need to be accessed at each level of attribute taxonomy. Then, for each level of attribute taxonomy, the method splits the attributes under different sensitive classes. From the attribute set generated, the method identifies the sensitive class, and the list of attributes the user can access/cannot access has been identified. The generated attribute set and access/nonaccess sets have been used to estimate the HSS measure in the next stage.

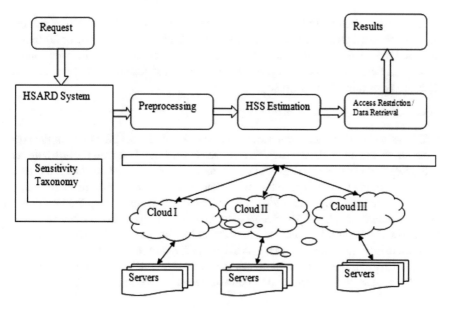

Fig. 1 Architecture of HSARD data retrieval system

Algorithm:
Intake: Request R, Attribute Taxonomy AT, Sensitivity Taxonomy ST
Output: Attribute Set Ats, Access Set Acs, Non Access Set NaS, Sensitive class Set Scs
Start

   Read R, AT, ST.

   Identify list of attributes needs to access.

   $$\text{Ats} = \int_{i=1}^{size(AT)} \sum AT(i) \rightarrow R$$

   For reach sensitive class Sc

       Identify list of attributes from ATs.

       $$\text{Scs}(i) = \int_{i=1}^{ST} \sum ATs(k) \in ST(i)$$

   End

   Identify the list of attributes the user has access ACs.

   $$\text{ACs} = \int_{i=1}^{size(AT)} \sum_{k=1}^{size(Ats)} Ats(k) == AT(i) \&\& AT(i). \text{UserName} == \text{User}$$

   Identify list of attributes the user has no access. Nas.

   $$\text{Nas} = \int_{i=1}^{size(AT)} \sum_{k=1}^{size(Ats)} Ats(k) \not\ni AT \&\& AT(i). \text{UserName}! = \text{User}$$

Stop

The above-discussed algorithm receives the user request and identifies the list of sensitive attributes in different classes or hierarchies. Then, the list of attributes accessed by the user has been identified and also the list of attributes the user cannot access has been identified. The generated and extracted features have been used to measure HSS value in the next stage.

HSS Estimation

In this stage, from the access/nonaccess set and sensitivity set generated in the previous stage, the method estimates various measures. First, based on the attribute set and access/nonaccess set, the method estimates the number of attributes the user has access in the current hierarchy level. Then, the total number of attributes being accessed in the current hierarchy level is estimated. Using this, the method estimates attribute access support AAS. Second, the method estimates the number of sensitive attributes the user has access in the current hierarchy and estimates the number of nonsensitive attributes the user has access in the current hierarchy. Using these two information, the method estimates the hierarchy sensitivity support measure. Estimated measure has been used to perform access restriction and data retrieval.

Algorithm:
Intake: Attribute Set As, Access Set Acs, Non Access Set NaS, Sensitivity Set Scs, Attribute Taxonomy AT.
Output: Hss
Start

  Read As, AcS, NaS, SCs.
  Identify list of attribute in current hierarchy CH.
  $NoHA = \int_{i=1}^{size(AT)} \sum AT(i) \in CH$
  Compute no of attributes access in current hierarchy.
  $NoA = \sum_{i=1}^{size(As)} As(i) \in NoHA.$
  Compute $AAS = {NOA}/{size(NOHA)}$
  Compute $NoSA = \sum_{i=1}^{size(As)} As(i) \in Scs.$
  Compute number of non sensitive attribute access NoNSA.
  $NoNSA = \sum_{i=1}^{size(As)} As(i) \in Scs.$
  Compute $HSS = \frac{NOSA}{NONSA} \times {NoSA}/{ASS}$
Stop

The above-discussed algorithm computes various measures on the attributes being identified. Identified sensitive and nonsensitive attributes are used to measure the hierarchical sensitive support measure.

HSARD Data Retrieval and Access Restriction

In this stage, the method receives the user request and performs preprocessing on the request. The method identifies the list of attributes being get affected by the request and identifies the list of attributes being get affected in each hierarchy. Based on the sensitive taxonomy and attribute taxonomy, the method computes the hierarchical sensitive support measured in each level of the taxonomy. Based on the HSS measure of different hierarchy levels, the method computes the cumulative data retrieval support measure. Based on the CRDS measure, the data have been retrieved to produce the result to the user.

Algorithm:
Input: User Request UR, Attribute Taxonomy AT, Sensitivity Taxonomy ST
Output: Result
Start

        Read UR.

        [Attribute Set, Access Set, Non Access Set] = preprocessing(UR, AT, ST)

        For each level

                Compute HSS = HSS-Estimation(UR, AT, ST)

        End

        Compute CDRS $= \frac{\Sigma HSS}{No\ of\ Levels} \times \Sigma HSS > Th$

        If CDRS> DTh then //DTh-data access threshold

                Result = Extract Data

        End

Stop

The above-discussed algorithm receives the user request and estimates the hierarchical sensitive support for each hierarchy level of the data. Finally, a cumulative data retrieval support is measured to produce the result to the user.

4 Results and Discussion

The proposed algorithm has been implemented and evaluated for its performance in various qualities of service parameters of cloud environment. The method has been implemented using advanced java and evaluated with varying scenarios. The method has produced efficient results on all the parameters considered. The method has produced the following results.

The proposed HSARD method has been validated for its performance in various factors according to the details given in Table 1. The performance of access restriction has been measured on various numbers of levels. The result has been compared with other methods and the proposed HSARD algorithm has produced higher performance in access restriction in all the conditions than other methods (Figs. 2 and 3).

The method has been evaluated for its efficiency in data retrieval and compared with other algorithms on the same. The result shows that the proposed HSARD algorithm has produced higher data retrieval efficiency in all the levels than other algorithms considered.

Table 1 Simulation details

Name	Value
Simulation tool	Advanced Java
Levels	200
Data points	10 Million
User size	5000

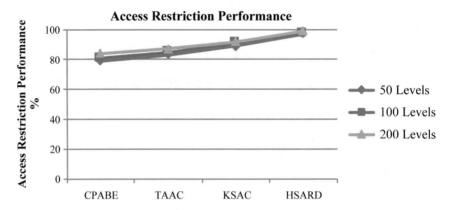

Fig. 2 Restriction performance versus no. of levels

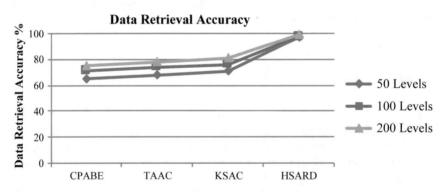

Fig. 3 Data retrieval efficiency versus no. of levels

5 Conclusion

In this paper, an efficient hierarchical sensitivity support-based data retrieval and access restriction algorithm is presented. The method receives the user request and identifies the list of attributes being get affected. Then, the method identifies the list of sensitive items that get affected and nonsensitive items that get affected. Using these information, the hierarchical sensitive support measure has been computed. The HSS value has been measured for all the hierarchical layer and using the HSS value, the method computes the CRDS measure. Based on the CRDS value, the method performs data retrieval. The method produces higher efficiency in data retrieval and access restriction.

References

1. W. Teng, G. Yang, Attribute-based access control with constant-size ciphertext in cloud computing. IEEE Cloud Comput. (99), 1 (2015). https://doi.org/10.1109/TCC.2015.2440247
2. J. Luo, A novel role-based access control model in cloud environments. IJCIS **9**(1) (2016)
3. L. Xie, C. Wang, Cloud multidomain access control model based on role and trust-degree. Hindawi Electr. Comput. Eng. **2016** (2016)
4. K. Yang, Time-domain attribute-based access control for cloud-based video content sharing: a cryptographic approach. IEEE Multimedia **18**(5) (2016)
5. J. Kim, S. Nepal, A cryptographically enforced access control with a flexible user revocation on untrusted cloud storage. Data Sci. Eng. **1**(3), 149–160 (2016)
6. M. Sookhaka, F. Richard Yua, Attribute-based data access control in mobile cloud computing Taxonomy and open issues. Future Gener. Comput. Syst. **72**(2017), 273–287 (2017), Elsevier
7. B. SrinivasaRao, A framework for predicate based access control policies in infrastructure as a service cloud. IJERA **6**(2), 36–44 (2016)
8. N. Paladi, Providing user security guarantees in public infrastructure clouds. IEEE Cloud Comput. **5**(3) (2017)
9. J. Luna, Quantitative reasoning about cloud security using service level agreements. IEEE Cloud Comput. **3**(5) (2017)
10. Z. Yan, Flexible data access control based on trust and reputation in cloud computing. IEEE Cloud Comput. **5**(3) (2017)
11. K. Fan, Privacy protection based access control scheme in cloud-based services. IEEE China Commun. **14**(3) (2017)
12. Y. Zhu, *Towards Temporal Access Control in Cloud Computing* (IEEE, INFOCOM, 2012)
13. Z. Zen, Keyword search with access control over encrypted cloud data. IEEE Sens. J. **17**(3) (2017)
14. J. Wei, Secure and efficient attribute-based access control for multi-authority cloud storage. IEEE Syst. J. (99) (2017). https://doi.org/10.1109/JSYST.2016.2633559
15. J. Li, Fine-grained data access control systems with user accountability in cloud computing. Cloud-Computing (2010)

Provision of Efficient Sentiment Analysis for Unstructured Data

C. Priya, K. Santhi and P. M. Durairaj Vincent

Abstract Applications on sentiment analysis via diverse context in analyzing the individual opinion on various issues such as political events, e-governance, and product reviews. Decision-making through the sentiment analysis improves the understanding of the public opinion. Opinion mining can be achieved by retrieving data through social network, microblogs, blogs, and search engines. Twitter tweets are an invaluable source of knowing the individual opinion from lots and lots of unique personality. However, the unstructured data to the huge volume and unwanted punctuation used in context and the emoticons used in the context is the main task to analyze the efficiency data and with greater accuracy. Most of the existing computational methods/algorithms identify sentiment on unstructured data via algorithms on machine learning like (BOW) bags of word approach. Here is the work of both the supervised and unsupervised approaches on various training datasets used. Automatic generation of the sentiment for the tweets extracted is provided by the unsupervised approach. And various algorithms in machine learning are used to determine the sentiment analysis they are as Maximum entropy (ME), Multinomial Naïve Bayes (MNB) and support vector machines (SVM) and are used to identify sentiment from the tweets. Here in this work I have achieved an accuracy of 87% unsupervised 97% in supervised approach. The ngram, unigram, bigram, and parts-of-speech (POS) were combined together to identify the hidden emotion and sentiment in the context that mentioned in the tweet that are all in an unstructured format. The lexicon-based approach 75.20% is achieved, based on the sentiment prediction the opinion is given.

Keywords Opinion · Unstructured data · Ngram · Bigram · Unigram
Supervised approach · Unsupervised approach · Decision-making

C. Priya (✉) · K. Santhi · P. M. Durairaj Vincent
Vellore Institute of Technology, Vellore, Tamil Nadu, India
e-mail: Priya.2017@vitstudent.ac.in

K. Santhi
e-mail: santhikrishnan@vit.ac.in

P. M. Durairaj Vincent
e-mail: pmvincent@vit.ac.in

© Springer Nature Singapore Pte Ltd. 2019
S. C. Satapathy et al. (eds.), *Information Systems Design and Intelligent Applications*,
Advances in Intelligent Systems and Computing 862,
https://doi.org/10.1007/978-981-13-3329-3_19

1 Introduction

As a human being we all need to know "what other people think" and we are curious in knowing other's opinion because this plays a vital role in decision-making. In the past decade, the World Wide Web was not used by everyone and they asked their friends to suggest a better product or to discuss on elections on whom they are voting for. However, now the World Wide Web is there in their hand itself, this Internet made it possible in providing the opinion from a wide people of unique personalities externally which is far from the one unique personality own and personal network where they are being connected to the wide range of unknown people. In that network each and every single person is providing their opinion on the particular product in vast numbers that are being published publically and unknown people are also able to access the opinion shared by the individual one. Opinion mining is also known as sentiment analysis, it analyses and identifies each individual's opinion, emotion, hidden sentiment in the context, uncover the attitude of the individual towards the product.

1.1 Literature Review

An opinion mining provides the aggregated opinion of one's individual belief and the individual's assessment or judgment or evaluation of the particular issue. The mean of a wide range of opinion from different individual on the particular issue/product provides greater impact and guidance to the other individual, MNC's, governance, and other social networking communities for making the decision. Effective decision-making generally made the basis of providing précised and timely information. It is a human nature to refer a friend, specialist, and relatives to gather a different kind of opinion, to know both the positive and the negative about the particular product or an issue. There are various challenges that have to be focused on sentiment analysis in the syntactic entity of the accent and semantically oriented language and engrossment of internal extraction of sentiment in the context and keep track on the emotions involved in the context. Hence the fine-grained approach is used to identify the sentiment and generate the automatic sentiment prediction the rule-based approach is that which extracts tweet's context and determines the hidden sentiment and emotion. Opinion mining and the sentiment analysis is known as "The set of evaluative text document D that contain emotion/sentiment context that which provide opinion about an object, sentiment analysis goal is to extract attributes via context and components of the object that have been commented on in each document $d \in D$ and to identify out whether the given comment under the particular section is positive, negative or neutral" according to Liu usually the opinion is conveyed from the opinion holder in the three perspectives of (happy, unhappy, neither happy, nor unhappy) on an issue or product Directly/indirectly globally wide range of people are interconnected to each other from one country to another country by social networks such as Facebook, twitter etc. Hence anyone can express their opinion in their own perspective without unveiling their true identity and without fear [1]. And another important challenge is to identify the false reviews that which lead to provide false opinion. And the other challenge is to identify the mixed feelings [2–7] and the news articles that which points out on the current news feed [8–13]. Providing the sentiment analysis on the informal language that is used in the day-to-day life by the human is being used in the comment section or microblogs, etc., is complicated. Lexicons are used in the process of identifying sentiment and the opinion for online reviews and the news articles. This research is keen on the informal linguistic that is the unstructured data used in the comment section (e.g., tweets).

2 Related Work

Most of the researchers in the recent years' work to generate automatic sentiment analysis which is more accurate and with the best efficiency. This research on sentiment analysis is to provide automatic generation of the sentiment and the opinion to the tweets that are retrieved from the twitter. Previously, most of the researcher's used the rule-based method and statistical approach in machine learning to determine the sentiment from the context and uncover the hidden emotion in the context. Here in Sect. 2 some of the algorithms, approaches, techniques, and their applications are being discussed briefly. Sentiment analysis and opinion mining used different techniques that are presented in the detailed survey and also it classify the context in the semantic-oriented phrases by using the unsupervised algorithms. In [14] Polarity is determined using the lexicon-based approach for the words or context provided in the document Esuli and Sebastiani [15]. The opinion strength for each and every term is calculated by SentiWordNet, and the reliability of the SentiWordNet for calculating the sentiment and the opinion for the certain data has been evaluated by Ohana [16]. Hamouda and Rohaim et al. [17] applied the online reviews and classified using the SentiWordNet. To find the aspect of the opinion holder the dictionary learning approach was proposed by Liu et al [18] in a way to consider the adjectives only. Lee and Pang used NB, ME, and SVM to classify the movie reviews. And by combining grams such as bigram, unigrams they showed that they are supporting the special features also to determine the sentiment and emotion in the context. In this paper, the supervised and unsupervised machine learning approaches to classify the tweets and perform the sentence level identification are used. The non-polar words are removed while preprocessing and polar words are considered to predict the emotion which is presented in the context, to increase the accuracy of the classifier. This proposed sentiment analysis with the unsupervised approach provided better performance for identifying and analyzing the sentiment over the tweets extracted and compared with the other methods which existed previously. And in the supervised approach the unigram, bigram and POS of the features were used to identify the sentiment and opinion of the unstructured context and found that MNB classifier with the unigram feature is more effective in classifying and providing the sentiment.

3 Data Preprocessing and Collection

3.1 Data Collection

Here in this research, the collected data are as tweets from the twitter by using the Twitter API. Two types of APIs are offered by Twitter, they are **Streaming and REST**. The Streaming API is the long-lived connection because it offers the data presented near and around tweets as much as the user wants. And in the REST API, it is contrast to Streaming API because the REST API is widely known as short-lived API, it is because a certain amount of the tweets can only be downloaded in per day by person (Table 1).

3.2 Training Dataset

Two types of the dataset are used for training the classifier—they are subjectivity data and the neutral data. The subjectivity dataset has the sentiment within the context where the neutral dataset does not have the sentiment over the context.

Subjective data
Subjective data is that which contain the sentiment over the context and provide the sentiment in two ways such as happy or unhappy. There are sufficient amounts of the negative data (tweets) and the positive data (tweets) for the two consecutive days collected to train the dataset by using the classifier.

Preprocessing the unstructured tweets
Twitter is the one where people are connected from one country to another country through the world, and share their own views as the tweets and those people are off with a different native language. English is the international language used all over the world widely.

Neutral tweets
Neutral tweets only contain the factual words about the particular issue or contain both the sentiment of positive and negative that is very confusing.

Table 1 Data by topics

Topic	#Positive	#Neutral	#Negative	#Irrelevant	Search term
Google	2480	4815	1616	5280	#Google
Apple	2218	6102	4070	1940	#Apple
Twitter	2537	5245	2112	6410	#Twitter
Microsoft	2250	4487	3203	5430	#Microsoft
Total	9485	20,649	11,001	19,060	60,195

3.3 Data Preprocessing Process

This preprocessing is the first step in the sentiment analysis and it is done before processing any other of the machine learning technique/algorithms and semantically oriented lexicon analysis.

4 Proposed Methodology

4.1 Unsupervised Approaches

In the previous research, the lexicons were majorly used to determine the sentiment over the tweet's context. A major problem in the lexicon analysis is domain-specific words that are not used in those lexicons; and determining the sentiments for those lexicons is the biggest challenge. The proposed work does not require training for the classifier to determine the sentiment for the domain-specific lexicons. In this research, two words are considered, namely excellent and poor. Excellent is considered to be highly positive context and poor is considered to be highly negative context. And score (t) is generated for each word by the following rule:

$$\text{prediction}(S) = \frac{\log(\text{hit}(s \wedge \text{good})\text{hit}(\text{worst})}{\text{hit}(s \wedge \text{worst})\text{hit}(\text{good}) - 1} \tag{1}$$

4.2 Labeled Training Data

Labeled training data are said to be supervised approach because this supervised approach is the one which maps the sample input with output. These unseen examples are mapped accurately by using this function. This supervised learning is a three-step process. Data preprocessing, feature extraction, training and testing.

Feature extraction
Machine learning approaches are represented in the (BOW) Bag of words to represent the feature. The feature is represented to the attribute which is used to map the pattern of the data. There are various features used to determine sentiment analysis, such as unigram, bigram, engram, POS tagger. In this work, the tweet is represented in BOW. This BOW is represented in two different stages such as attribute selection and instance representation.

Training and Testing
The features such as the unigram and the bigram are used in the proposed method and the feature will be considered for each tweet and each post. Then the three machine learning approaches are used to classify the sentiment. MNB multinomial

Naive Bayes algorithm is the classifier which uncovers the underlying meaning of the context presented in the tweets.

5 Evaluation and Result

To determine the accuracy and performance of each classifier following rule is (Figs. 1, 2, 3 and 4; Table 2):

Prediction accuracy

$$\text{Prediction accuracy} = \frac{\text{LT}}{\text{TT}} \tag{2}$$

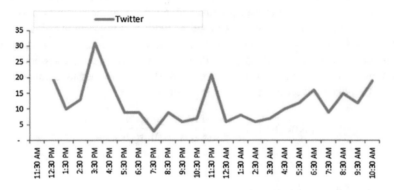

Fig. 1 Snapshot on tweets with respect to time of #anger

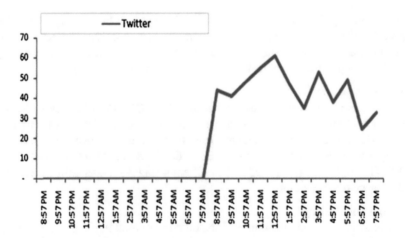

Fig. 2 Snapshot on tweet with respect to time of #Sad

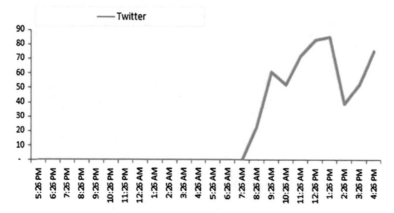

Fig. 3 Snapshot on tweet with respect to time of #Surprise

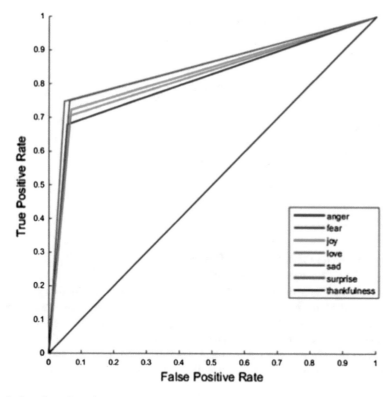

Fig. 4 Snapshot of emotion rate

Table 2 Results on unstructured data sentiment analysis

Testing Multinomial NB and Unigram Learning time 0.7746265928955078 s		Predicting time 0.20416146774292 s	
	Negative	Neutral	Positive
F1 score	0.28949672	0.25072993	0.81369782
Precision	0.85408163	0.48431373	0.85906623
Recall	0.84099617	0.42150171	0.87820513
Accuracy	0.8795943454210207		

Refined measure on tweets-precision

$$\text{Refined measure for positive data} = \frac{PT}{TP + FP} \tag{3}$$

$$\text{Refined measure for negative dat} = \frac{NT}{TN + FN} \tag{4}$$

Recall

$$RC = \frac{PT}{TP + FN} \tag{5}$$

F-measure

$$F-\text{measure} = \frac{2 * \text{Refined measure} * RC}{\text{Refined measure} + RC} \tag{6}$$

6 Conclusion

This system makes the human to think individually with number of opinion and help him to make a decision. Another challenge is that a machine does not know what human emotion is hence making the machine to learn the human emotion via context. Training the machine in a way to learn the structured format and unstructured format of data and train the system to predict the emotion with less false rate is a challenging issue. In this research, those challenges were overcome and achieved efficient results. Future work aims to increase the accuracy than the present accuracy and also explore the potential in reviewing the spam detection and intelligent recommender system.

References

1. N. Jindal, B. Liu, Mining comparative sentences and relations, in *Proceedings of the 21st National Conference on Artificial Intelligence*, vol 2, pp. 1331–1336 (2006)
2. E. Cambria, B. Schuller, Y. Xia, C. Havasi, New avenues in opinion mining and sentiment analysis. IEEE Intell. Syst. **28**(2), 15–21 (2013)
3. R. Chen, W. Xu, The determinants of online customer ratings: a combined domain ontology and topic text analytics approach. Electron. Commer. Res. (2016)
4. R. Feldman, M. Fresco, J. Goldenberg, O. Netzer, L. Ungar, Extracting product comparisons from discussion boards, in *Seventh IEEE International Conference on Data Mining (ICDM 2007)*, pp. 469–474 (2007)
5. Y. Li, Z. Qin, W. Xu, J. Guo, A holistic model of mining product aspects and associated sentiments from online reviews. Multimedia Tools Appl. **74**(23), 10177–10194 (2015). https://doi.org/10.1007/s11042-014-2158-0
6. B. Liu, Opinion mining and sentiment analysis, in *Web Data Mining: Exploring Hyperlinks, Contents, and Usage Data* (pp. 459–526) (2011). https://doi.org/10.1007/978-3-642-19460-3_11
7. Y. Ma, G. Chen, Q. Wei, Finding users preferences from large-scale online reviews for personalized recommendation. Electron. Commer. Res. **17**(1), 3–29 (2017). https://doi.org/10.1007/s10660016-9240-9
8. N. Godbole, M. Srinivasaiah, S. Skiena, Large-scale sentiment analysis for news and blogs. Proc. Int. Conf. Weblogs Soc. Media (ICWSM) **7**(21), 219–222 (2007)
9. M. Van de Kauter, D. Breesch, V. Hoste, Fine-grained analysis of explicit and implicit sentiment in financial news articles. Expert Syst. Appl. **42**(11), 4999–5010 (2015). https://doi.org/10.1016/j.eswa.2015.02.007
10. J. Li, S. Fong, Y. Zhuang, R. Khoury, Hierarchical classification in text mining for sentiment analysis of online news. Soft Comput. **20**, 3411–3420 (2015). https://doi.org/10.1007/s00500-015-1812-4
11. P. Liu, J.A. Gulla, L. Zhang, Dynamic topic-based sentiment analysis of large-scale online news, in *Proceedings of the 17th International Conference on Web Information Systems Engineering* (pp. 3–18) (2016)
12. S.Y.K. Mo, A. Liu, S.Y. Yang, News sentiment to market impact and its feedback effect. Environ. Syst. Decisions **36**(2), 158–166 (2016). https://doi.org/10.1007/s10669-016-9590-9
13. A.K. Nassirtoussi, S. Aghabozorgi, T.Y. Wah, D.C.L. Ngo, Text mining of newsheadlines for forex market prediction: A multi-layer dimension reduction algorithm with semantics and sentiment. Expert Syst. Appl. **42**(1), 306–324 (2015). https://doi.org/10.1016/j.eswa.2014.08.004
14. X. Ding, B. Liu, P.S. Yu, A holistic lexicon-based approach to opinion mining, in *Proceedings of the 2008 international conference on web search and data mining*, pp. 231–240 (2008). https://doi.org/10.1145/1341531.1341561
15. A. Esuli, F. Sebastiani, Sentiwordnet: a publicly available lexical resource for opinion mining, in *Proceedings of 5th language resources and evaluation*, vol 6, pp. 417–422 (2006)
16. B. Ohana, Opinion mining with the sentiwordnet lexical resource. M.Sc. dissertation, Dublin Institute of Technology (2009)
17. A. Hamouda, M. Rohaim, Reviews classification using sentiwordnet lexicon, in *World Congress on Computer Science and Information Technology* (2011)
18. G. Fei, B. Liu, M. Hsu, M. Castellanos, R. Ghosh, A dictionary-based approach to identifying aspects implied by adjectives for opinion mining, in *Proceedings of 24th International Conference on Computational Linguistics*, p. 309 (2012)

Automated Shopping Experience Using Real-Time IoT

A. Krishnamoorthy, V. Vijayarajan and R. Sapthagiri

Abstract Automated shopping experience based on embedded sensors using Raspberry Pi with ultrasonic sensor is introduced. Such a system is well suitable in places like grocery stores, where it can reduce the employee's requirements to manage the whole place. The main objective is to help consumers to checkout items in real time. In addition, by processing all the information on the go will help to acquire and analyze the needs of the users and user interest on specific items. Instead of making the user to wait in queue for making the bill, our system will automate the process by detecting the item when it is picked out of the shelf and added to the user virtual cart. It makes the system attractive to both seller and buyer perspective. When the user left the store, those who picked the items will be added up and charged using their connected bank account to the application. To detect who is entering and leaving the store is monitored with the help of a camera implemented along with the facial recognition algorithm called Kairos. And, all the actions are updated in real time using a database called Google's Firebase as a backend. The system design and experimental setup are presented in this paper.

Keywords Internet of things · Raspberry Pi · Ultrasonic · Firebase · Kairos
Virtual cart · Distributed network

1 Introduction

In the future, the reality between the commercial and personal world is going to be the same. The integration between these two spaces is going to be tightly bonded. Just imagine a case, where users can pick anything from a commercial space, and

A. Krishnamoorthy · V. Vijayarajan (✉) · R. Sapthagiri
School of Computer Science & Engineering, VIT, Vellore, Tamil Nadu, India
e-mail: vijayarajan.v@vit.ac.in

A. Krishnamoorthy
e-mail: krishnamoorthy.arasu@vit.ac.in

R. Sapthagiri
e-mail: sapthagiri.r2016@vitstudent.ac.in

© Springer Nature Singapore Pte Ltd. 2019   209
S. C. Satapathy et al. (eds.), *Information Systems Design and Intelligent Applications*,
Advances in Intelligent Systems and Computing 862,
https://doi.org/10.1007/978-981-13-3329-3_20

simply leave the store. The traditional processes like checkout, billing, inventory reports have been carried out without the human intervention. This seamless process involves lots of technological aspects. Nowadays, everything moving toward the world of automation, convenience, and time is the key factor to the user.

Automated checkout in the brick-and-mortar spaces will improve the overall sales of the store and ease for the customer to move from commercial to personal space by replacing the conventional method. It involves a lot of key technologies to validate the vision, which is evolving at a really faster pace because of the need and usability has been grown to many areas.

There are various constraints involved in the effects of technologies and miniaturizing of hardware, sensors for reading, scavenging, and energy-saving components. In this new era of evolving technology, all this technology finds a place in every aspect of fields such as healthcare, industries, transport, and stores, etc. The major constraints in automated intelligent systems may include false picking, energy consumption and cost management. Though most of the sensors have the broad area of sensing capabilities, the system needs to focus on the scope such as placements of sensors in the area of interest.

Communication among the embedded nodes will be updated on the real time to the database, in order to maintain the inventory, cart, user behavior, etc. For this paper, we will be taking the case of grocery stores, where the designed system will improve the overall experience of the consumer and retailers. Major issues faced in the traditional stores are:

1. Long waiting time for checkout and payment.
2. It requires a lot of human power to serve all the demanding user needs.
3. Inefficiency in the checkout process leads to variable quality of service to consumers.
4. It lacks personalized and retailer's insights over the user needs.

To solve these issues, we proposed a system to automate the shopping experience which will improve the way consumer and retailers do their tasks. Though it has a lot of its own problems but still, we try to implement a solution in order to solve the problem addressed.

1. It will improve the shopping experience of the user by reducing the time of checkout.
2. It will eliminate the need for checkout counters and no need for human power.
3. The virtual cart can easily handle the consumer billing process.
4. No waiting time for checkout and all the process will happen on the go.
5. Retailers can track the interest of the consumers.

In this paper, the designed system is considerably optimizing the problem stated. Hardware involvement and software integration needs to be robust enough to handle the data in real-time manner. All the actions happen in the systems needed to be updated on to the database. And, it communicated between different nodes of embedded sensors, user application, manager applications, and cameras to maintain the state of process to solve the problem. Every tray in the store is equipped with the

Fig. 1 Basic overview of architecture

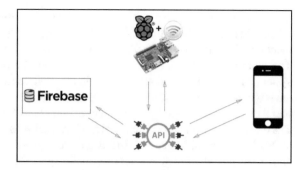

sensor embedded on to it along with the processing system Raspberry Pi for computation and transfer of data to the real-time database. And, a camera fitted on to the top of the store sealing will be monitoring the actions performed by the consumers. In order to track who is picking what item out of the store tray, every consumer is identified when they enter into the store and assigned a unique code for them to track their actions within the store. All the actions are shared among the managing applications of the store that will keep track of who is inside the store and what are all the items that are picked out of the store and added into their virtual cart with respect to the user. Once the customer is done with shopping, he/she can leave the store without the need of checkout. All the billing and paying process will be carried out automatically (Fig. 1).

2 Existing System

As the need for efficient system to quicken the process of checkout is increased that leads to rise in more number of counters, many researchers have tried to innovate to improve the existing model to be efficient. RFID-based smart shopping system which is incorporated along the shopping trolley, whenever user adds something into the trolley will be automatically billed to that user [1] is by proposed Swadhin et al., and it is efficient enough to solve the problem. Innovation build on top of the smart trolley system moved to wireless sensor networks where it can communicate between each other to create a mesh around network [2] is proposed by Udita et al. The proposed Outayo et al. [3] model of RFID based billing system is a cost effective one, but it needs to be further improved to satisfy the current technologies. To make the billing process even simpler, Menuka et al. [4] proposed a glove model with an inbuilt RFID Scanner which will automatically adds the shoping items to the user bill account.

Again, a similar model of RFID-based system is proposed by Umar et al. It is a simple billing system with a reader that detects the item using the RFID [5]. Some stores have adapted to self-checkout system to improve the user experience, and it will help to reduce labor involvement for the stores. But, it involves high user checking out time that will bother subsequent user in the queue. To even improve the electronic checkout, researchers have implemented RFID into items that will automatically get billed when it is detected by the RFID reader which is linked along with the POS machine. By using RFID, it will increase the speed of scanning the item for billing. There are several factors involved in influencing the operation time for checkout system like time involved to scan the items, update in POS, and packing. In [6], envisage the increase in lifetime of nodes by properly selecting the Cluster Head (CH) based on the residual energy state and total number of frames transmitted to the sink. The role of the sensor node is modeled as Finite State Machine (FSM) and it is realized as Markov process. In general, IoT is a rapidly growing system of physical sensor and its peripherals, which enables information gathering and monitoring aspects [7]. The proposed work [8] discussed about the data acquisition and habitat monitoring in WSN based water quality management system with IoT.

3   Proposed Architectural Overview

This section describes the architecture of the proposed system and algorithm used in the system. It has various modules that need to communicate and coordinate to make the process flow of the system.

3.1   Overview

In this proposed system, while the user enters into the store will be detected by the entry cameras placed in the storefront which is integrated with the backend to process the facial algorithm to detect the face and mark the entry into the database in real time. Once the user ID is generated and state changed to In-Store will create a virtual cart for the user. Consumer can check-in their own status through the user application installed on their smartphone. Now, the user has the ability to pick the items out of the tray, and those items will be automatically added to their virtual cart by listening the changes happening in the tray with the help of embedded sensors integrated into it. In this case, we will be using ultrasonic sensor that will monitor the variation in the distance changes.

Those changes will trigger the listener to monitor who is picking the item at that prime moment will be captured by the tray facing cameras placed on top of the shelves. And, all the interactions is recorded on the real-time database to make appropriate entries like change in inventory of that particular item, updating of virtual cart, amount calculation, etc. The item of interest is the basket which will hold the items that are interested by the user but not added into the cart. It also needs to monitor by the system.

Once the user finished shopping, he/she can leave the store without the need of checkout counter. Virtual cart will add up the items and credit it from the user's bank

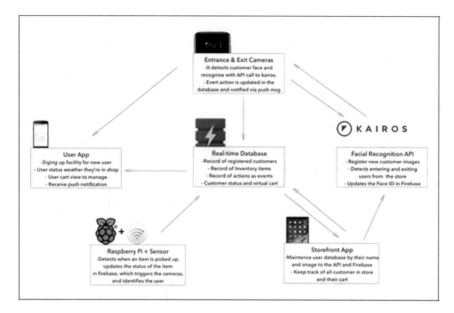

Fig. 2 Architectural overview of proposed system

account integrated through the application. User will get the notification for every action performed by them inside the store.

3.2 Architecture

Basically, it consists of various modules communicate between each other and make decisions among them. All this action is stored in the central real-time database called Firebase. The interaction between the modules are shown in Fig. 2 which is the proposed overall structure of the system.

3.2.1 Entry/Exit Cameras

These cameras are the most important entity of this whole system because it detects and identifies the entering and exiting users from the store. And, it is backed up with facial recognition algorithm called Kairos. It basically works on the principal of API calls made on to it whenever there is some action triggered from the camera. Here, we are using Android phone cameras for our purpose.

Whenever someone walks into the store will be detected and verified against the database which contains the unique ID for each and every user who registered with the shop. Once the user got verified, it will set the status of the user to In-Store mode which means the users is inside the store for shopping. All the actions are stored and maintained in the database and communicated over notifications with user and manager application.

3.2.2 Real-Time Database

In our proposed architecture, we will be using Firebase for the backend data storage. It is a non-relational database based on JSON. It will collect and process the data in real time. Every component needs to interact within real time. Firebase allows customized listeners to be added and changed within the database. It is responsible for all the customer data, inventory management, and user carts.

3.2.3 Kairos API

It is an industry grade facial recognition algorithm, and it has many use cases. It is based on API Calls, we will be primarily using two API available in Kairos, namely enroll and recognize.

- Enrolling: It can be done via API/Enroll and the image will be sent to the server and returns a unique ID.
- Recognize: It can be done via API/Recognize once the image is called along with recognize will return all possible matches, sort by highest confidence level.

3.2.4 User Application

It will allow the new user to register by uploading or taking a picture of them, name and payment. Once registered, they can go shopping and storefront will automatically detect them with the help of facial recognition. The process behind registering the user involves calling of an API called Kairos Enroll call, it will register the user into the real-time database by mean of unique ID generated for that user. And whenever the user enters into the shop will call a Verify API to identify the user who enters the store. And, the whole data is stored in a real-time storage called Firebase provided by Google. Enrolled images are stored in firestone service provided by Google. The data model used for register is Name, Image URL, and Payment Gateway. Every user ID is generated with cart in their data structure which is used when they are shopping and checking out. Users will receive a push notification for every action done by them like adding to cart, enter/exit the store.

Fig. 3 Item tray components

3.2.5 Item Tray Embedded Sensor

This whole setup is built for the item tray, sensors used are ultrasonic and light sensor, and Raspberry is used for processing the item and updating information on demand to firebase. Camera is used to detect who is picking the item from the tray. The process behind the whole setup is items are placed in an inclined position over the sensor, whenever there is difference happens in the sensor reading, it will trigger the Raspberry Pi to send an update signal to the inventory database. This event will simultaneously call the camera event to capture who is picking the item and updates the virtual cart for that user.

For this proposed system, we will be opting to Raspberry Pi3, ultrasonic sensor for detecting the distance Pi3 will be running the script for computing the sensor and accessing the database system. Figure 3 shows the components used in the system.

3.3 Algorithm

In the following paragraphs, we will be explaining about the algorithm used in the system.

Item cart detection

The main objective of this algorithm is to detect the items in the tray and update the changes in the tray inventory. If the tray is triggered, then it should process the variation and detect how many quantities have been taken and whom taken it. And update it correspondingly to the inventory and virtual cart.

Algorithm 1 Pseudo-code of Item cart detection
1: **while** True **do**   -Till the end of shopping.
2:    **if** item Tray varies **then**
3:       Calculate the distance of variation
4:       Detect the no of item picked using distance
5:       Update the Inventory and trigger camera to detect user
6:       Add the item into the virtual cart

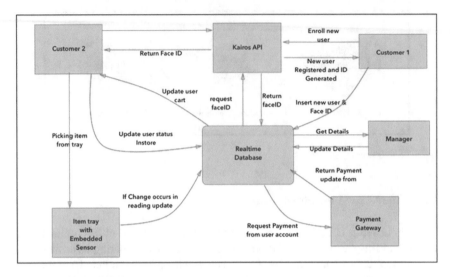

Fig. 4 Data flow diagram

As shown in algorithm 1, it will loop all the time checking for the change happen in the item tray. If change happens, it will trigger the distance from the ultrasonic sensor. And that distance is used to calculate the item taken. Item inventory is updated in the database. That will trigger the camera action to find who took the item out of the tray. Later, it will be added to the corresponding user's virtual cart.

3.3.1 Processes Overview

In this section, we explain the setup of our proposed model which has various entities interlinked to communicate between them. Here, the embedded sensor is placed in the item tray to detect the variation happens whenever there is some user actions performed. The data flow between the user entity and store entity has been shown the Fig. 4. Every action is based out of trigger events that will be stored in the database.

1. *Customer Actions*

The proposed prototype uses a mobile application for all user interaction between user and the storefront. Basic actions by the customer are new user registration which makes the user to get enrolled in the database by uploading or taking a picture. Other features are mostly done through the notification for actions like entering into store, picking items out of tray will be added into cart, billing, payment, and leaving the store is maintained by the user application. Figure 5 refers the user application.

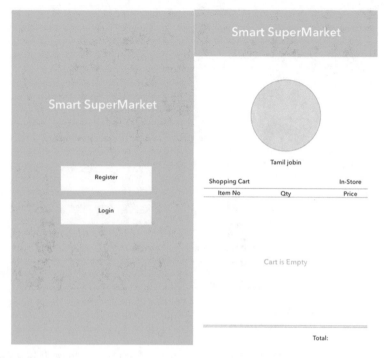

Fig. 5 User application

2. Storefront Actions

It is responsible for maintaining the state of the customers and inventory management in the store. All the entry, exit, and tray cameras are built and integrated with the storefront application. And, it monitors all the actions performed by the users. Every module communicates in real-time manner to respond to actions to set the state of the store. Storefront application is an Android application combined with the rest API calls to Kairos and Firebase.

3. Firebase

It is the central component of the whole purposed model and it connects all the integrated modules in the system where data communication needs to the updated on the real-time basis. Firebase allows the custom listener to be set on top the actions made through the different modules. Simple architecture of firebase along with the endpoints open for CRUD operations is shown in Fig. 6.

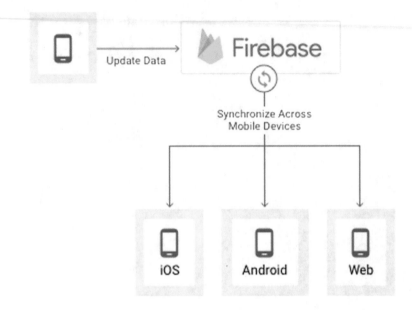

Fig. 6 Firebase overall flow

The data stored in firebase will be in the form of JSON structure. It is a simple schema that contains array of different items and users. Figure 7 shows the schema.

4. Ultrasonic Sensor embedded tray

Ultrasonic sensor is positioned inside the item tray and it is connected along with the Raspberry Pi that will be processing the reading by the program embedded inside the controller. Item tray is kept in the sliding position in order to make item roll down like queued manner that will create the space on the item tray that will make the sensor to trigger some action.

4   Results and Testing

The experimental setup is tested with a single shopping item with entry/exit face detection placed around the store. Our proposed model is built to handle multiple customers within the store with the help of face recognition. All this information is maintained with central firebase database. Most of the time, it will result in proper outcomes for all the cases except some error happening in the item tray and tray camera module. This is because the object recognition using the ultrasonic sensor, the logic behind calculating the distance to measure the no of items picked. Next, we observe the time taken to detect the face might take time to recognize. To solve the issue, we opted for a revolving door to control the customer entry into the store. And

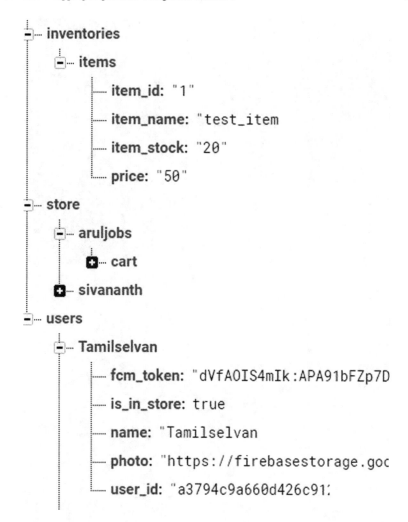

Fig. 7 JSON schema

the entire process to takes places with respect to placement and distances of sensors and cameras. All this required to decide upon the design phase of store. To assess a distributed environment, it requires proper placement of item tray with sensors and cameras. The fault tolerance of failing to add the item into cart depends on the detection of user by the face recognition.

We can say, it is similar to mote-to-mote communication where all sensors interact with each other with the central database in real time. Finally, it should push the cart at the time of user checking out from the store.

The item tray (Fig. 8) is positioned in a inclined position and an ultrasonic sensor is attached at the end of the row. The ultrasonic sensor is connected to the pi which

Fig. 8 Item tray prototype

process the data fetched from the ultrasonic sensor. Reading will be the distance for the closest object. It is done with the help of a python code loaded into our Raspberry Pi. When the item is removed out from the tray, the change will be manipulated automatically by triggering the inventory in our database at realtime.

Fig. 9 Test user app

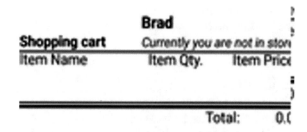

By this action, it will trigger the smartphone placed on top of the tray and camera will detect the face and attempts to add it to the recognized user cart.

For the usability and ease of access, we opted for a smartphone-based access camera. Android phone will be running our facial recognition using its camera. All the possible operations are performed by the smartphone like taking picture, uploading, and processing of images.

Cameras will be placed on the entry and exit doors of the store, whenever someone enters into the store through the revolving door, revolving doors will be used to control the user entry into the store. All the features are implemented heavily with the help of API calls to Kairos API, and Face API provided by Google. The overall architecture is a pipelined native detecting system for human faces.

Whenever someone enters into the store, they will be detected and that will trigger a verify call to Kairos, that will return a subject ID. Subject ID will be matched in our database and manipulate the status of the user to In-Store. Face will be captures at some particular distance from the camera.

The same process will be carried away when the customer exits the store, they will be detected by the facial recognition. And in the user-facing app (Fig. 9), they will be prompted with a push notification by saying the total amount. It can be done by retrieving the data from the database, and calculated the total amount to the customer spent, then pushed to the user with the help of FCM.

5 Conclusion and Future Enhancement

This work shows the successful demonstration of the possibilities of incorporating the usability of embedded hardware to make a smart automated shopping experience. This proposed system benefits when compared to the traditional system in use. However, it is not suitable for the large-scale environment with multiple carts and large number of user scenarios. By adapting distributed environment into this system will handle such situation better. Besides, it changes the whole experience of the customers as well as the seller.

Furthermore, we can use the data collected from other embedded sensors to predict the user needs to recommend on the go. To improve the item identification, we can generate a confidence score from multiple sensors and user history for confirming the user. Our current proposal does not explain about the setup and infrastructure of the sensors and cameras inside the store.

References

1. S. Pradhan, E. Chai, K. Sundaresan, S. Rangarajan, L. Qiu, Konark: a RFID based system for enhancing in-store shopping experience, in *Proceedings of the 4th International on Workshop on Physical Analytics* (ACM, USA, 2017), pp. 19–24
2. U. Gangwal, S. Roy, J. Bapat, Smart shopping cart for automated billing purpose using wireless sensor networks, in *SENSORCOMM 2013: The Seventh International Conference on Sensor Technologies and Applications*, vol. 7, pp. 168–172 (2013)
3. O. Boyinbode, O. Akinyede, A RFID baed inventory control system for Nigerian supermarket. Int. J. Comput. Appl. **116**(7) (2015)
4. M.G. Senevirathna, S. Harshini, Hands free POS automated RFID scanning glove to reduce waiting time of store checkout lines, in *2016 IEEE International Conference on Information and Automation for Sustainability (ICIAfS)* (2016)
5. M.S. Umar, J.A. Ansari, M.Q. Rafiq, Automated retail store based on RFID, in *2013 International Conference on Communication Systems and Network Technologies (CSNT)* (IEEE, 2013), pp. 17–21
6. A. Krishnamoorthy, V. Vijayarajan, Energy aware routing technique based on Markov model in wireless sensor network, Int. J. Comput. Appl. (2017). https://doi.org/10.1080/1206212x.2017.1396423
7. J. Boman, J. Taylor, A.H. Ngu, Flexible IoT middleware for integration of things and applications, in *2014 International Conference on Collaborative Computing: Networking, Applications and Worksharing (CollaborateCom)* (IEEE, 2014), pp. 481–488
8. C. Encinas, E. Ruiz, J. Cortez, A. Espinoza, Design and implementation of a distributed IoT system for the monitoring of water quality in aquaculture, in *Wireless Telecommunications Symposium (WTS)* (IEEE, 2017), pp. 1–7

A Comparative Survey on Three-Dimensional Reconstruction of Medical Modalities Based on Various Approaches

Sushitha Susan Joseph and D. Aju

Abstract The area of three-dimensional reconstructions has made advances in the recent years. Image reconstruction is the mathematical process which converts the signals obtained from the scanning machine into an image. Particularly in the medical field, the reconstructed images aid in the surgery and research. This survey provides an overview of three-dimensional reconstruction techniques in medical images using various imaging modalities like MRI, CT, biplanar radiography, and light microscopy along with the related disease. The reconstruction techniques such as Marching Cubes, Delaunay's Triangulation, Outlier Removal, Edge Enhancement and Binarization, False Positive Pruning, Contours, Support Vector Machines, Poisson Surface Reconstruction, Dictionary Learning, and Parametric Models are briefly described. The advantages and disadvantages of each technique are discussed and some possible future directions are suggested.

Keywords 3D reconstruction · Marching cubes · Delaunay's triangulation · PSR Dictionary learning · Parametric models

1 Introduction

Revolutions in the medical field take place with the advancement of imaging technology. Recent advances in imaging technology have had a huge impact on the diagnosis of disease as well as its treatment. The use of imaging techniques is undoubtedly on the rise. The physicians and surgeons seek computer assistance for more than a decade. The scanning devices provide images of high quality in two dimensions. The two-dimensional digitized information needs to be interpreted in three dimensions

S. S. Joseph (✉) · D. Aju
SCOPE, Vellore Institute of Technology, Vellore, India
e-mail: sushithasusan.joseph2017@vitstudent.ac.in

D. Aju
e-mail: daju@vit.ac.in

© Springer Nature Singapore Pte Ltd. 2019
S. C. Satapathy et al. (eds.), *Information Systems Design and Intelligent Applications*,
Advances in Intelligent Systems and Computing 862,
https://doi.org/10.1007/978-981-13-3329-3_21

223

in order to visualize the intricate anatomical morphology [1]. The diagnosis and treatment procedures provide better results when it is guided with three-dimensional images or models [2]. Reconstruction is the process through which these models can be obtained. The three-dimensional reconstructed model from patient-specific data helps the surgeons to visualize the underlying anatomical structures which are distinctive and particular to the patient [3]. Lamade et al. [4] were able to show that the impact of using 3D reconstructions in liver surgery is the increased precision in tumor localization and operation planning. It helps in identifying the areas of risk so that the resection can be precisely calculated for the improved operation planning [5, 6]. Inexperienced surgeons can use the reconstructed model for their studies [7].

2   Reconstruction Techniques

This survey is a study pertaining to three-dimensional reconstruction techniques that are applied to medical images which were acquired from different scanning devices. It summarizes the techniques used along with the disease for which the reconstruction is associated with. The two-dimensional images are acquired through scanning devices such as MRI, CT, CTA, and stereo radiography. MRI generates images using radio waves and magnetic field [8]. MRI does not produce radiations which are harmful [9] when compared to CT scan and X-rays. It is widely used noninvasive [10] technique in research to measure the soft tissues in brain. When compared to CT, high-resolution images of soft tissues can be produced by MRI. Computed Tomography (CT) is an advanced extension of X-ray [11] that uses X-rays and computer to create images of the body. CT is the standard technique for scanning bones. However, CT and MRI are not suitable for diagnosing the deformities in lower limbs and spine [12]. It underestimates the actual deformations of spine as they are obtained from the supine position of patients [13]. The images of spine and lower limbs are taken in standing position so that they can be examined with weight bearing condition using stereo radiography [14]. Computed Tomography Angiography (CTA) is a noninvasive test for the diagnosis of coronary artery disease which is increasingly utilized and highly accurate [15]. It is used to view the arteries and veins.

2.1   Marching Cubes (MC)

Marching cubes is a high-resolution algorithm for 3D surface reconstruction proposed by Lorensen and Cline [16]. A polygonal mesh is generated by proceeding through the input data. Isosurface creation is heavily utilized in medical visualization. Isosurfaces recreate the digitized images taken by CT, MRI scans, and CTA. Arakeri and Reddy [17] used MC algorithm for the reconstruction of brain tumor. The volume of tumor is computed on the basis of slice thickness, gap between slices and the area of tumor on each abnormal slice. Gnonnou and Smaoui [18] successfully reconstructed breast using marching cubes algorithm from MRI images.

Nugroho et al. [19] have presented an approach that uses both marching squares and marching cubes for the reconstruction of heart. Kigka et al. [20] applied Marching Cubes algorithm in three-dimensional reconstruction of coronary arteries. The accumulation of atheromatic plaques in the coronary arteries leads to Coronary Artery Disease. It is one of the common types of heart disease and is the leading cause of mortality [21]. Coronary Artery Disease is investigated using Computed Tomography Angiography. The reconstruction methodology consists of the following steps. Preprocessing of the images, centerline extraction of vessels, weight function estimation for outer wall, lumen and calcified plaque, segmentation of lumen, outer wall and plaque, and 3D surface reconstruction. From a three-dimensional discrete field, marching cubes obtains polygon mesh of isosurfaces computed. A triangulation approach is used. For each CTA image, three-dimensional models are constructed by connecting the identified border points.

2.2 Delaunay's Triangulation

It is the most popular and most commonly used methods in problems where mesh generation is involved. It is a triangulation such that for every d-simplex, the circumsphere is empty which means its interior does not contain any of the given points. Bharathi and Manimegalai [22] used Delaunay's triangulation for reconstructing brain tumor. The images from MRI scans are segmented. Sobel edge detection and morphology operations such as erosion and dilation are used. Then, the reconstruction is performed using Delaunay's triangulation.

2.3 Outlier Removal, Edge Enhancement and Binarization, False Positive Pruning (OEBFP)

Dendritic spines are the key structures in the central nervous system that regulate the neural activity [23]. Morphological changes in the shape, size, or number of dendritic spine leads to brain and neurological disorders [24]. Reberger et al. [25] did research on dendritic spines. They were successful in providing the visualization through the three-dimensional reconstruction procedure. The samples are taken from human brain and light microscopy is used for data acquisition. The neuron is observed through the z-axis. The reconstruction is performed by removing the outliers, enhancement of edges, binarization, and pruning of false positives.

2.4 Contours

De-xin [26] proposed the idea of triangle meshes in the three-dimensional reconstruction of models. The surface of the brain was reconstructed from MRI images.
Two successive slices with triangular patches are selected and each set of contours
from them are linked for estimating the external surface. Zhang et al. [27] proposed
a method for the three-dimensional reconstruction of spine, in which the landmarks
need not be identified. It was done using contour matching with the Hough transform technique. Angelopoulou et al. [28] present growing neural gas which is a
self-organizing network for automatically extracting landmarks. Reconstruction of
brain ventricles is performed by representing the contours on them using growing
neural gas.

2.5 One Class Support Vector Machines (OCSVM)

Support vector machine which is a concept in statistical learning theory is used for
solving classification as well as regression problems. A hyperplane separates the
set of samples into two classes such that there is maximum distance between the
hyperplane and closest vector in an optimal manner [29]. They are used for the
reconstruction of encephalic tissues in brain and scoliosis.

When the SVM handles the training with positive information only, it is called
one class support vector machines. The data is mapped into the feature space using
kernel functions and the dataset is described using a hypersphere. Wang et al. [30]
were the first to apply OCSVM into 3D reconstruction. The seven encephalic tissues
from MRI scans were reconstructed. The spine reconstruction is done for scoliotic
patients. Statistical shape models relying on Gaussian distribution were commonly
is used for this. Lecron et al. [31] took the advantage of OCSVM to build the spine
model. They built a shape model which is kernel based apart from Gaussian distribution. The reconstruction algorithm is formulated as an optimization problem for the
correspondence score of the shape of the spine with respect to the OCSVM defined
hyperplane.

2.6 Immune Sphere-Shaped Support Vector Machine
     (ISSSVM)

Sphere shaped SVMs solved the classification problems with class imbalance. Guo
et al. [32] extended this concept to three-dimensional reconstruction. They found that
this method can reconstruct any arbitrarily irregular surfaces and the prior information
regarding the object surface is not needed for reconstruction. For selecting the optimal
parameters of kernel function, Immune algorithm has been used. They performed the

three-dimensional reconstruction of encephalic tissues by combining SSSVM and Immune algorithm resulting in Immune Sphere-Shaped Support Vector Machine (ISSSVM).

2.7 Poisson Surface Reconstruction (PSR)

The concept of Poisson Surface Reconstruction for reconstructing the kidney is presented by Leonardi et al. [33]. Their purpose is to obtain a solid three-dimensional model of the kidney. Post treatment of final mesh is done in order to obtain a kidney model. This can be used as a real-time model in the medical environment like surgery. The penetration of coronary artery into the heart has to be identified earlier for a successful surgery. Khaleel et al. [34] proposed an approach to reconstruct coronary artery from oriented cloud points. The Poisson surface reconstruction technique was applied on the oriented cloud points. With the aim of using it as a tool in surgery, they measured the curvatures and compared it to the standard depth so that a warning can be given to the doctors regarding penetration. Palomara et al. [35] successfully reconstructed the models of liver parenchyma which was patient specific using PSR to be useful in liver resection procedures. Gradient fields were computed for cloud point orientation.

2.8 Dictionary Learning

Dictionary is a method for the sparse representation of the image. Zhang et al. [36] successfully reconstructed 3D bone model from CT images using dictionary learning. The initial dictionary was created using vertex of mesh triangles obtained using random sampling. The triangular mesh and point cloud have a projection relationship. This relationship is used to construct the sparse coefficient matrix. Then, iterative updation of dictionary and matrix using the objective function should be performed for the reconstruction.

2.9 Parametric Models

Pomero et al. [37] have introduced the concept of parametric models that uses transversal inferences for the semiautomatic reconstruction of spine. For a given vertebra, the relationship between the geometric descriptors provides the transversal inference. Humbert et al. [38] proposed an approach that uses longitudinal inference together with transversal inference. The parametric spine model for longitudinal inference was explained by the length of a curve which passes through the spinal curve, depthwise, widthwise and also along the spinal curve for each vertebral endplate. The parametric vertebrae model is determined by the depth and width of each endplate and the anterior, posterior, left and right heights of the vertebral body along with the 3D coordinate of 19 anatomical points.

For the reconstruction of cranial vault, the determination of its thickness is of great significance as it varies from one person to another [39–41]. Sommer et al. [42] use superquadric modeling for the approximation of external surface but could not evaluate the thickness as the internal surface was not taken into consideration. Another approach which considers internal surface was proposed by Laurent et al. [43]. In this method, landmarks and contours were identified on Digital Reconstructed Radiographs, which was generated for the parametric description of cranial vault. Through iterative deformation, the cranial vault is reconstructed. The combination of statistical and parametric model is another approach introduced by Chaibi et al. [44] for the fast reconstruction of lower limb. The Simplified Parametric Model of lower limb is represented by geometric parameters. The regression between these geometrical parameters is described by statistical inference. Quijano et al. [45] further explored this concept to improve the initial solution by reducing the corrections done manually. They proposed extra-digitization which helps to enrich the statistical model. The focus of this research was to provide a step further to automate the three-dimensional reconstruction of lower limbs. The orientation and angle of vertebrae in the biplanar radiographs were used by Kumar et al. [46] to reconstruct spine. The vertebral endplate midpoints are detected and the spine midline is constructed. Comparison of the angles in the projected model and images is performed and the generic model is iteratively deformed till matching occurs. This results in the personalized reconstruction of three-dimensional spine model.

2.10 Statistical Shape Models (SSM)

Kadoury et al. [47] proposed a hybrid model for the reconstruction of scoliotic spine. They used statistical model along with the model based on the image. Coronal vertebrae and sagittal vertebrae images were used and the geometrical parameters were included using the transformation algorithm. Dworzak et al. [48] proposed Statistical Shape Models (SSM) to reconstruct human rib cage. The purpose was to aid the interval studies which help to understand the changes in the course of a particular disease in a patient. Two-dimensional projection images are obtained and SSM is adapted to it. Distance measures based on Silhouette are used for this adaptation.

The culmination of various 3D reconstruction techniques based upon its advantages and disadvantages are summarized in Table 1. Also, the applicability of the reconstruction methods that could be applied and manipulated on different human organs are shown in the below table.

Table 1 Advantages and disadvantages of techniques for reconstruction in medical images

S. No.	Reconstruction technique	Reconstructed organ	Advantage	Disadvantage
1	Poisson surface reconstruction	Kidney, Liver	(i) Creates smooth surfaces (ii) Robust to noise (iii) Fast	The input data is in the form of 3D scalar fields which has to be converted to oriented cloud points
2	Marching cubes	Brain, Breast, Coronary Artery	(i) Simple (ii) Efficient (iii) Robust	(i) Irregularities in topology (ii) Cracks in adaptive resolution (iii) Sharp features of edges are not protected
3	Delaunay's triangulation	Brain	(i) Efficient in storing and manipulating data (ii) Can easily describe the surfaces even though the resolution changes	(i) The quality of tetrahedral mesh of Delaunay's triangle are brilliant in 2D but it diminishes as it comes to 3D (ii) Post processing is required to improve it
4	OCSVM	Brain, Spine	(i) Efficient and accurate (ii) Irregular surfaces can be reconstructed easily (iii) Presence of outliers are detected	Selection of appropriate kernel function is difficult
5	ISSSVM	Brain	(i) No prior information regarding the shape of the object is needed (ii) Irregular surfaces can be reconstructed easily	(i) Selection of appropriate SSSVM parameters and kernel function is difficult (ii) Processing time is high

(continued)

Table 1 (continued)

S. No.	Reconstruction technique	Reconstructed organ	Advantage	Disadvantage
6	Dictionary learning	Bone	(i) Efficient (ii) Robust (iii) Easily adaptable (iv) Accurate	The presence of noise reduces the smoothness of the reconstructed surface
7	Parametric models	Spine	Reduces the number of anatomical landmarks to be digitalized	Interaction of an experienced operator is required for adjusting control points
8	Statistical shape models	Spine	(i) Accurate	(i) The model cannot represent variations of shape (ii) Outliers are not detected
9	OEBFP	Dendritic spine	No need of expensive hardware and software	Processing time is high
10	Contours	Brain, Spine	Less computational complexity	Execution time is large

3   Discussion and Future Directions

The survey reveals that there are many reconstruction techniques that can be applied for medical images. The surfaces of any organ can be successfully reconstructed using marching cubes with some post processing done for obtaining the sharp edges. The Poisson surface reconstruction can be applied only for organs with smooth surfaces such as kidney, liver, and spleen. It is suitable only when the input is in the form of oriented cloud points. Delaunay's triangulation can be applied for reconstruction of soft tissues but requires post processing for improving the mesh quality. OCSVM are appropriate for reconstructing surface with high irregularity like spine but the selection of appropriate kernel function is a difficult task. ISSSVM has the advantage of the fact that no prior knowledge is required but the selection of kernel function determines the quality of the reconstructed surface. Dictionary learning can be used data with high dimensions such as bones but the presence of noise reduces the smoothness of reconstructed surface. Parametric models are suitable for the reconstruction of spine but it requires the help of an experienced operator. SSM is commonly used for the reconstruction of the spine. Various approaches are introduced to make it more efficient. OEBFP is the technique that requires no costly hardware or software. The survey concludes that the selection of reconstruction technique depends on the structure of the organ. The parameters like smoothness of surface, presence of outliers, and representation of input data are the parameters that determine the processing time of reconstruction and the quality of reconstructed images. Since the application is in medical field, processing time and quality of reconstruction are important. There are some future directions which can be applied in the reconstruction of medical images. (i) The dictionary learning method can be improved by learning dictionaries in a greedy manner. Deep learning incorporated into dictionary learning can be used. (ii) The Power Crust algorithm which is a robust algorithm that constructs the surfaces without hole filling mechanism can be used. (iii) The concept of parallel implementation of Ball pivoting algorithm is also a good technique that can be used in the medical image reconstruction.

References

1. C.A. Landes, F. Weichert, P. Geis, F. Helga, M. Wagner, Evaluation of two 3D virtual computer reconstructions for comparison of cleft lip and palate to normal fetal microanatomy. Anat. Rec. **288**, 248–262 (2006)
2. C.S. Wang, W.H. Wang, M.C. Lin, STL rapid prototyping bio-CAD model for CT medical image segmentation. Comput. Ind. **61**, 187–197 (2010)
3. L.T. De Paolis, M. Pulimeno, G. Aloisio, Visualization and interaction system for surgical planning, in ed. by V. Luzar-Stiffler, I. Jarec, Z. Bekic, *Proceedings of the 32nd International Conference on Information Technology Interfaces (ITI 2010)*, Cavtat/Dubrovnik, Croatia, June 21–24, 1st Ed. Zagreb, Croatia: University of Zagreb, University Computing Centre, pp. 269–274 (2010)
4. W. Lamadé, G. Glombitza, L. Fischer, P. Chiu, C.E. Cárdenas, M. Thorn, H.P. Meinzer, L. Grenacher, H. Bauer, T. Lehnert, C. Herfarth, The impact of 3-dimensional reconstructions on operation planning in liver surgery. Arch. Surg. (Chicago, Ill.: 1960) **135**(11), 1256–1261 (2000)
5. H. Lang, A. Radtke, M. Hindennach, T. Schroeder, N.R. Frühauf, M. Malagó, H. Bourquain, H.-O. Peitgen, K.J. Oldhafer, C.E. Broelsch, Impact of virtual tumor resection and computer-assisted risk analysis on operation planning and intraoperative strategy in major hepatic resection. Arch. Surg. (Chicago, Ill.: 1960) **140**(7), 629–638 (2005)
6. C. Hansen, S. Zidowitz, B. Preim, Impact of model-based risk analysis for liver surgery planning. Int. J. Comput. Assist. Radiol. Surg. **9**(3), 473–480 (2014)
7. P. Lamata, F. Lamata, V. Sojar, P. Makowski, L. Massoptier, S. Casciaro, W. Ali, T. Stüdeli, J. Declerck, O.J. Elle, B. Edwin, Use of the resection map system as guidance during hepatectomy. Surg. Endosc. **24**(9), 2327–2337 (2010)
8. M.S. Nizam, B.J.J. Abdullah, N. Kwan-Hoong, A.C. Ahmad, Magnetic resonance imaging: Health effects and safety, in *Proceedings of the International Conference on NonIonizing Radiation at UNITEN ICNIR2003, (ICNIR2003)*, 20 Oct (2003)
9. A. Berger, Science editor Article on Magn. Reson. Imaging BMJ **324**(7328), 35. PMCID: PMC1121941, US National Library of Medicine National Institutes of Health (2002)
10. M. Zhao, D.A. Beauregard, L. Loizou, B. Davletov, K.M. Brindle, Non-invasive detection of apoptosis using magnetic resonance imaging and a targeted contrast agent, MRC Laboratory of Molecular Biology, Hills Road, Cambridge, UK. Nat. Med. **7**, 1241–1244 (2001)
11. G.T. Herman, *Fundamentals of Computerized Tomography: Image Reconstruction from Projections* (Springer Publishing Company, Incorporated, 2009)

12. P. Gamage, S.Q. Xie, P. Delmas, Diagnostic radiograph based bone reconstruction framework: application to the femur. Comp. Med. Imag. Graph. **35**, 427–437 (2011)
13. M. Yazici, E.R. Acaroglu, A. Alanay, V. Deviren, A. Cila, A. Surat, Measurement of vertebral rotation in standing versus supine position in adolescent idiopathic scoliosis. J. Pediatr. Orthop. **21**(2), 252–256 (2001)
14. E. Melhem, A. Assi, R. El Rachkidi, I. Ghanem, EOS® biplanar X-ray imaging: concept, developments, benefits, and limitations. J. Child. Orthop. **10**(1), 1–14 (2016)
15. K.M. Woods, C. Fischer, M.K. Cheezum, E.A. Hulten, B. Nguyen, T.C. Villines, The prognostic significance of Coronary CT angiography. Curr. Cardiol. Rep. **14**(1), 7–16 (2012)
16. W. Lorensen, H. Cline, Marching cubes: a high resolution 3D surface construction algorithm. Comput. Graph. **21**(4), 163–169 (1987)
17. M.P. Arakeri, G. Ram Mohana Reddy, An effective and efficient approach to 3D reconstruction and quantification of brain tumor on magnetic resonance images. Int. J. Signal Process. Image Process. Pattern Recogn. **6**, 135–142 (2013)
18. C. Gnonnou, N. Smaoui, Segmentation and 3D reconstruction of MRI images for breast cancer detection, in *International Image Processing Applications and Systems Conference* (2014)
19. P.A. Nugroho, D.K. Basuki, R. Sigit, 3D heart image reconstruction and visualization with marching cubes algorithm, in *Knowledge Creation and Intelligent Computing*, pp. 35–41 (2016)
20. V.I. Kigka, G. Rigas, A. Sakellarios, P. Siogkas, I.O. Andrikos, T.P. Exarchos, D. Loggitsi, C.O. Parodi, Dimitrios I. Fotiadis.: 3D reconstruction of coronary arteries and atherosclerotic plaques based on computed tomography angiography images. Biomed. Signal Process. Control **40**, 286–294 (2018)
21. M.J. Budoff, Assessment of coronary artery disease by cardiac computed tomography a scientific statement from the American heart association committee on cardiovascular imaging and intervention, council on cardiovascular radiology and intervention, and committee on cardiac imaging, council on clinical cardiology. Circulation **114**(16), 1761–1791 (2006)
22. A.S. Bharathi, D. Manimegalai, 3D Digital reconstruction of brain tumor from MRI scans using Delaunay triangulation and patches. ARPN J. Eng. Appl. Sci. **10**, 9227–9232 (2015)
23. P. García-López, V. García-Marín, M. Freire, Dendritic spines and development: towards a unifying model of spinogenesis—a present day review of Cajal's histological slides and drawings. Neural Plas. **2010**, 1–29 (2010)
24. P. Penzes, M.E. Cahill, K.A. Jones, J.-E. VanLeeuwen, K.M. Woolfrey, Dendritic spine pathology in neuropsychiatric disorders. Nat. Neurosci. **14**, 285–293 (2011)
25. R. Reberger, A. Dall'Oglio, C.R. Jung, A.A. Rasia-Filho, Structure and diversity of human dendritic spines evidenced by a new three-dimensional reconstruction procedure for Golgi staining and light microscopy. J. Neurosci. Methods **293**, 27–36 (2018)
26. Z. De-xin, A method for brain 3D surface reconstruction from MR images. Optoelectron. Lett. **10**(5), 383–386 (2014)
27. J. Zhang, L. Lv, X. Shi, Y. Wang, F. Guo, Y. Zhang, H. Li, 3-D Reconstruction of the Spine from biplanar radiographs based on contour matching using the hough transform. IEEE Trans. Biomed. Eng. **60**(7), 1954–1962 (2013)
28. A. Angelopoulou, A. Psarrou, J. Garcia-Rodriguez, S. Orts-Escolano, J.A. Lopez, K. Revett, 3D reconstruction of medical images from slices automatically landmarked with growing neural Models. Neurocomputing **150**, 16–25 (2015)
29. L. Manevitz, M. Yousef, One-class svms for document classification. J. Mach. Learn. Res. **2**, 139–154 (2001)
30. L. Wang, G. Xu, L. Guo, X. Liu, S. Yang, 3D reconstruction of head MRI based on one class support vector machine with immune algorithm, in *Proceedings of 29th Annual International Conference of IEEE Engineering in Medicine and Biology Society*, pp. 6015–6018 (2007)
31. F. Lecron, J. Boisvert, S. Mahmoudi, H. Labelle, M. Benjelloun, Three-dimensional spine model reconstruction using one-class SVM regularization. IEEE Trans. Biomed. Eng. **60**(11), 3256–3264 (2013)
32. L. Guo, Y. Li, D. Miao, L. Zhao, W. Yan, X. Shen, 3-D Reconstruction of Encephalic Tissue in MR Images Using Immune Sphere-Shaped SVMs. IEEE Trans. Magn. **47**(5), 870–873 (2011)

33. V. Leonardi, V. Vidal, J. Mari, M. Daniel, 3D reconstruction from CT-scan volume dataset application to kidney modeling, in *Proceedings of the 27th Spring Conference on Computer Graphics*, vol D, pp. 111–120 (2011)
34. H.H. Khaleel, R.O.K. Rahmat, D.M. Zamrin, R. Mahmod, N. Mustapha, 3D surface reconstruction of coronary artery trees for vessel locations' detection. Arab J. Sci. Eng. **39**, 1749–1773 (2014)
35. R. Palomara, F.A. Cheikh, B. Edwin, A. Beghdadhi, O.J. Elle, Surface reconstruction for planning and navigation of liver resections. Comput. Med. Imaging Graph. **53**, 30–42 (2016)
36. B. Zhang, X. Wang, X. Liang, J. Zheng, 3D Reconstruction of human bones based on dictionary learning. Med. Eng. Phys. **49**, 163–170 (2017)
37. D.M. Pomero, S. Laporte, J. A. de Guise, W. Skalli, Fast accurate stereo-radiographic 3D-reconstruction of the spine using a combined geometric and statistic model. Clin. Biomech. **19**, 240–247 (2004)
38. L. Humbert, J.A. De Guise, B. Aubert, B. Godbout, W. Skalli, 3D reconstruction of the spine from biplanar X-rays using parametric models based on transversal and longitudinal inferences. Med. Eng. Phys. **31**, 681–687 (2009)
39. Y.S. Jung, H.J. Kim, S.W. Choi, J.W. Kang, I.H. Cha, Regional thickness of parietal bone in Korean adults. Int. J. Ora. Maxillofac Surg. **32**, 638–641 (2003)
40. N. Lynnerup, Cranial thickness in relation to age, sex and general body build in a Danish forensic sample. Forensic Sci. Int. **117**, 45–51 (2001)
41. J.H. Kidder, A.C. Durband, A re-evaluation of the metric diversity within Homoerectus. J. Hum. Evol. **46**, 299–315 (2004)
42. H.J. Sommer, R.B. Eckhardt, T.Y. Shiang, Superquadric modeling of cranial and cerebral shape and asymmetry. Am. J. Phys. Anthropol. **129**, 189–195 (2006)
43. C.P. Laurent, E. Jolivet, J. Hodel, P. Decq, W. Skalli, New method for 3D reconstruction of the human cranial vault from CT-scan data. Med. Eng. Phys. **33**, 1270–1275 (2011)
44. Y. Chaibi, T. Cresson, B. Aubert, J. Hausselle, P. Neyret, O. Hauger, J. A. de Guise, W. Skalli, Three-dimensional reconstruction of the lower limb from biplanar calibrated radiographs. Med. Eng. Phys. **35**, 1703–1712 (2013)
45. S. Quijano, A. Serrurier, B. Aubert, S. Laporte, P. Thoreux, W. Skalli, Three-dimensional reconstruction of the lower limb from biplanar calibrated radiographs. Med. Eng. Phys. **35**, 1703–1712 (2013)
46. S. Kumar, K. Prabhakar Nayak, K.S. Hareesha, Improving visibility of stereo-radiographic spine reconstruction with geometric inferences. J. Digit Imaging **29**, 226–234 (2016)
47. S. Kadoury, F. Cheriet, H. Labelle, Personalized X-Ray 3-D reconstruction of the scoliotic spine from hybrid statistical and image-based models. IEEE Trans. Med. Imaging **28**(9), 1422–1435 (2009)
48. J. Dworzak, H. Lamecker, J. von Berg, T. Klinder, C. Lorenz, D. Kainmüller, 3D reconstruction of the human rib cage from 2D projection images using a statistical shape model. Int. J. Comput. Assist. Radiol. Surg. **5**, 111–124 (2010)

Estimating the Distance of a Human from an Object Using 3D Image Reconstruction

R. M. Swarna Priya, C. Gunavathi and S. L. Aarthy

Abstract Obstacle detection, pedestrian detection, human motion detection, etc., are the recent technologies which are booming high in the computer vision industry. These technologies play a major role in various applications like surveillance, autonomous cars, driverless vehicles, etc. Our focus is on identifying and calculating the distance of a human from an object using 3D image reconstruction. The object can be a video camera or a sensor in case of surveillance or driverless vehicle. Using this distance, intelligent decisions could be taken. This methodology can help the computer vision researchers to detect any obstacle in their region of interest and also their distance from the particular point. This work could also detect the frame in which the person or the obstacle is detected along with the distance. The above said methodology could be incorporated in traffic monitoring system for identifying or detecting the pedestrians so that accidents could be avoided.

Keywords 3D reconstruction · Depth estimation · Obstacle detection

1 Introduction

The distance measurement of an object from a particular point has become an important aspect of mobile autonomous systems nowadays. In case of autonomous systems like self-driving cars, accurately recognizing the surroundings and in lesser time is the most important and also a challenging task. There are two methods for measuring the distance, namely, active method and passive method. In active method, some

R. M. Swarna Priya (✉) · C. Gunavathi (✉) · S. L. Aarthy
School of Information Technology and Engineering, Vellore Institute
of Technology, Vellore, Tamil Nadu, India
e-mail: swarnapriya.rm@vit.ac.in

C. Gunavathi
e-mail: gunavathi.cm@vit.ac.in

S. L. Aarthy
e-mail: aarthy.sl@vit.ac.in

© Springer Nature Singapore Pte Ltd. 2019
S. C. Satapathy et al. (eds.), *Information Systems Design and Intelligent Applications*,
Advances in Intelligent Systems and Computing 862,
https://doi.org/10.1007/978-981-13-3329-3_22

235

sorts of signals are sent to the object from which the distance has to be measured. For example, various kinds of systems which use sensors like light sensors, laser light sensors, ultrasonic sensors, etc., could be used for sending the signals and also to scan the environment of the object. These systems could identify the environment better when compared to the human. The major drawback of this method is that the initial deployment cost required is higher. Hence, nowadays researchers are focusing on reducing this cost by preparing some methodologies that could be used by the cameras.

The passive methods are those techniques which try to measure the position of the object by using the information received from the object and few passive measurements. The recent research trend is to use image processing techniques in the computer vision domain for measuring the distance of the object. This has a variety of applications like robotics, remote sensing, virtual reality and industrial automation. Moreover, the passive methods have a major advantage of being efficient in measuring the distance even in various atmospheric conditions like lightning with the help of low-cost cameras.

This paper focuses on identifying a human (or) an object and then calculating the distance between them using 3D image reconstruction. The methodology which we have discussed here could be used for any kind of application like obstacle detection in case of driverless vehicles, pedestrian detection in case of traffic monitoring system, etc. We have tried to identify the obstacle (or) the object (or) the human from a video captured by a remote camera and then also find the distance between them.

The rest of the paper is organized as follows. Section 2 discusses the related work carried out with respect to this research domain. Then Sect. 3 discusses the system model and dataset used. Section 4 elaborates the algorithm that is designed. Section 5 discusses the results obtained as a result of the experiment conducted. Finally, the paper is concluded.

2   Related Work

The recent focus of researchers, academicians and industrialists is towards the autonomous vehicles especially cars that could drive by themselves without human intervention. These autonomous vehicles are expected to be used in various applications like military, transport and production. The practitioners claim that these cars would be replacing the human intervention in all the above-mentioned domains [14]. The autonomous driving cars are built based on computer systems which help human beings to control the parts of the vehicle automatically. These control parts vary in a wide frame ranging from competency to anti-lock braking systems and cruise control which is the base need for full automation of driving [1].

Though this has become the most recent trend in automotive engineering and electronics, the expectations are too high as well as the safety plays a major role due to the risk in roads. The drivers need to be relieved completely or to be partially relieved using these autonomous cars [9]. Moreover, this automation needs to reduce the accidents that occur on roads and not increase them due to low-quality automation [7]. The obstacles between the drive are to be detected accurately by these autonomous cars, which is another major challenge. This challenge is being attended using various stereo-based vision and 2D/3D sensors [3]. The usage of single camera for detecting the obstacles is the widely used technique in the existing traditional approach [4] of 3D reconstruction [6], which has a lower accuracy [5]. The reason behind this less accuracy is that the scene is reconstructed using an assumption when a single camera is used for acquiring the image [18]. The reconstruction accuracy still goes down when spatial data is used. This problem of lower accuracy could be reduced when stereo cameras [8] are used for acquired the scene [6]. Few researchers showed an increased accuracy with the help of multiple vision sensors for calculating the distance of the object from the car and also to measure their size [10]. Kuhnert [11] and Stommel [12] proposed a new framework by fusing the two images that were acquired by the stereo camera and PMD camera. The reason for increased accuracy in this methodology is that the disparities are removed or nullified when two images are fused by completely different camera parameters. Santoro, AlRegib and Altunbasak [19] provided a methodology for reducing the depth errors which are the outcome of the camera shift process. They tried to reduce the errors using the estimated motion for correcting the errors due to the alignment of the stereo cameras [2]. All the alignment errors like translation, rotation, etc., were taken care of.

Lee et al. [13] conducted a survey of the various visual ego-motion estimation algorithms that could be used in most of the self-driving cars which use multiple cameras. They also compared their outcomes with many real-time data collected from cars with multi-cameras.

Mahammed et al. [15], Mrovlie et al. [16] and Murray et al. [17] showed interest in estimating the distance of an obstacle from the sensor which captured the object using two web cameras. But this had a lower accuracy than the available techniques. Generally, MATLAB is used by almost all practitioners for estimating the distance of the object. But they had a limitation that they could be used only for maximum distance of 1.2 m.

Though there are numerous methodologies for estimating the distance of an object from the acquiring camera, the camera parameters, lenses distance, etc., are omitted. A revised method using 3D image reconstruction for distance measurement is proposed in this paper.

3   System Model and Setup of the System

The methodology which is developed in this work gets the input as a video file which is captured using the system model as shown in Fig. 1. A camera is fixed at the junction of the road where pedestrians cross over the pedestrian line. The camera records the scenario and stores the video every five minutes time stamp. The methodology identifies if a person is available in the video and if available then calculates the distance at which the person is walking from the camera. The video

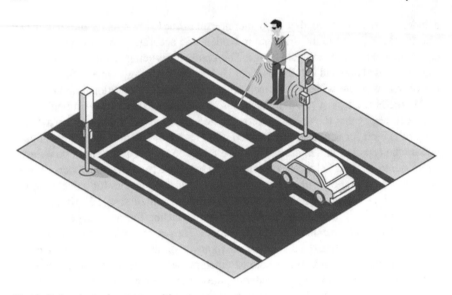

Fig. 1 Pedestrian and camera position

file which is acquired will be stored in the extension .mp4 or .avi. The methodology developed will accept the video file in any format.

3.1 Description of Dataset

The dataset that is used for analysing the performance of the methodology discussed in this paper is the widely used UCF crowd dataset. This dataset consists of videos of people walking on roads, traffic videos and other high moving objects. The videos available in the dataset are collected from the BBC Motion Gallery and Getty Images. The video file could be taken and can be fed into the methodology as raw video file. No preprocessing is required. Hence, the time required for preprocessing the raw video file is omitted.

4   Algorithm for Estimating the Distance from Camera

The algorithm which is proposed for detecting the presence of human being and then the distance of the human from the camera in metres is discussed below.

Algorithm: Person Detection and Distance Estimation (PDDE)
Input: Video file taken in traffic.
Output: Person, distance and 3D view.

1. Initiate the recognition by loading the stereo parameters of the camera.
2. The video file is read.
3. The frames are subdivided.
4. The disparity between the right and left camera are rectified by reconstructing the entire scene in 3D view.
5. Evaluate the disparity.
6. Using the disparity value, the 3D view of the scene is reconstructed.
7. From the 3D view, the human is detected.
8. The distance of the human from the camera is evaluated.
9. Steps 4 to 8 are repeated for the entire set of frames.

The video file can be of any format and length. The dataset consists of videos varying from 1 min to 1 h duration. Each and every camera has various parameters which are called as stereo parameters (or) calibration parameters. The algorithm developed uses two cameras with the geometrical value as given below:

(i) Rotation of camera: 3×3.
(ii) Translation of camera: $-121.5110, 0.0939, -0.3216$.
(iii) Fundamental matrix size: 3×3.
(iv) Essential matrix size: 3×3.

The calibration parameters are as given below:

(i) Number of patterns: 12.
(ii) World points: 48×2 with measurement in 'mm'.

The system provides an error rate of 22.35% for reproduction. These parameters are initially loaded. The user can change the parameter values as per their camera resolution. The video file is read from the stored location, and the video is played in the background by the player. While playing, the entire video is subdivided into frames each of size 1080×1920 pixels.

Since there are two cameras: one facing towards the right side and the other facing towards the left side, there is a high disparity in the acquired scene. To rectify this, we try to reconstruct the entire scene into 3D view. The generated 3D view of the scene has the following properties:

(i) The x, y, z position of the view angle is $-0.9671, -5.4502, -5.6191$, respectively.
(ii) The camera view angle is $39°$.
(iii) The view position is $-10.0625, -23.4824$.
(iv) The projection followed is orthographic projection.

After the disparity is reduced, the current disparity value is calculated for each pixel in the left image by computing the distance to the corresponding pixel in the right image, and it is proportional to the distance of the corresponding world point from the camera. Then the disparity map is generated. The 3D world coordinate points of each pixel are reconstructed from the disparity map. The world coordinate points are converted into the distance in metres using which the point-cloud objects are framed and virtualized. Based upon the points, the human is detected. Once the human is detected, compute the distance from the centroid to the camera in metres and find the 3D world coordinates of each detected human being. The list of steps which are discussed above is repeated until the entire sets of subdivided frames are processed. Finally, the total number of humans detected and their corresponding distances will be given.

5 Experimental Results

The algorithm which is proposed is implemented using MATLAB R2017a. The results are tested using pedestrian video in .mp4 format. The video runs for 8 s where two humans cross the pedestrian. Initially, the video is subdivided into eight frames of 1080×1920 pixel size, i.e. each second in the video is subdivided and eight frames are formed and a sample frame is shown below (Fig. 2).

The frame image is converted into greyscale where one image is the left camera image and the other is right camera image. Once both the images are converted then the disparity between the image is identified. The disparity is calculated by finding the distance between each pixel in the left image and the pixel in the right image.

Fig. 2 Sample frame

Table 1 Disparity map

$-3.40E+38$	$-3.40E+38$	$-3.40E+38$	$-3.40E+38$

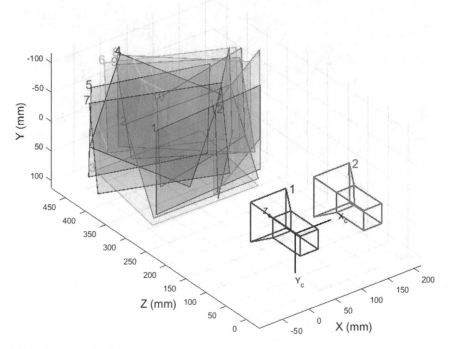

Fig. 3 The point cloud

Table 2 Centroid values computed for the video frame

X	Y	Z
2.334024	0.012572	1.732762

The distance is proportional to the distance of the corresponding world point from the camera. The distance is recorded as disparity map and a sample is shown above (Table 1).

Using the disparity map, the image is reconstructed to three-dimensional scene by mapping each pixels disparity. The final output will be the reconstructed 3D image in the form of point cloud as shown above (Fig. 3).

The 3D reconstructed image is used to detect the person/human in the video for each frame by creating a people detector object.

The centroid has been identified for each detected person, and the distance is computed from the centroid of each person to the camera in metres. The centroid calculated for the input video is tabulated in Table 2.

Fig. 4 Detected person and distance

Finally, the output of the detected person/human along with the distance of the person from the camera is displayed for each frame in the video as shown in the figure above (Fig. 4).

The accuracy of the estimated distance of the person from the camera is estimated using the mean projection error rate, which is evaluated as given below.

Mean re-projection error	0.2235

6 Conclusion

The proposed method for object–distance measurement designed purely relies on 3D image reconstruction. This method is able to aid in measuring the distance between the cars and the objects to determine the safe driving distance. The method developed in this work uses only two cameras to capture the view on the roadside. Then 3D image reconstruction technique is used to identify the centroid, point cloud and disparity map. Using these, the final 3D cloud is reconstructed. The person is detected using an opaque angle and then the distance from the camera is estimated and displayed. The methodology can be extended to identify any object or obstacle on the road. Moreover, the error could be still reduced by modifying the way the image is reconstructed.

References

1. Administration, N. H. T. S., *Preliminary Statement of Policy Concerning Automated Vehicles*, pp. 1–14, Washington, DC, (2013)
2. P. Alizadeh, M. Zeinali, *A Real-Time Object Distance Measurement Using A Monocular Camera* (2013)
3. N. Appiah, N. Bandaru, *Obstacle detection Using Stereo Vision for Self-driving Cars* (2011)
4. G. Calin, V. Roda, Real-time disparity map extraction in a dual head stereo vision system. Latin Am. Appl. Res. **37**(1), 21–24 (2007)
5. Carnicelli, J. (2005). Stereo vision: measuring object distance using pixel offset
6. H. Fathi, I. Brilakis, Multistep explicit stereo camera calibration approach to improve euclidean accuracy of large-scale 3D reconstruction. J. Comput. Civ. Eng. **30**(1), 04014120 (2014)
7. N. Goodall, Ethical decision making during automated vehicle crashes. Transp. Res. Rec.: J. Transp. Res. Board **2424**, 58–65 (2014)
8. C. Häne, T. Sattler, M. Pollefeys, Obstacle detection for self-driving cars using only monocular cameras and wheel odometry. Paper Presented at the Intelligent Robots and Systems (IROS), 2015 IEEE/RSJ International Conference on Intelligent Robots and Systems (IROS) (2015)
9. A. Hohm, F. Lotz, O. Fochler, S. Lueke, H. Winner, automated driving in real traffic: from current technical approaches towards architectural perspectives. SAE Technical Paper (2014)
10. T.-S. Hsu, T.-C. Wang, An improvement stereo vision images processing for object distance measurement. Int. J. Autom. Smart Technol. **5**(2), 85–90 (2015)
11. C. Ilas, Electronic sensing technologies for autonomous ground vehicles: a review. Paper presented at the 2013 8th International Symposium on Advanced Topics in Electrical Engineering (ATEE) (2013)
12. K.-D. Kuhnert, M. Stommel, Fusion of stereo-camera and pmd-camera data for real-time suited precise 3d environment reconstruction. Paper presented at the 2006 IEEE/RSJ International Conference on Intelligent Robots and Systems (2006)
13. G.H. Lee, F. Faundorfer, M. Pollefeys, Motion estimation for self-driving cars with a generalized camera. Paper presented at the Proceedings of the IEEE Conference on Computer Vision and Pattern Recognition (2013)
14. C. Li, J. Wang, X. Wang, Y. Zhang, A model based path planning algorithm for self-driving cars in dynamic environment. Paper presented at the Chinese Automation Congress (CAC) (2015)
15. M.A. Mahammed, A.I. Melhum, F.A. Kochery, Object distance measurement by stereo vision. Int. J. Sci. Appl. Inf. Technol. (IJSAIT) **2**(2), 05–08 (2013)
16. J. Mrovlje, D. Vrancic, Distance measuring based on stereoscopic pictures. Paper presented at the 9th International Ph.D. Workshop on Systems and Control: Young Generation Viewpoint (2008)
17. D. Murray, J.J. Little, Using real-time stereo vision for mobile robot navigation. Auton. Rob. **8**(2), 161–171 (2000)
18. M. Pollefeys, D. Nistér, J.-M. Frahm, A. Akbarzadeh, P. Mordohai, B. Clipp, P. Merrell, Detailed real-time urban 3d reconstruction from video. Int. J. Comput. Vis. **78**(2–3), 143–167 (2008)
19. M. Santoro, G. AlRegib, Y. Altunbasak, Misalignment correction for depth estimation using stereoscopic 3-d cameras. 2012 IEEE 14th International Workshop on Paper presented at the Multimedia Signal Processing (MMSP) (2012)
20. H. Walcher, *Position Sensing: Angle and Distance Measurement for Engineers* (Elsevier, Amsterdam, 2014)

Adaptive Routing Mechanism in SDN to Limit Congestion

Anusha Kannan, Sumathi Vijayan, Manikandan Narayanan and Monesh Reddiar

Abstract A network whose behavior can be dynamically controlled, changed, and managed through various interfaces is called a Software Defined Network. The flow of packets in the network is determined by a communication protocol called open flow. In SDN, rules are implemented in control plane and rules are processed in data plane. With the increase in use of Internet and the emergence of IOT, data sent or received over a network is increasing each day and it is testing the network capabilities to deal with congestion. To resolve the specified issue, in this paper, we will control congestion in SDN by splitting traffic dynamically by analyzing the statistics gathered by each switch in the network. The traffic split will take place in such a way that when a flow is rerouted to another path, the controller does check in advance that this action does not lead to congestion in the new path. The main aim during the above implementation of the work will be to reduce overutilized links and decrease packet loss.

Keywords SDN · Congestion · Adaptive routing · Congestion management
Load calculation · Open flow

A. Kannan (✉) · M. Reddiar
School of Computing Science and Engineering,
VIT Chennai Campus, Chennai, Tamil Nadu, India
e-mail: Anusha.k@vit.ac.in

M. Reddiar
e-mail: moneshm.reddiar2016@vitstudent.ac.in

S. Vijayan
School of Electrical and Electronics Engineering,
VIT Chennai Campus, Chennai, Tamil Nadu, India
e-mail: vsumathi@vit.ac.in

M. Narayanan
School of Information Technology and Engineering,
VIT Vellore Campus, Vellore, Tamil Nadu, India
e-mail: Manikandan.n@vit.ac.in

© Springer Nature Singapore Pte Ltd. 2019
S. C. Satapathy et al. (eds.), *Information Systems Design and Intelligent Applications*,
Advances in Intelligent Systems and Computing 862,
https://doi.org/10.1007/978-981-13-3329-3_23

1 Introduction

SDN is a network where an existing infrastructure is monitored by a central control panel, and this allows various companies to monitor and control their network, so the demand for this type of network is increasing. SDN is used by CDN (Content Delivery Network) service providers to efficiently direct flow of packet to the high demand areas and various companies use it while creating cloud framework. SDN is basically divided into three categories, i.e., Application Plane, Control Plane, and Data Plane. The application plane consists of various programs that either analyze the network conditions or generate a mechanism for switches to modify current flow of network. Control Plane in SDN interacts with the Application Plane as well as the Data Plane, it collects the statistics of network and generates routing table according to the request of application and sends the table to respective switches. The Data Plane is the collection of physical switches in the network. Figure 1 shows the abstract view of SDN architecture, and the NBI in it stands for northbound interface and SBI stands for southbound interface.

OpenFlow is a protocol specifically designed for SDN to allow Control Plane to define the route of the packets. It is located above Transmission Control Protocol (Fig. 2).

A network link is said to be congested if the queue of packets to be routed through switch exceeds the limit and new incoming packets start to drop out of the queue resulting in retransmission of the packet. Congestion occurs with the increase of communication flow in the network. It can be controlled by rerouting the packets through other path according to the existing scenario.

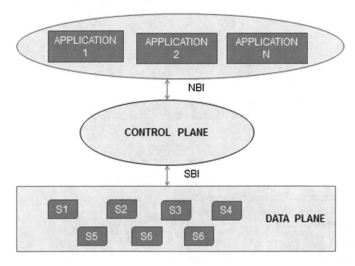

Fig. 1 SDN architecture

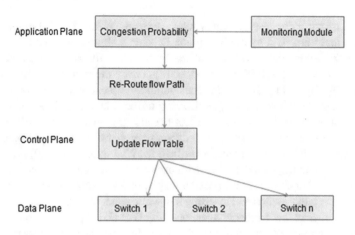

Fig. 2 Adaptive routing architecture

In SDN congestion occurs in two areas—the first one is between switches and control plane while requesting flow route by the switches and the second is in the data plane, we will be dealing with the second type here.

The problem with the existing methods is that the rerouted link has high probability of congestion therefore the percentage of packet drops remains the same therefore we will dynamically route the flow to solve the mentioned problem.

In this paper, we have proposed a way to limit congestion through adaptive routing mechanism.

2 Literature Survey

It is always a difficult task to reduce congestion in a large network due to the dynamic traffic generation. Various kinds of congestion management techniques for software-defined networks have been referred in this literature.

Kamisinski et al. [1] proposed two algorithms that reroute flow based on path overload probability. The first algorithm will reroute flow by calculating the estimated congestion probability of other links whereas the second algorithm uses number if links in the path to reroute flow to minimize congestion. Mendiola et al. [2] survey the involvement of SDN in traffic engineering and focus on the benefits of the abstraction provided by the interface of SDN. Munasinghe et al. [3] suggest a three-tier architecture for LTE networks to unload packets at the first phase, i.e., before the packets reach the backbone network to increase throughput of the network in case of congestion. Pang et al. [4] proposed an architecture for SDN based DCN (Data Center Networks) that finds the flow path by the use of traffic scheduling, multiple path selection, and segment routing. This ensures the collision of packets does not take place due to unplanned selection of paths. Pham and Hwang [5] survey methods to control congestion over wireless network. Wang et al. [6] implement an algorithm that finds path by modifying the Dijkstra's algorithm and adding various probabilistic factor related to status of the links to find the cost of each path does result in efficient path selection. Shu et al. [7] convey the importance of managing real-time flow by taking QOS under consideration. Tuncer et al. [8] propose a way to manage congestion by flexibly redirecting traffic. Flexible redirection can have a positive effect as well as negative effect because the uncertainty while choosing random path through sub masking may decrease probability of congestion on other links but will be difficult to predict the traffic flow. Wichtlxuber et al. [9] proposed an approach a combining CDN and ISP with use of SDN switches to handle real-time data efficiently. Xu et al. [10] proposed a way to defend against new flow attack by identifying the target switch and inserting it in white list or blacklist. This is important to deal with attacks that trigger congestion. Wojcik et al. [11] analyze the FAMTAR approach of adaptive routing that focuses on minimizing the updates carried out on routing table by using a max–min threshold.

From the related works analyzed, we come to know that all the path evaluation and flow routing depend on the using current status of the links and rerouting is done once congestion occurs. Therefore, designing a method to predict the congestion in advance will decrease the packet loss caused by congestion relatively.

3   Proposed Work

In this section, we propose a way to find congestion probability of each switch and redirect the flow accordingly.

The proposed method consists of three stages

Stage 1—Network Monitoring
Stage 2—Calculating Congestion Probability
Stage 3—Redirecting flow.

3.1   Network Monitor

There are two things to take under consideration while analyzing the capacity of network to handle flow, i.e., link capacity (LC) and switch capacity (SC). Link capacity is the maximum amount of packets a link can hold. Switch capacity is the maximum number of packets the switch can process.

In this phase, the Monitoring Module will analyze the number of packets processed by the switch within a time limit in regular intervals and store it.

3.2 Calculating Congestion Probability

To determine in advance the possibility of congestion in the network the CP is calculated. CP is determined using the data stored by monitoring module. The formula proposed in this paper to calculate CP is

$$CP = (S1 * S2 \ldots * SN)/LC * SC * N \tag{1}$$

In the above formula, the S are the number of packets the switch has processed according to monitoring module in regular intervals of $1 \ldots N$ and LC is link capacity and SC is switch capacity. This formula will quantify the statistics collected in a range between 0 and 1.

3.3 Redirecting Flow

In this phase, the flow is rerouted to other path if the CP crosses a predefined threshold. Whenever a reroute is initiated, the Estimated Congestion Probability (ECP) of the other possible paths is calculated by using the following equation:

$$ECP = \sum_{1}^{n} CP * DJ(cost) \tag{2}$$

The above equation a links cost and CP of each link is combined for all links in a path to obtain an ECP value. The ECP with the least value will be taken as a path for rerouting. The flow table of switches involved in rerouting is updated through control plane.

Algorithm for redirecting flow

Step 1-If CP of a switch exceeds threshold do,
Step 2-Calculate ECP of all other links
Step 3-Sort the Path based on ECP in incremental order
Step 4-For each new flow do,
Step 5-Check Source and Destination of flow and allot it to the path with least ECP

3.4 Selection of Threshold

The threshold selection is an important aspect in the above-proposed model as the level of congestion tolerance depends on the threshold and if the threshold is higher, the delay to update flow will be high whereas in case of lower threshold value, high amount of rerouting will take place.

4 Implementation and Results

From Figs. 3, 4, 5, and 6, we limit congestion and choose the efficient path from the possible path to send packets.

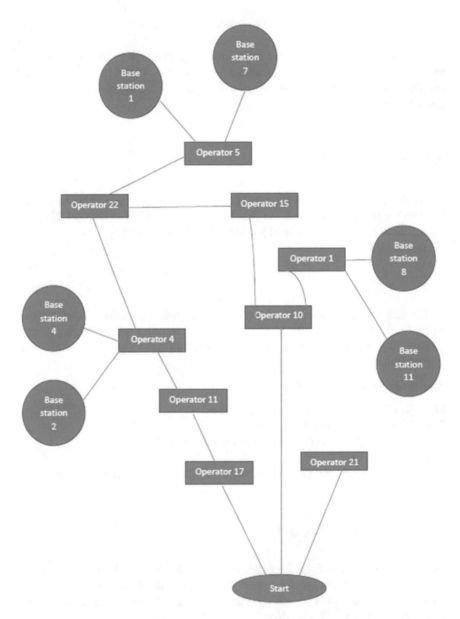

Fig. 3 Controller and nodes setup

```
--controller=remote,ip=10.0.2.15,port=6633
*** Creating network
*** Adding controller
*** Adding hosts:
h1 h2 h3 h4 h5 h6 h7 h8
*** Adding switches:
s1 s2 s3 s4 s10 s11 s17 s18 s21 s22
*** Adding links:
(h1, s1) (h2, s1) (h3, s2) (h4, s2) (h5, s3) (h7, s4) (h8, s4) (s1, s10
) (s1, s21) (s2, s10) (s2, s21) (s3, s11) (s3, s22) (s4, s11) (s4, s22) (s10, s1
8) (s11, s17) (s17, s21) (s18, s22)
*** Configuring hosts
h1 h2 h3 h4 h5 h6 h7 h8
*** Starting controller
*** Starting 10 switches
s1 s2 s3 s4 s10 s11 s17 s18 s21 s22
*** Starting CLI:
mininet> pingall
*** Ping: testing ping reachability
h1 -> X X h4 h5 h6 h7 h8
h2 -> h1 h3 h4 h5 h6 h7 h8
h3 -> h1 ^C
Interrupt
mininet> pingall
*** Ping: testing ping reachability
h1 -> h2 h3 h4 h5 h6 h7 h8
h2 -> h1 h3 h4 h5 h6 h7 h8
h3 -> h1 h2 h4 h5 h6 h7 h8
h4 -> h1 h2 h3 h5 h6 h7 h8
h5 -> h1 h2 h3 h4 h6 h7 h8
h6 -> h1 h2 h3 h4 h5 h7 h8
h7 -> h1 h2 h3 h4 h5 h6 h8
h8 -> h1 h2 h3 h4 h5 h6 h7
*** Results: 0% dropped (56/56 received)
```

Fig. 4 Open daylight topology

```
Cleaning up...
monesh@monesh-VirtualBox:~/cong$ python odl.py
Enter Host 1
1

Enter Host 2
4

Enter Host 3 (H2's Neighbour)
5

Device IP & MAC

{'10.0.0.8': '72:8f:7a:be:31:60', '10.0.0.5': '5a:be:46:92:f5:82', '10.0.0.4': '
76:99:14:87:65:68', '10.0.0.7': '8a:c6:d1:f5:df:05', '10.0.0.6': '2a:c6:31:0b:34
:b6', '10.0.0.1': '92:22:d9:e3:31:a3', '10.0.0.3': '9a:46:c9:6e:ce:9f', '10.0.0.
2': 'ca:0f:6f:5c:2e:3f'}

Switch:Device Mapping

{'10.0.0.8': 'openflow:4', '10.0.0.5': 'openflow:3', '10.0.0.4': 'openflow:2', '
10.0.0.7': 'openflow:4', '10.0.0.6': 'openflow:3', '10.0.0.1': 'openflow:1', '10
.0.0.3': 'openflow:2', '10.0.0.2': 'openflow:1'}

Host:Port Mapping To Switch

{'10.0.0.8': '2', '10.0.0.5': '1', '10.0.0.4': '2', '10.0.0.7': '1', '10.0.0.6':
 '2', '10.0.0.1': '1', '10.0.0.3': '1', '10.0.0.2': '2'}

Switch:Switch Port:Port Mapping

{'11::17': '3::2', '10::18': '3::1', '18::10': '1::3', '4::22': '3::1', '22::4':
 '1::3', '22::3': '2::4', '1::10': '4::1', '2::21': '3::2', '22::18': '3::2', '1
7::21': '1::3', '18::22': '2::3', '4::11': '4::2', '2::10': '4::2', '3::22': '4:
:2', '3::11': '3::1', '1::21': '3::1', '11::3': '1::3', '21::17': '3::1', '21::1
```

Fig. 5 Host and controller configuration

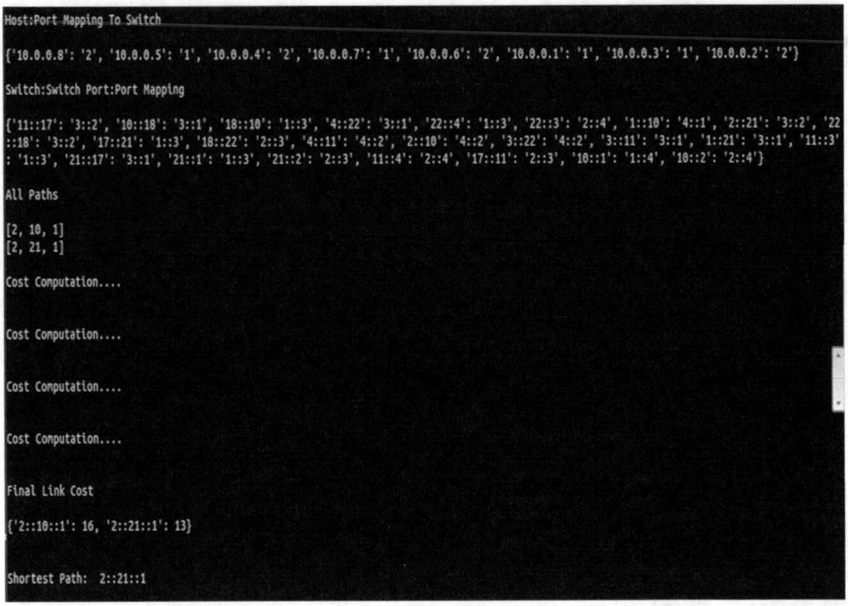

Fig. 6 Cost computation

5 Conclusion

In this paper, we have proposed a way to reduce packet loss in case of congestion by detecting the congestion in advance. In our design the flow is rerouted in a dynamic way, i.e., an adaptive approach by utilizing the past data is followed to get the best result. The proposed model can be extended by implementing a random forest regression while calculating ECP to increase the efficiency of the module. The proposed model whenever simulated in terms of reliability will show a positive result. The simulated model will show less packet loss while handling congestion.

References

1. A. Kamisinski, J. Domzal, R. Wojcik, A. Jajszczyk, Two rerouting-based congestion control algorithms for centrally managed flow-oriented networks. IEEE Commun. Lett. **20**(10), 1963–1966 (2016)
2. A. Mendiola, J. Astorga, E. Jacob, M. Higuero, A survey on the contributions of software-defined networking to traffic engineering. IEEE Commun. Surv. Tutorials **19**(2), 918–953 (2017)
3. K. Munasinghe, I. Elgendi, A. Jamalipour, D. Sharma, Traffic offloading 3-tiered SDN architecture for DenseNets. IEEE Netw. **31**(3), 56–62 (2017)
4. J. Pang, G. Xu, X. Fu, SDN-based data center networking with collaboration of multipath TCP and segment routing. IEEE Access **5**, 9764–9773 (2017)

5. Q. Pham, W. Hwang, Network utility maximization-based congestion control over wireless networks: a survey and potential directives. IEEE Commun. Surv. Tutorials **19**(2), 1173–1200 (2017)
6. M. Wang, J. Liu, J. Mao, RouteGuardian: Constructing secure routing paths in software-defined networking. Tsinghua Sci. Technol. **22**(4), 400–412 (2017)
7. Z. Shu, J. Wan, J. Lin, S. Wang, D. Li, S. Rho, C. Yang, Traffic engineering in software-defined networking: measurement and management. IEEE Access **4**, 3246–3256 (2016)
8. D. Tuncer, M. Charalambides, S. Clayman, G. Pavlou, Flexible traffic splitting in OpenFlow networks. IEEE Trans. Netw. Serv. Manage. **13**(3), 407–420 (2016)
9. M. Wichtlxhuber, R. Reinecke, D. Hausheer, An SDN-based CDN/ISP collaboration architecture for managing high-volume flows. IEEE Trans. Netw. Serv. Manage. **12**(1), 48–60 (2015)
10. T. Xu, D. Gao, P. Dong, H. Zhang, C. Foh, H. Chao, Defending against new-flow attack in SDN-based internet of things. IEEE Access **5**, 3431–3443 (2017)
11. R. Wojcik, J. Domzał, Z. Dulinski, P. Gawłowicz, D. Kowalczyk, Performance evaluation of flow-aware multi-topology adaptive routing, in *2014 IEEE International Workshop Technical Committee on Communications Quality and Reliability (CQR)* (2014)

Optimization and Decision-Making in Relation to Rainfall for Crop Management Techniques

K. Lavanya, Anand Vardhan Jain and Harsh Vardhan Jain

Abstract Disease prediction has a high degree of uncertainty which is due to the complex and imperfect nature of symptoms that are used in diagnosis. Diseases continue to be a threat to crop yield and investments even though technological advancements have been made in agricultural sector. Clinically screened database has been taken as knowledge base for weather and crop symptoms. The present work highlights the degree of effect of different weather parameters on rainfall and then further utilizes the findings for decision-making on crop production including disease detection and crop selection. Parameters used in this experiment are Wind Speed (WS), Relative Humidity (RH) and Temperature (T). Taguchi Orthogonal arrays (OA) will be used for data optimization. Parameter optimization is done with the help of ANOVA (Analysis of Variance).

Keywords Data optimization · Taguchi · ANOVA · Fuzzy · Disease detection

1 Introduction

India is the largest producer of paddy crops globally. It is the second biggest exporter of rice worldwide. In Tamil Nadu, majority of farmers belong to minimal category. Although farmers use cultural and chemical methods for controlling pests, it leads to higher production costs. The excessive use of chemicals results in lower yield hence the profits in rice production. A major factor which results in crop failure is pest damage. This, in turn, leads to an increase in the price of the crop. A pathologist is able to draw conclusion with the information available to him regarding crop

K. Lavanya (✉) · A. V. Jain · H. V. Jain
School of Computer Science and Engineering, VIT University, Vellore, India
e-mail: lavanya.k@vit.ac.in

A. V. Jain
e-mail: anand.vardhanjain2013@vit.ac.in

H. V. Jain
e-mail: harshvardhanjain95@gmail.com

© Springer Nature Singapore Pte Ltd. 2019
S. C. Satapathy et al. (eds.), *Information Systems Design and Intelligent Applications*,
Advances in Intelligent Systems and Computing 862,
https://doi.org/10.1007/978-981-13-3329-3_24

symptoms, which is uncertain and highly environment specific. The prediction of a disease using this data is difficult for a human expert. Finding subsets of this data allows supervised learning to introduce accuracy concepts. Taguchi models are predefined models used for optimization of large data Sets to improve the quality of data available. We have used L-32 Taguchi model. It divides each parameter value into ranges (levels). Our model consists of a mixed level design rather than a full factorial design. ANFIS, as the name implies, is a combination of both fuzzy Inference system and neural networks hence having a single framework that includes advantages of both the areas. System used for this paper is a four-layered feed-forward neural network. Input climatological factors are represented by the first layer. The second layer (hidden layer) comprises of the fuzzy rules. The input specific result will be represented by the third layer. The last layer would identify the diseases which are diagnosed by the system. Literature survey and background survey have been highlighted in Sects. 2 and 3. Section 4 explains the proposed system and Sect. 5 discusses the experiment conducted and its results.

2 Related Works

In medicine, disease diagnosis is efficiently achieved using decision algorithm modelling. It has led to considerable progression in the comprehension of clinical data, conversion of clinical understanding into models. These models are converted into experimental programs. ANFIS was used to diagnose liver disease because with technology advancing doctors have developed a tendency towards useful applications to reach better and accurate diagnosis. Gaussian and triangular membership functions for three membership function times for each input and output. 80% of the data was chosen for the training and 10% for testing which managed to reduce the error [1]. Omid Sojoodi Shijani, Majid Ghonji Feshki proposed PSO algorithm along with feed-forward neural network improving the diagnosis of heart disease. Specific dataset was worked on providing efficient criteria for heart disease diagnosis. It also concluded that the above-mentioned methods have the most accurate criteria [2]. Data mining techniques were incorporated to reduce large Datasets to usable data which in turn would aid in decision-making and diagnosis of heart diseases [3]. Used ANN, ANFIS and Genetic Algorithm processes for forecasting rainfall and then compared it with linear multiple regression models to show the benefits of Artificial Intelligence models [4]. Endoscopic images were acquired from volunteers. First, extraction of image features was conducted followed by preforming diagnostic task with the help of a neuro-fuzzy scheme. Clustering Algorithm is used for creating clusters and the results help to create fuzzy base rules. Multiple classifier approaches provided encouraging results [5]. This paper shows the Taguchi's parameter design approach, and how it has been applied for optimizing cylindrical grinding machine parameters. Depth of cut and cutting speed were evaluated as parameters here to analyze the effect of these parameters, an orthogonal array, signal-to-noise (S/N) ratio and analysis of variance (ANOVA) are employed. Planning of the experiment is done by using an orthogonal array. Analysis of variance (ANOVA) is used to find which input parameters affect the output parameter. This procedure eliminates the need for repeated experiments and helps in saving time and resources [6]. In this paper, LM6Al/5%SiC composites were machined by ECM process. The effect of different process parameters on important machining parameter, i.e. MRR was studied. Effect of the process parameters was found using the design method of Taguchi. Analysis of the effect required an application of S/N ratio, ANOVA and orthogonal array. The result of this paper for metal removal rate concludes the successful Taguchi parameter design for cutting parameters [7]. In this paper, Taguchi method is used to find process parameters and their interactions which have an effect on thermo-catalytic degradation of polypropylene into liquid fuel. The amount of liquid fuel yielded in the process is a machining parameter which was affected by process parameters. To analyse the effect of these parameters, an orthogonal array, signal-to-noise (S/N) ratio and analysis of variance (ANOVA) are employed. An orthogonal array has been used to plan the experiments. For investigative purposes, ANOVA is used to determine what parameters affect the quality characteristic. The result concluded that the yield of liquid fuel in this process was highly dependent on temperature followed by acidity of catalyst and catalyst concentration [8].

3 Background Study

3.1 ANFIS

ANFIS is adaptive neuro-fuzzy inference system, with inputs and weights consisting of real value [0, 1]. Fuzzy systems incorporate learning ability hence leading to intelligent decision-making. System used for this paper is a fourth layer feed-forward neural network. Input variables are represented by the first layer which includes the different combinations of parameters. The second layer (hidden layer) comprises of the fuzzy rules. The produced result will be represented by the third layer. Output of the data set would be described by the fourth layer.

3.2 Taguchi

Taguchi methods are statistical methods or sometimes called robust design methods, developed by Genichi Taguchi to improve the quality of manufactured goods, and more recently also applied to engineering, biotechnology and advertising. To choose a process/product which in the operating environment works more consistently, Taguchi designed experiment is used. Taguchi designs identify the control factors. During experimentation, the noise parameter is manipulated to allow variability to occur and then find optimal process parameter values that make the product resistant to variations from noise factors. Orthogonal arrays are used in Taguchi designs to evaluate the effects of parameters on response mean and variation. Orthogonal array is required for a balanced design. This helps in assessing each parameter (factor) independently, hence not affecting another factor's estimation thus reducing costing and time associated with the experimentation.

4 Proposed Methodolgy

Disease prediction leads to a better understanding of crop origins and their risk interactions resulting in better yields. Models like Multiple Regression and Rolling Average results in disease prediction. But if not validated, these can lead to uninterrupted results. Optimization of Taguchi outputs by artificial neural networks results in an optimal solution. Product of weight vector and its corresponding input vector results in ANN output, adding the values and applying activation function to the sum. The approach for neural network in predicting disease is as following:

Non-linear model is advantageous due to complex data set of disease and the difficulty to model it.

Neural network requires a humungous data set to recognize a disease more efficiently (Fig. 1).

Empirical Data:

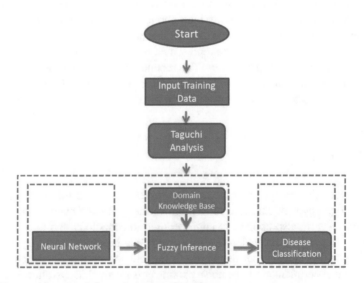

Fig. 1 Proposed framework for optimization of weather parameters

Table 1 Notation representation of diseased table

Paddy diseases	Notation	Input linguistic variable
Rice blast	D1	Spores
Sheath rot	D2	Parts of plant
Brown leaf spot	D3	Dwarfing
Bacterial leaf blight	D4	Border colour

Crop fields in Vellore district area were surveyed for disease occurrence. 27 locations were considered which widely covered varieties of soil, types of cultivators, climatic conditions and methods of cultivation. The extremities of the diseases were then tabulated, compiling a huge dataset as the target. Incomplete and Useless datasets were eliminated thus ensuring consistency. The knowledge collected in these steps helped us in understanding the above elements for diagnosis of diseases in paddy (Table 1).

4.1 Taguchi Optimization and ANOVA Analysis

The study conducted in this paper includes three parameters, namely Wind Speed (WS), Relative Humidity (RH) and Temperature (T) for predicting the amount of rainfall. Specified ranges of the attributes are provided based on the district chosen for the experiment. WS has been given two levels, RH and Temperature both include three levels each, covering all domain values. These levels explain different scenarios of the parameters and different combinations of levels of all the parameters indicate the amount of rainfall.

4.1.1 Data Optimization Using Taguchi Models

Taguchi models are predefined models used for reduction of large data sets to improve the quality of data available. Optimization of data leads to decrease in runtime while covering all the domain values. It also saves a lot of space and is more economical. We divided each parameter value into ranges (levels) for its use in the Taguchi model. Table 2 shows the above-mentioned levels and their range values. Our model consists of a mixed level design rather than a full factorial design. We have used L-32 Taguchi model.

4.1.2 Use of Simulation Software

The software used for Taguchi analysis and ANOVA simulation is Minitab. Minitab has an inbuilt feature to create mixed level designs in addition to full factorial design.

Table 2 Parameter ranges

Parameters	Level 1	Level 2	Level 3
Wind speed	1–7 m/s	7–9 m/s	–
Relative humidity (%)	≥70	60–70	25–60
Temperature	Normal	Warm	Cold

Table 3 Taguchi combinations

Exp. no.	WS	Temp.	RH	Rainfall
1.	1	1	1	93.758
2.	1	2	2	54.569
3.	1	3	3	0.000
4.	1	1	2	84.064
5.	1	2	3	54.125
6.	1	3	1	104.750
7	2	1	2	46.600
8.	2	2	3	70.300
9.	2	3	1	0.000
10.	2	1	3	0.000
11.	2	2	1	0.000
12.	2	3	2	49.600

Table 4 ANOVA analysis

Source	DOF	Adj. SS	Adj. MS	p-value
WS	2	2342.2	1171.4	0.289
RH	2	8488.7	4244.4	0.017
Temp.	1	11,627.6	11,627.6	0.001
Error	30	27,167.5	905.6	
Total	35	49,626.6		

According to process parameters and their corresponding level a mixed level design is chosen. Minitab produces 36 combinations representing different levels of parameters with respect to each other. Below shown are twelve of those 36 combinations (Table 3).

4.1.3   ANOVA Analysis for Rainfall

ANOVA analysis on the above-mentioned data given in Table 4 indicating the degree of freedom for each parameter along with adjusted SS/MS, p-value and error which in turn confirm the level of effect of each parameter on rainfall.

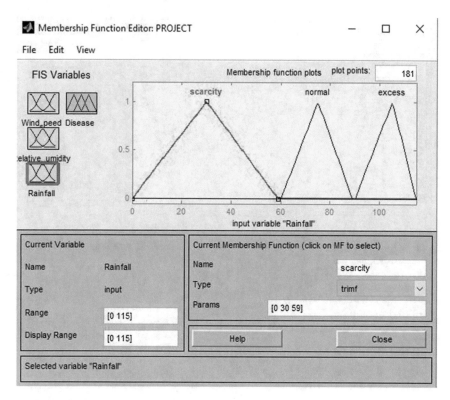

Fig. 2 Membership function plot of rainfall

4.2 ANFIS

4.2.1 Predicting Disease by Neuro-fuzzy System

Here, the model used for disease diagnosis is back propagation neural network system. Each input gets a corresponding output which in result train weights by the gradient descent on the error E. The model below has four stages.

4.2.2 Describing Output and Input Variables

The reason behind the application of fuzzy logic to the input can be the result of the fact that behaviour of the disease symptoms cannot be specifically quantified as the attributes which affect them are physiological and behavioural in nature.

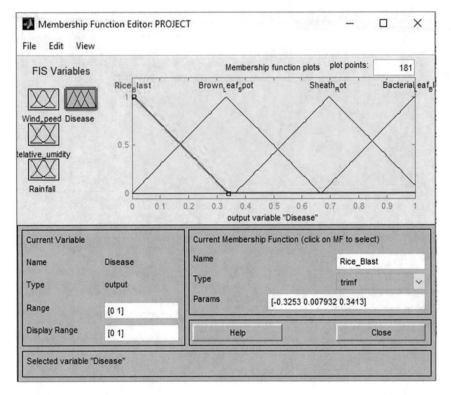

Fig. 3 Membership function plot of disease

Table 5 Membership functions for weather parameters

Inputs	Class	Fuzzification method
Temperature	Numerical	Triangular
Wind speed	Numerical	Triangular
Relative humidity	Numerical	Triangular

4.2.3 Define Membership Functions

The membership value in a fuzzy set is between [0, 1] for every element. A fuzzy set of crop symptoms is given in Fig. 2—(for graphs of ANFIS) with following form $A = \{(\mu_A(a), \gamma_A(a))/a \, \varepsilon \, A\}$; where $\mu_A(a)$, $\gamma_A(a)$ represent membership and non-membership of a in A, respectively, with the condition $0 \le \mu_A(a) + \gamma_A(a) \le 1$, for all $a \, \varepsilon \, A$ (Fig. 3; Table 5).

Table 6 Evaluation table

	Sheath rot	BLB	Rice blast	BLS
TP	11	14	16	12
TN	7	6	8	10
FP	7	4	4	2
FN	5	6	2	6
Accuracy	0.60	0.66	0.80	0.733
Specificity	0.5	0.6	0.66	0.837
Sensitivity	0.916	0.70	0.88	0.666

4.2.4 Determining Fuzzy Rules

The disease classifying rules, in ANFIS are made using domain-specific knowledge and observation. 'If-Then' rules are used to describe fuzzy logic based models. The fact after 'If' is called a premise or hypothesis or antecedent. Based on this premise, we can infer another fact which is called a conclusion. Fuzzy sets are interpreted based on the data of factors in the hypothesis. Table 6 describes disease prediction rules:

1. If (Wind Speed is 0–7 m/s) and (Temperature is in the Normal range) and (Relative Humidity is Normal) and (Rainfall is 93.5) then the disease is RICE BLAST.
2. If (Wind Speed is 0–7 m/s) and (Temperature is in the Warm range) and (Relative Humidity is 77–83%) and (Rainfall is 54.569) then the disease is BROWN LEAF SPOT.
3. If (Wind Speed is 0–7 m/s) and (Temperature is in the Cold range) and (Relative Humidity is 74–79) and (Rainfall is 0) then the disease is SHEATH ROT.
4. If (Wind Speed is 0–7 m/s) and (Temperature is in the Normal range) and (Relative Humidity is 60–73%) and (Rainfall is 84.064) then the disease is BROWN LEAF SPOT.
5. If (Wind Speed is 0–7 m/s) and (Temperature is in the Warm range) and (Relative Humidity is 74–87%) and (Rainfall is 54.125) then the disease is RICE BLAST.
6. If (Wind Speed is 0–7 m/s) and (Temperature is in the Cold range) and (Relative Humidity is $\geq$86%) and (Rainfall is 104.75) then the disease is BACTERIAL LEAF BLIGHT (Fig. 4).

4.2.5 Defuzzified Output

Paddy Disease prediction solution has to be converted to a discrete value for decision-making. Fuzzy set is summarized to a single value through Defuzzification. In deriving centroid of the shape of the output fuzzy set, we obtain discrete value in centroid method.

The defuzzified result:

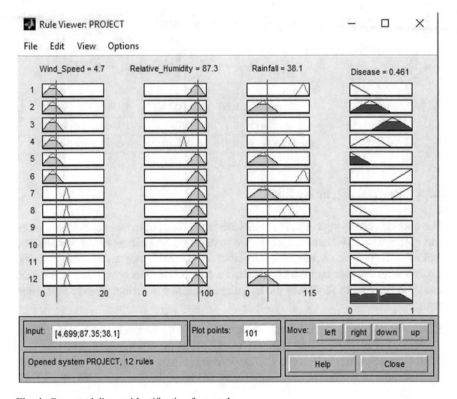

Fig. 4 Generated disease identification fuzzy rules

$$\frac{\sum_{i=1}^{n} x_i \mu(x_i)}{\sum_{i=1}^{n} \mu(x_i)}$$

5 Experimental Details for Fuzzy

5.1 Performance Metrics

Accuracy is determined as follows: Accuracy = (TP + TN)/(TP + FP + TN + FN) where TP denotes true positives referring to the positive tuples that were correctly labelled by the classifier. True negatives are denoted by TN, are the negative tuples that were correctly labelled by the classifier. FP denotes false positives which include the negative tuples that were incorrectly labelled by the classifier. FN denotes false negatives. False negatives are the positive tuples that were incorrectly labelled by the classifier. The resulted values of TP, TN, FP and FN are computed by training the Crop Disease Dataset by using Artificial Neural Approach for computation of Accuracy.

5.2 *Experimental Results*

Root mean square error (RMSE)

$$\sqrt{\sum_{i=1}^{n} (Y(t) - \overline{Y}(t))^2 / n}$$

Mean absolute error (MAE)

$$\sum_{i=1}^{n} Y(t) - \overline{Y}(t)/n$$

Error measure (E)

$$Y(t) - \overline{Y}(t)$$

$Y(t)$ is an indicator of expected output, $\overline{Y}(t)$ indicates true output in t time and n is the total number of target data (Table 7).

6 Conclusion

On the analysis of historical data, this paper explains the relationship between various symptoms and disease diagnosis. Above, RMSE and MAE values have been evaluated for all the possible diseases. The research techniques used here were successfully able to conduct the given experiment with more precision and efficiency. In future, the scope of this experiment can be increased to multiple districts and regions including more historical data leading to more accurate results.

Table 7 Evaluation of RMSE and MAE

	Sheath rot	BLB	Rice blast	BLS
RMSE	0.816	1.824	2.449	1.966
MAE	0.2	0.333	0.6	0.466

References

1. M.R. Farokhzad, L. Ebrahimi, A novel adaptive neuro fuzzy inference system for the diagnosis of liver disease. **1**, 2476–7638 (2016)
2. M.G. Feshki, O.S. Shijani, Improving the heart disease diagnosis by evolutionary algorithm of PSO and feed forward neural network, in *2016 Artificial Intelligence and Robotics (IRANOPEN)* (IEEE 2016, April), pp. 48–53
3. B. Kaur, W. Singh, Review on heart disease prediction system using data mining techniques. Int. J. Recent Innov. Trends Comput. Commun. **2**(10), 3003–3008 (2014)
4. S. Banik, F.H. Chanchary, K. Khan, R.A. Rouf, M. Anwer, Neural network and genetic algorithm approaches for forecasting bangladeshi monsoon rainfall, in *2008 11th International Conference on Computer and Information Technology* (2008, Khulna), pp. 735–740
5. V.S. Kodogiannis, Computer-aided diagnosis in clinical endoscopy using neuro-fuzzy systems, in *2004 IEEE International Conference on Fuzzy Systems (IEEE Cat. No. 04CH37542)*, vol. 3 (2004, Budapest, Hungary), pp. 1425–1429
6. R. Rudrapati, P.K. Pal, A. Bandyopadhyay, Modeling and optimization of machining parameters in cylindrical grinding process. Int. J. Adv. Manuf. Technol. **82**(9–12), 2167–2182 (2016)
7. S.R. Rao, G. Padmanabhan, Optimization of machining parameters in ECM of Al/B_4C composites using Taguchi method. Int. J. Appl. Sci. Eng. **12**(2), 87–97 (2014)
8. A.K. Pandaa, R. Singhb, Optimization of process parameters by Taguchi method: catalytic degradation of polypropylene to liquid fuel. Int. J. Multi. Curr. Res. 50–54 (2013)

Evaluation of Features to Identify a Phishing Website Using Data Analysis Techniques

Amalanathan Geetha Mary

Abstract With the growth in the present digital era, the Internet is the prime source of knowledge. This situation is depleted by phishers and they have drafted various websites which steals user's information and misuse it. Though it is hard to locate a phishing site, various features of the phishing site helps in uncovering its mask. This paper discusses several features to identify a phishing site. Using data mining techniques like classification and association rule mining many explorations are performed to prove the notion. Similarly, the impact of various features considered for analysis is studied too.

Keywords Phishing site · Legitimate site · CART algorithm · Classification
Association rule mining

1 Introduction

WWW is part of lives owing to the evolution of computing and architecture. Almost 90% of the transactions are happening as wired in some countries, which paves way for the phishers to steal information from the people. Though online services make user's life easy and comfortable, it may also make it cumbersome, and they have to take precautions. Phishing is done by blending social engineering and website spoofing, for instance, phisher sends several emails to random users, asking to update their personal details in the bank and mentioning a wrong website address similar to the bank website address. The fraudulent website is a mere representation of an actual bank website, with the same color, icon, links, etc., and in whole control of the phisher.

A. Geetha Mary (✉)
School of Computer Science and Engineering, Vellore Institute of Technology, Vellore, Tamil Nadu, India
e-mail: geethamary.a@gmail.com

© Springer Nature Singapore Pte Ltd. 2019

S. C. Satapathy et al. (eds.), *Information Systems Design and Intelligent Applications*,
Advances in Intelligent Systems and Computing 862,
https://doi.org/10.1007/978-981-13-3329-3_25

When the user attempts to login the forged bank website using the actual credentials like login id and password, then the phisher gets the complete control of the victim's account. Though lots of researches have been conducted on phishing, most of them have dealt it in email level, i.e., detecting the spam emails [1]. Regrettably, many anti-spam tools depend on regular training, so phishers break the tool easily. Consequently, many browser-related researchers have found effective and various tools that were developed, for instance, SpoofGuard, sitehound, Netcraft, AZProtect, EarthLink, and FirePhish. Comparisons of most of these tools are done by Abbasi and Chen [2] based on its performance during different intervals in a day. Zahedi et al. [3] have analysed various tools to detect phishing tools to detect fake websites and presented the impact of performance and quality measures in choosing a tool. Mohammad et al. [4] have used 17 attributes to detect a phishing site and they have presented rule for each feature to classify a phishing site. Data Mining is a process of extracting patterns from data. Data Mining is applied over many real world applications like health care, security, academics etc. [5–8]. In recent times, data mining is used to detect a phishing site using several techniques like classification [9, 10] and associative classification [11].

2 Analysis

Mohammad et al. [9] have provided a training data set with 30 conditional attributes and one decision attribute. The dataset has 11,055 data records which help in deciding whether a website is legitimate or phishing. Table 1 gives insight on the features considered to decide legitimacy of the website, the possible values the attributes can take and the criteria based on which the values are assigned.

Table 1 Attributes considered for analysis

S. No.	Feature	Values	Description
1	having_IP_Address	−1	Domain name does not have an IP address
		1	Domain name has an IP address
2	URL_Length	−1	Length of URL is less than 54 characters
		0	Length of URL is greater than or equal to 54 and less than or equal to 75 characters
		1	Length of URL is greater than 75 characters
3	Shortening_Service	−1	Did not use the shortening service
		1	Have used shortening service
4	having_At_Symbol	−1	Not having "@" symbol
		1	Having "@" symbol

(continued)

Table 1 (continued)

S. No.	Feature	Values	Description
5	double_slash_redirecting	−1	Last occurrence of "//" is in or before 7th position
		1	Last occurrence of "//" is after 7th position
6	Prefix_Suffix	−1	No prefix or suffix of "-"
		1	Prefix or suffix of "-"
7	having_Sub_Domain	1	If one or less than one subdomain
		0	2 subdomains
		−1	More than 2 subdomains
8	SSLfinal_State	−1	HTTPS certificate issued before 2 years
		0	Has HTTPS certificate but issuer is not trusted
		1	Otherwise
9	Domain_registeration_length	1	Domain expires in less than a year
		−1	Otherwise
10	Favicon	−1	Favicon included from the same website
		1	Otherwise
11	Port	−1	Necessary ports are closed
		1	Otherwise
12	HTTPS_token	1	HTTPS token is not part of the domain of the URL
		−1	Otherwise
13	Request_URL	−1	Percentage of the URL request <22%
		0	Percentage of the URL request ≥22 and 61%
		1	Otherwise
14	URL_of_Anchor	−1	Percentage of anchors of URLs <31%
		0	Percentage of anchors of URLs ≥31 and ≤67%
		1	Otherwise
15	Links_in_tags	−1	% of links <17%
		0	% of links ≥17 and ≤81%
		1	otherwise
16	SFH	−1	SFH is "about: blank\" or empty
		0	SFH refers to a different domain
		1	Otherwise
17	Submitting_to_email	1	"mail()" or "mailto:" function is used to submit information
		−1	Otherwise

(continued)

Table 1 (continued)

S. No.	Feature	Values	Description
18	Abnormal_URL	−1	URL includes host name
		1	Otherwise
19	Redirect	−1	No. of page redirects ≤1
		0	No. of page redirects >1 and <4
		1	No. of page redirects ≥4
20	on_mouseover	1	onMouseOver changes status bar
		−1	Otherwise
21	RightClick	1	Right click is disabled
		−1	Otherwise
22	popUpWindow	1	Popup window contains text fields
		−1	Otherwise
23	Iframe	1	Using iframe
		−1	Otherwise
24	age_of_domain	−1	Age of domain ≥6 months
		1	Otherwise
25	DNSRecord	1	DNS record is not found for the domain
		−1	Otherwise
26	web_traffic	−1	Rank of the website ≤100,000
		0	Rank of the website >100,000
		1	Otherwise
27	Page_Rank	1	Page rank <0.2
		−1	Otherwise
28	Google_Index	−1	Webpage indexed in google
		1	otherwise
29	Links_pointing_to_page	−1	Number of webpage pointing to the page = 0
		0	Number of webpage pointing to the webpage >0 and ≤2
		1	Otherwise
30	Statistical_report	1	Webpage is listed in top phishing IPs or domains
		−1	Otherwise
31	Result	−1	Website is legitimate
		1	Website is phishing

To locate whether the website is legitimate or phishing, the following attributes were considered:

- having_IP_Address: If the URL has a IP address, then chances of it being a phishing website is high.
- URL_Length: In the study done by Mohammad et al. [9], they have concluded that the if the URL length is more than 75, then the website being a phishing website is high and most of the legitimate users website addresses length is less than 54. So URL length takes three labels.
- Shortening_Service: URL may be shortened using any of the services like TinyURL.com. So, ultimately the URL may be longer, but by using the shortening feature, the URL is shortened. Though the present URL is shortened, it will lead to the required webpage which may have a longer URL, which is referred as "HTTP Redirect".
- having_At_Symbol: The browser will ignore the text preceding "@" symbol in the URL. So chances of redirecting the page to the real site, i.e., the phishing site is high
- double_slash_redirecting: Consider the URL, http://www.legitimate.com//http://www.phishing.com. Browser will ignore the text before "//" and redirect the page to the web address after "//". In this case, the page will be directed to http://phishing.com. In a legitimate web address, double slash would occur in 6th position if it follows http or in 7th position if it follows https.
- Prefix_Suffix: A legitimate website address will not have a "-". For example, https://www.indianbank.net.in is a legitimate address and www.indian-bank.com is a phishing website.
- having_Sub_Domain: Consider a website www.vit.ac.in, in which "in" stands for India, country code top-level domains (ccTLD), other than this a legitimate user may have one subdomain. Here it is "ac" which stands for academic institution. So, the number of subdomains contributes toward deciding whether a web address is legitimate or not.
- SSLfinal_State: HTTPS assure website legitimacy, but while scrutinizing, it was clear that the certificate issuer and the age of certificate plays a major role in deciding the legitimacy of the website
- Domain_registeration_length: Phishing website operates for a short spell of time. Legitimate domains pay recurrently and some pay before ahead.
- Favicon: Favicon is an icon (image) specific for the website. If the favicon is loaded from the same website then chances for it being a legitimate website is high.
- Port: Network Address Translation (NAT), firewalls and proxy servers block unnecessary port other than the port required by the website., i.e., HTTP and HTTPS. So, a legitimate website would not try to open other ports.
- HTTPS_token: Phishing websites may add HTTPS as part of the domain name in the URL, for example, http://www.https-indianbank.net.in.
- Request_URL: In most of the legitimate websites, objects like web pages, images, videos are loaded from the same domain. So, according to the percentage of the contents from other domains, legitimacy of the website can be decided.

- URL_of_Anchor: Anchor is an element used to connect to other web pages. Phishing websites include an anchor element which does not link to other webpage but may run a javascript file. Number of links that are connecting to other websites decides the legitimacy of the website.
- Links_in_tags: Tags like <Meta> is used to add description about webpage, <Link> is used to link the webpage to other webpage and <Script> to run a program to process. In a legitimate website, all of these links would be pointing to the same domain name. So the percentage of the links pointing to the same domain name also contributes in legitimacy of website.
- SFH: Filled in forms from the browser is processed in the server. Server Form Handler (SFH) carrying an empty string or "about:blank" is suspicious.
- Submitting_to_email: Filled in forms from the browser can be forwarded to mail of a phisher instead of submitting it to the server, for which mail() or mailto: function can be used.
- Abnormal_URL: whois is a database which has information on all the domain names. If the domain name is not a part of the URL then the possibility of the website being phishing is high.
- Redirect: A legitimate website is redirected once, but if the redirect is more than one then website is likely to be a phishing site.
- on_mouseover: javascript can be used to show a fake URL in the status bar. To verify a webpage's legitimacy, the webpage's source code has to be checked with the option "onMouseOver" event and test for any changes.
- RightClick: Phishers used to disable right click option in a web page, else the user may inspect the source code of the page. Hence, legitimacy of the website could be checked using this option.
- popUpWindow: Legitimate websites never asks its user to enter personal information through a popup window.
- Iframe: Phishers use iframe tag to inbuilt a frame into the page and where the legitimacy checks were done only in a webpage and not in iframe, so checking an iframe is mandatory too.
- age_of_domain: Most of the phishing websites are short-lived and hence their age would be less than 6 months. Age of the domain can be checked in https://who.is database and legitimacy of the website can be ensured.
- DNSRecord: If the whois database does not recognize the claimed identity of a website or no DNS record found about the website then probability of it being a phishing site is more [12].
- web_traffic: Alexa database (Alexa the Web Information Company) maintains the number of visits for a website and ranks it. Low rank implies the high legitimacy of a website. If the site is not ranked then it is a phishing site.
- Page_Rank: Page rank signifies the importance of a site. The value ranges between 0 and 1. Most of the phishing site has a page rank of zero.
- Google_Index: A website is checked against Google's index. If a website is not indexed in google then it may be a phishing website.

```
        SSLfinal_State in [-1.000 0.000] [Mode: Legitimate] ⇨ Legitimate
    ⊟  SSLfinal_State in [1.000] [Mode: Phishing]
        URL_of_Anchor in [-1.000] [Mode: Legitimate] ⇨ Legitimate
    ⊟  URL_of_Anchor in [0.000 1.000] [Mode: Phishing]
        ⊟  web_traffic in [0.000] [Mode: Phishing]
            ⊟  URL_of_Anchor in [0.000] [Mode: Phishing]
                having_Sub_Domain in [-1.000 0.000] [Mode: Legitimate] ⇨ Legitimate
                having_Sub_Domain in [1.000] [Mode: Phishing] ⇨ Phishing
            URL_of_Anchor in [1.000] [Mode: Phishing] ⇨ Phishing
        web_traffic in [-1.000 1.000] [Mode: Phishing] ⇨ Phishing
```

Fig. 1 Decision tree: a tree which is generated by applying CART algorithm

- Links_pointing_to_page: Due to short lifespan phishing websites does not have any links from external websites. Thus, number of links pointing to a website determines legitimacy of the website.
- Statistical_report: From Phishtank (PhishTank Stats, 2010–2012) and Stopbadware (StopBadware, 2010–2012) database, the phishing websites were considered from top 10 and 50, respectively.

All these attributes help in deciding whether a website is a legitimate or phishing site, in addition, few more analysis can also be done on the data set. Clementine 11.1.1, now called as IBM SPSS Modeler, is used to perform required explorations. First 30 attributes can be used to decide whether a website is legitimate or not, which is referred to as conditional attributes. Using the 11,055 data records a model is trained. I have used CART algorithm to generate a classification tree. The generated classification tree is presented in Fig. 1. From the 30 attributes, the algorithm has chosen four attributes from which legitimacy of a website can be decided. For instance, deciding whether a website is a phishing site or not, we have to check, SSL-final_state, if it is −1 (HTTPS certificate issued before 2 years) or 0 (Has HTTPS certificate but issuer is not trusted), then it's a legitimate site. If SSLfinal_state is 1 then URL_of_Anchor has to be examined, if it is −1 (% of URL of Anchor <31%) then again it's a legitimate site, else verify web_traffic.

From Fig. 1, major attributes for deciding the legitimacy of the website is SSL-final_State, URL_of_Anchor, Web_traffic, and having_sub_Domain. Here examine the impact of these attributes individually over the deciding attribute using the distribution graphs. Figure 2 gives the distribution of each value across Result attribute. From Fig. 2, it is clear that among 11,055 data records, whichever has SSLfinal_state value as 0, is legitimate and if its −1, then most of the sites were legitimate and if it is 1, then most of the sites were phishing.

The distribution graphs for the attributes URL_of_Anchor, Web_traffic and having_sub_Domain is given in Figs. 3, 4 and 5, respectively.

Figure 6 displays the association rules generated from the 31 attributes and 11,055 data records of the data set. Relationship among these 30 attributes in deciding the legitimacy of a website is set by specifying all the 30 attributes as antecedents and decision attribute, Result as consequent. Maximum of 5 attributes are allowed as antecedent. Measures, support and confidence are set as 17% and 90%, respectively.

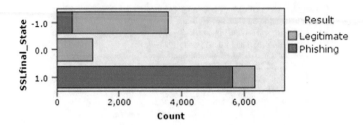

Fig. 2 Distribution of the attribute SSLfinal_state over Result attribute

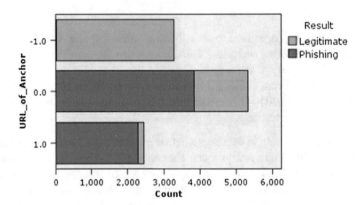

Fig. 3 Distribution of the attribute URL_of_Anchor over Result attribute

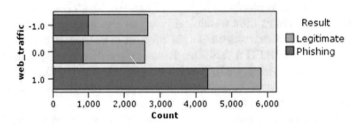

Fig. 4 Distribution of the attribute Web_traffic over Result attribute

Association rules can also be used as decision rules, for instance, from Fig. 6, it is clear that if URL_of_Anchor is −1 and SSLfinal_State is −1, then Result is legitimate, where the confidence is 100%.

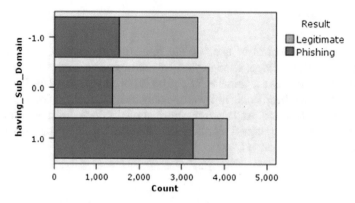

Fig. 5 Distribution of the attribute having_sub_Domain over Result attribute

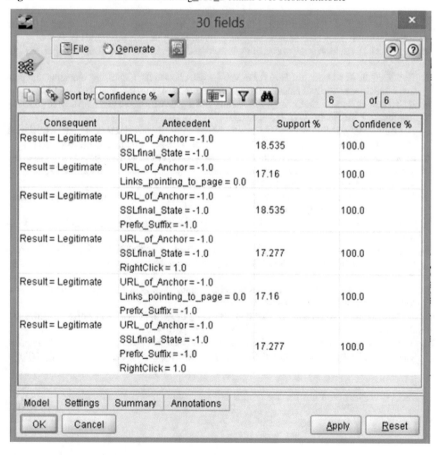

Fig. 6 Association rules generated from the data set

3 Conclusion

In this digital era, most of us use various tools to communicate and share data, information, and knowledge. Though phishers follow various methods to collect user data, mostly they fake as a legitimate site and try to obtain sensitive data. In this paper, various analysis has been done, noteworthy, only using four attributes legitimacy of a website can be deducted. And also the contributions of these 4 attributes for finding a phishing website is examined. Furthermore, associations of 30 attributes to locate a phishing site is also derived.

References

1. M. Crawford, T.M. Khoshgoftaar, J.D. Prusa, A.N. Richter, Najada H. Al, Survey of review spam detection using machine learning techniques. J. Big Data **2**(1), 23 (2015)
2. A. Abbasi, H. Chen, A comparison of tools for detecting fake websites. IEEE Comput. **42**(10), 78–86 (2009)
3. F.M. Zahedi, A. Abbasi, Y. Chen, Fake-website detection tools: identifying elements that promote individuals' use and enhance their performance. J. Assoc. Inf. Syst. **16**(6), 448 (2015)
4. R.M. Mohammad, L. McCluskey, F. Thabtah, Intelligent rule based phishing websites classification. IET Inf. Secur. **8**(3), 153–160 (2014)
5. A. Geetha Mary, D.P. Acharjya, N.Ch.S.N. Iyengar, Privacy preservation in fuzzy association rules using rough computing and DSR. Cybern. Inf. Technol. **14**(1), 52–71 (2014)
6. M.A. Geetha, D.P. Acharjya, N.Ch. Sriman Narayana Iyengar, Privacy preservation in fuzzy association rules using rough set on intuitionistic fuzzy approximation spaces and DSR. Int. J. Auton. Adapt. Commun. Syst. **10**(1), 67 (2017)
7. M.A. Geetha, Fuzzy-based random perturbation for real world medical datasets. Int. J. Clin. Pract. Suppl. **1**(2), 111 (2015)
8. A. Geetha Mary, *Privacy Preservation using Intelligent Techniques* (LAP LAMBERT Academic Publishing, Germany, 2016), p. 176
9. R.M. Mohammad, L. McCluskey, F. Thabtah (University of California, School of Information and Computer Science, Irvine, CA, 2015), https://archive.icsuci.edu/ml/datasets/Phishing+Websites
10. R. Jabri, B. Ibrahim, Phishing websites detection using data mining classification model. Trans. Mach. Learn. Artif. Intell. **3**(4), 42–51 (2015)
11. N. Abdelhamida, A. Ayesha, F. Thabtahb, Phishing detection based associative classification data mining. Expert Syst. Appl. **41**(13), 5948–5959 (2014)
12. Y. Pan, X. Ding, Anomaly based web phishing page detection. Paper presented at 22nd computer security applications conference, IEEE Computer Society, Los Alamitos, CA, USA (2006)

Prediction of Heart Disease Using Long Short-Term Memory Based Network

K. S. Umadevi

Abstract Technology has become one of the greatest needs of humankind and, in every field, has made lives better. In the field of medicine, technology has helped to treat, diagnose, and cure the diseases far better than the traditional method. Atherosclerosis is a condition where the cholesterol in the body gets deposited on the walls of blood vessels, thus narrowing them and reducing the blood flow to the organs. If the rate of atherosclerosis is greater than 50%, the patient stands at a high risk of heart diseases. The healthcare industry collects a large amount of data from the patients, but the potential of these data remains untapped. It retains various relationships and patterns and helps to determine the relationship between clinical parameters like blood pressure, cholesterol, and their association with heart diseases. The objective of the proposed work is to predict the rate of atherosclerosis for a patient based on the clinical parameters like blood pressure and cholesterol and predict the rate using various machine learning classification algorithms. This method will help the medical practitioner to determine the need for an operative procedure based on the outcome. This system predicts the percentage of narrowing down of blood vessels as the output.

Keywords Atherosclerosis · Heart attack · Long short-term memory · Prediction

1 Introduction

Heart is the most diligent part of the human body. The cardiovascular unit which comprises the heart and veins is responsible for circulating blood through the body. By the age of 70, the human heart would have pumped an excess of 2.5 billion liters of blood. The heart pumps around 2000 gal of blood everyday. A cardiovascular system is essential for providing the body with oxygen and supplements.

K. S. Umadevi (✉)
School of Computer Science and Engineering, VIT, Vellore, India
e-mail: umadeviks@vit.ac.in

© Springer Nature Singapore Pte Ltd. 2019
S. C. Satapathy et al. (eds.), *Information Systems Design and Intelligent Applications*,
Advances in Intelligent Systems and Computing 862,
https://doi.org/10.1007/978-981-13-3329-3_26

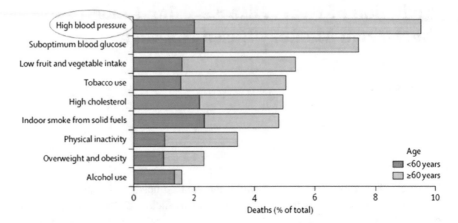

Fig. 1 Causes of coronary heart diseases for a particular age group

Modernization has led to a vast change in lifestyles of the current generation. Improper food habits, smoking, consumption of alcohol, stress, lack of sleep, diabetes, physical inactivity, and so on have a negative impact on an individual's health. This leads to cardiovascular diseases, obesity, malnutrition, hypertension, and even decrease in life expectancy of human beings. Coronary heart disease is a disease in which small blood vessels used to supply blood and oxygen to the heart narrows down. World Health Organization has estimated that 17.5 million people died from cardiovascular diseases in 2012, representing 31% of all global deaths. Out of these, 7.4 million were due to coronary heart disease and 6.7 million were due to stroke. World Health Organization (WHO) estimated by 2030 that almost 23.6 million people will die due to heart disease [1], Fig. 1.

Atherosclerosis is a condition where cholesterol deposition occurs inside the arteries thereby decreasing the amount of blood supply to various arteries in the body resulting in cramps, blood clotting, and even organ damage. Major causes of atherosclerosis include high blood pressure, high cholesterol levels, diabetes, smoking, alcohol consumption, and even family history. Common symptoms of atherosclerosis include chest pain, pain in hand or leg due to blocked artery, fatigue, shortness of breath, etc.

India is experiencing fast epidemiological progress as a result of financial and social change, and cardiovascular disease is turning into an inevitably critical reason for death. India's disease pattern has experienced a noteworthy move over the previous decade. According to WHO report, currently out of 10 deaths in India, 8 are caused by noncommunicable diseases (NCD), for example, cardiovascular ailments and diabetes in urban India. In India, 6 out of 10 deaths are caused by NCDs [2].

The proposed system is developed to integrate the latest technology in the day-to-day lives, so as to collect various information from a human being and predict the rate of narrowing of blood vessels within them, and thus help the doctor determine the need of an operative procedure. This reduces the cost of unnecessary operative procedures performed on the patient and thus, risking their lives.

2 Literature Survey

Since the heart disease is life threatening, many researchers tried to find solution through various data mining techniques like Naïve Bayes, neural networks, and decision trees to predict the heart attack in order to save human lives. It uses clinical parameters like age, blood cholesterol levels, blood pressure levels, etc., to predict the likelihood of patients getting heart disease. This system can support complex "what if" queries that other systems cannot support. This model uses the CRISP-DM method [3]. It is divided into six phases: business understanding, data understanding phase which focuses on understanding the objectives and requirements from a business perspective, converting this knowledge into data mining problem, and finally designing a plan to achieve the objectives. Data mining extension tools are used for accessing the models' contents and data. The database used here is a Cleveland database with 15 medical attributes like sex, blood pressure, cholesterol, ECG readings, etc. To validate the effectiveness of this model, they have used two methods, namely: classification matrix and Lift method [4]. This research focuses on using data mining techniques to predict heart diseases using data from patients' transthoracic echocardiography and applied classification models for large size dataset.

The dataset was cleaned by replacing the outliers with mean values and missing values were replaced by the most probable values. The model was tested using recall, f-measure, and ROC area measurement. The model efficiency was compared using all attributes and using selected attributes. The results have shown that the decision tree algorithm outperforms the other two algorithms. Thus, this study shows that data mining algorithms can be efficiently used to model and predict heart disease cases. To develop an efficient algorithm and predict the heart diseases for a patient, various algorithms like Naïve Bayes, weighted association classifier (WAC), and apriori algorithm are used. To further improve the accuracy and reduce the error rate of the solution presented, prediction module data is tested and predicted through ID3 (Iterative Dichotomiser 3) algorithm.

Recurrent neural network systems have been customized to function using a set of parameters in an arranged structures, by preferring appropriate advanced methods and parallel processing the results can be further improvised. The accuracy of prediction for heart disease data using j48 is 4% higher than Naïve Bayes and the accuracy has been improved by using NN.J48.

Using the information collected from the patient, there is a huge demand to find out an effective solution for all day-to-day problems. With the increase in mortality rate due to heart diseases, it is necessary to create awareness about healthy lifestyle among people. In lieu of the current scenario, it was necessary to create a system that predicts the probability of having a heart disease based on certain clinical factors like blood pressure and cholesterol.

3 Proposed System

Heart rate and cholesterol have been identified to be the major factors of atherosclerosis and thus choosing these values as the attributes while using the classification algorithm. The purpose of the research is to have used the following classification algorithms: Logistic Regression, K-Nearest Neighbors, Support Vector Machine, Kernel Support Vector Machine, Naïve Bayes, Random Forest, and Decision Tree classification. Kernel SVM is a classification technique used for nonlinearly separable data. It is an advanced version of SVM. The data is projected into higher dimensional space and the data can be classified using various kernels. There are three major types of kernels: Gaussian kernel, sigmoid kernel, and polynomial kernel. Decision tree algorithm is a classification algorithm for categorical and continuous value dependent variables. Each node in a decision tree contains a test on a condition and its branches hold the outcome.

Recurrent Neural Networks (RNNs) are connection models that capture the progression of arrangements by means of cycles among the connected nodes. Dissimilar to feedforward neural systems, repetitive systems hold in a state that can speak to data from a subjectively long setting window. Recurrent neural systems have been customized to function using a set of parameters in an arranged structures, by preferring appropriate advanced methods and parallel processing the results can be further improvised. As of late, frameworks in view of long short-term memory (LSTM) and bidirectional (BRNN) models have exhibited weighty execution on errands as changed as image, languages, and recognition [5].

One of the most successful RNN models for sequence learning as of now from 1997 is Long Short-Term Memory introduced by Hochreiter and Schmidhuber [6]. It consists of a memory cell and a unit of calculation that replaces conventional procedures used neurons in the hidden layer of the network. Using these memory cells, network overcomes a few challenges that are faced during the training phase. Next, Bidirectional Recurrent Neural Networks by Schuster and Paliwal present the BRNN architecture in which data from both the future and the past are utilized to decide the output at any time t. Instead, the neurons in the neural network are replaced by memory cells. It is used to rectify the gradient vanishing problem across other RNNs. The RNN cannot remember longer sequence and instead have short dependencies and are trained by a separate set of weights for remembering and forgetting outputs (Fig. 2).

An LSTM unit reads an input x_t and depends on prior output h_{t-1} and results in an output h_t. It has a memory cell c_t, an input gate i_t, an output gate o_t, and a forget gate f_t. Each LSTM cell performs the following functions:

1. Use the current input x_t and the previous hidden state (h_{t-1}) to decide data to be deleted from the memory vector (c_{t-1}), represented as:
   $f_t = \text{func}(w_f(h_{t-1}\ x) + b_f)$ where b_f is a bias and w_f is a set of weights.
2. Using x_t and h_{t-1}, a matrix is constructed that permits a specific information to be updated in c_{t-1}.
   $i_t = \text{func}(w_i.(h_{t-1}\ x) + b_i)$

Fig. 2 A RNN cell

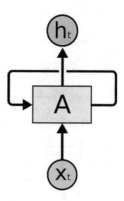

3. Use x_t and h_{t-1} to gather information that should be included.
   $$c_t = \text{func}(w_c.(h_{t-1}\ x) + b_c)$$
4. Finally, merge the new information and the old information
   $$c_t = f_t.c_{t-1} + i_t.c_t.$$

It can be clearly seen that by using stochastic gradient descent, this model will be used to train so that it can differentiate the information to be forgotten, preserved, or retained.

4 Results and Discussion

In the recent past, data mining has played a pivotal role in heart diseases research. To locate the concealed restorative data from the diverse and wide range of chronic ill people in the existed clinical information is an observable and effective approach in the investigation. Coronary illness grouping gives the basic premise to the treatment of patients. Measurements and machine learning are two principle approaches which have been connected to foresee the status of coronary illness in view of the outflow of the clinical information.

The finding of coronary illness, as a rule, relies upon an unpredictable blend of clinical and neurotic information. Due to this many-sided quality, there exists a lot of enthusiasm among clinical experts and specialists with regard to the effective and exact forecast of heart disease. According to the statistical data from WHO, one-third population worldwide died from heart disease; heart disease is found to be the leading cause of death in developing countries by 2010. The disclosure of biomarkers in coronary illness is one of the key commitments utilizing computational science. This procedure includes the improvement of a prescient model and the mix of various sorts of information and learning for symptomatic purposes. Besides, this procedure requires the plan and blend of various approaches from measurable investigation and information mining (Fig. 3).

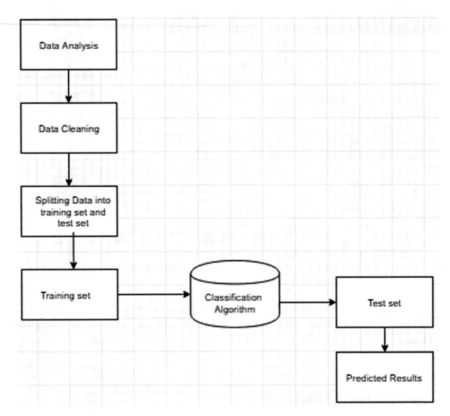

Fig. 3 Framework of the proposed system

The following are the categories of blood pressure and cholesterol levels. These play a very important part in deciding whether the patient suffers from atherosclerosis. The normal values are the optimal values of the blood pressure readings. These readings imply that the heart is functioning normally. The hypotension signifies low blood pressure readings. This means that the blood pressure readings are less than the optimal values and signifies some irregularity in the proper functioning of the heart. The high normal values signify that the values of the blood pressure are slightly higher than optimal readings but there is no risk. The stage 1 hypertension signifies that the values are higher than that of optimal values. The medical practitioner may suggest some lifestyle changes and some medications stabilize the high blood pressure. It is the threshold by which the condition can be treated by medication. Stage 2 hypertension implies that the blood pressure is much higher than the optimal readings. The medications should be started immediately and frequent checkups should be performed. And lastly, if the blood pressure is higher than 180 mm of Hg (systolic), the patient should be immediately rushed to the hospital (Table 1).

Table 1 Blood pressure levels

Blood pressure category	Systolic reading	Diastolic reading
Normal	90–119 mm of Hg	60–79 mm of Hg
Hypotension	<90 mm of Hg	<60 mm of Hg
High normal	120–129 mm of Hg	<80 mm of Hg
Stage 1 hypertension	130–139 mm of Hg	80–89 mm of Hg
Stage 2 hypertension	>140 mm of Hg	>90 mm of Hg
Hypertension	>180 mm of Hg	–

Table 2 Cholesterol levels

Cholesterol level	Cholesterol reading (mg/dl)
Good	<200
Borderline	200–239
High	>240

Table 3 Results

Classification algorithm	Accuracy (%)
Long short-term memory	92
Decision tree	76
Kernel SVM	90
Naive Bayes	69
Support vector machine	86

The cholesterol consumed by individuals is divided into two categories: the low-density lipoprotein cholesterol (Bad cholesterol) and the high-density lipoprotein cholesterol (Good cholesterol). This low-density lipoprotein cholesterol has a tendency to deposit on the interior of blood vessels and cause atherosclerosis. The high-density cholesterol protects the heart from heart diseases by keeping the bad cholesterol from building on the walls of blood vessels (Table 2).

Classification algorithms are a class of supervised learning and they are used to predict the values of categorical variables. Classification algorithms here predict whether the patient's arteries have narrowed the percentage based on the values of clinical parameters like blood pressure and cholesterol. The output variable used here is categorical variable with two values namely: 1 and 0. Value $= 1$ signifies that the arteries have narrowed by 50% or more and value $= 0$ signifies that the arteries have narrowed by less than 50%. This model is extremely helpful as it will help the medical practitioner determine the need for an angiography by using the predicted values (Table 3).

The proposed system was simulated using keras module from tensorflow package. The data set is used from the UCI repository for heart diseases [7, 8]. The system uses 950 samples for training and 50 samples for testing. From the results achieved, LSTM gives the most accurate results. Further, decision tree and logistic regression provide better results than other algorithms. This is because there was no linear way to divide the data points into two. Kernel SVM goes to one higher dimension to divide the data points into two parts.

5 Conclusion

To conclude, LSTM provides the most accurate results in this study. The accuracy of the all the algorithms is considerably low as seen from the above table. The results of this study can be improved by working on improving the accuracy of the algorithms and also applying dimensionality reduction so as to decrease the number of variables required to make a prediction.

References

1. D. Mozaffarian, E.J. Benjamin, A.S., Go, D.K., Arnett, M.J., Blaha, M. Cushman, ... M.D. Huffman, Executive summary: heart disease and stroke statistics-2015 update: a report from the American Heart Association. Circulation **131**(4), 434–441(2015)
2. R.P. Upadhyay, An overview of the burden of non-communicable diseases in India. Iran. J. Public Health **41**(3), 1 (2012)
3. H.D. Masethe, M.A. Masethe, Prediction of heart disease using classification algorithms, in *Proceedings of the World Congress on Engineering and Computer Science*, vol. 2 (2014), pp. 22–24
4. A. Taneja, Heart disease prediction system using data mining techniques. Orient. J. Comput. Sci. Technol. **6**(4), 457–466 (2013)
5. G. Purusothaman, P. Krishnakumari, A survey of data mining techniques on risk prediction: heart disease. Indian J. Sci. Technol. **8**(12) (2015)
6. Z.C. Lipton, J. Berkowitz, C. Elkan, A critical review of recurrent neural networks for sequence learning. arXiv preprint arXiv:1506.00019 (2015)
7. F.A. Gers, J. Schmidhuber, F. Cummins, Learning to forget: continual prediction with LSTM, in *Proceedings of 9th International conference on Artificial Neural Networks ICANN'99*, pp. 850–855 (1999)
8. R. Das, I. Turkoglu, A. Sengur, Effective diagnosis of heart disease through neural networks ensembles. Expert Syst. Appl. **36**(4), 7675–7680 (2009)

Oceanographic Mapping and Analysis

Tejas Mohlah and Thirugnanam Tamizharasi

Abstract Global warming is a threat the world faces as a whole, while there are many facets to this problem; the most significant is the rise in ocean temperature. One of the causes of this is eutrophication, which is the explosive growth of algae in any water body. Eutrophication occurs due to excessive amounts of nutrients in the water. These nutrients are, by extension, the reason for the rise in temperature and alkalinity, and reduction of oxygen availability in ocean water. Data for various factors like phosphates, nitrates, alkalinity, oxygen levels, etc. are collected from the national oceanic and atmospheric administration. This analysis is expected to be carried out initially on RapidMiner. Before the data is analyzed, the best clustering methods will be analyzed on various metrics of performance. Then, analysis is done using clustering. Analysis will also be done using correlations between factors including nutrients. These correlations and analysis can shed insight into the trends of eutrophication in ocean water.

Keywords Eutrophication of ocean water · Clustering
Data preprocessing phosphates and nitrates

1 Introduction

Due to the large quantities of data used, data preprocessing methods were necessary for handling missing and noisy data. The preprocessing methods used are

- binning,
- clustering, and
- normalization.

T. Mohlah · T. Tamizharasi (✉)
SCOPE, Vellore Institute of Technology, Vellore 632014, India
e-mail: tamizharasi.t@vit.ac.in

T. Mohlah
e-mail: tejazzmohlah@yahoo.co.in

© Springer Nature Singapore Pte Ltd. 2019
S. C. Satapathy et al. (eds.), *Information Systems Design and Intelligent Applications*,
Advances in Intelligent Systems and Computing 862,
https://doi.org/10.1007/978-981-13-3329-3_27

Clustering algorithms used

K-Means Clustering

K-Means is iterative in nature. It works by finding the local maxima for each iteration. It works in the following five steps:

1. State the number of clusters desired, K.
2. Assign each point to a cluster randomly.
3. Compute centroids of the clusters.
4. Determine the closest cluster centroid and reassign each point to it.
5. Compute the centroids of the clusters again.
6. Repeat steps 4 and 5 until no further change is possible. Repeat the 4th and 5th steps until stability has been reached. Once it has been reached, then termination of the algorithm takes place when no further switching of data points from one data set to the nth K-means clustering.

K-medoids Algorithm

It is a clustering algorithm that is related to two algorithms: the k-means algorithm and the medoid shift algorithm. K-medoids minimizes the total dissimilarities between points in a cluster and the point that has been selected as the center of that particular cluster.

K-medoids chooses data points as centers known as medoids. It is also a partitioning technique that clusters the data set containing n objects into k clusters. Silhouette is used to determine K which is known beforehand. The medoid of a finite data set is a data point from the same data set with the condition that the average dissimilarity of this point to all the other data points is minimal. This point has to be the most centrally located point in the set.

Density-based clustering algorithm

Density-based clustering algorithm plays an important role in finding density-based, nonlinear-shaped structure. The most widely used density-based algorithm is Density-Based Spatial Clustering of Applications with Noise (DBSCAN); the main concept used is density connectivity and reachability.

Conversion Ratios Used

Phosphate Phosphorus (PO_4-P)

MW PO_4 = 94.971482
MW P = 30.973762
1 μg PO_4/l = 1/MW PO_4 μg = 0.010529 μmol PO_4/l
1 μg PO_4/l = MW P/MW PO_4 = 0.326138 μg P/l
1 μg P/l = 1/MW P = 0.032285 μmol P/l

Nitrate Nitrogen (NO_3-N)

MW NO_3 = 62.005010

MW N = 14.006720 1 μg NO$_3$/l = 1/MW NO$_3$ μg/l = 0.016128 μmol NO$_3$/l
1 μg NO$_3$/l = MW N/MW
NO$_3$ = 0.225897 μg N/l
1 μg N/l = 1/MW N = 0.071394 μmol N/l

Nitrite Nitrogen (NO$_2$-N)

MW NO$_2$ = 46.005580
MW N = 14.006720 1 μg NO$_2$/l = 1/MW NO$_2$ = 0.021736 μmol/l
1 μg NO$_2$/l = MW N/MW NO$_2$ = 0.304457 μg/l N
1 μg N/l = 1/MW N = 0.071394 μmol N/l

Silicate Silicon (SiO$_3$-Si)

MW SiO$_3$ = 76.083820
MW Si = 28.085530 1 μg SiO$_3$/l = 1/ MW SiO$_3$ = 0.013143 μmol
SiO$_3$/l1 μg SiO$_3$/l = MW
Si/MW SiO$_3$ = 0.369139 μg Si/l
1 μg Si/l = 1/MW Si = 0.035606 μmol Si/l

2 Literature Survey and Related Work

Multiple researchers, analysts, and scientists have identified and discussed the effects of excess of silicate, phosphates, and nitrates on coastal and sea-growing microorganisms and how they play an important part in impedimenting the growth and propagation of not just individual organisms but growth of colonies of the same.

There is a delicate balance that needs to be preserved between a nutrient-rich environment and a nutrient-polluted one. Nutrient pollution is a rising threat to localized water bodies and the respective ecosystems. Phosphates and nitrates increase the amount of waterborne plants and algae, which increase the biochemical oxygen demand, which in turn decreases the oxygen content in water and increases the temperature of the same [1]. This increase in temperature and decrease in dissolved oxygen levels make it almost impossible for aquatic life to exist rather co-exist with the excessive algae duckweed and waterborne plant and animal life like phytoplankton [2]. Microorganisms such as diatom *Thalassiosira pseudonana* (a species of marine centric diatoms) are very important to the ecosystem cycles (like sulfur cycles) have been estimated to contribute to up to 40% of the global oceanic primary production [3] and to play a major role in exporting organic carbon from upper layers to the deep ocean [4] and are directly affected by overnutrition [5]. Nutrient pollution also leads to a drop in the pH level of water and an increase in alkalinity. The culmination of all of this is known as eutrophication.

Eutrophication gives rise to excess organic matter production and harmful algal blossoms which are increased exponentially due to the abundance of nutrients like nitrates, phosphates, and silicates [6]. Human activity has an enormous influence on the global cycling of nutrients. These cycles are dominated by human intervention and a certain one-way topology, wherein phosphate cycle generally goes from the land to the sea, where the concentration depends on the amount of human involvement in the environment. Similarly, the effect of human activity on the global cycling of nitrogen (N) is equally immense, and the rate of change in the pattern of use is much greater [7]. The single largest global change in the N cycle comes from increased reliance on synthetic inorganic fertilizers, which accounts for more than half of the human alteration of the N cycle [8]. These nutrients come from sources such as garbage dumping into rivers which enter oceans, after polluted rivers runoff into oceans, excess agricultural fertilizer runoff into rivers that flow into oceans, oil and chemical spills in oceans from cargo ships, direct ocean pollution from garbage dumps, etc [9]. All studies conducted on this subject have conclusively proved that when nutrients, namely, nitrates, phosphates, and silicates, cross a certain threshold, an explosive growth in biomass ensues, which in turn causes an increase in the biochemical oxygen demand and a decrease in the oxygen content in water, a drop in the pH level of water, and an increase in alkalinity. Even though the concentration of nitrates and phosphates can change on the basis of the stages of their respective cycles [10], it does not play a part in affected areas due to the extent of excessive nutrients in the system already.

Buoy-based systems are ideal for the usage of long-term systems and their implementation. This is mainly due to the fact that they are low maintenance since do not have overtly sensitive parts, have high durability, and have the ability to reliably collect data for long periods of time without significant maintenance. These properties make it perfect for collection of data over a long period of time, easier if they are networked together for faster more accurate data collection [11].

3   Implementation

System Architecture (Img. 1)

Description

The system architecture consists of

(1)   NOAA data warehouse,
(2)   NOAA data set,
(3)   local repository,
(4)   local repository to RapidMiner,
(5)   data cleaning,
(6)   normalization,
(7)   set role, and
(8)   analysis.

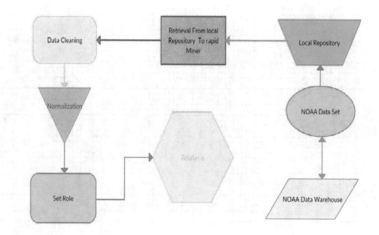

Img. 1 Implemented system architecture

The current implementation is done by retrieving a data set from the National Oceanic and Atmospheric Administration (NOAA) with the given coordinates for the respective region and after the specification of attributes that are required. The data set then is cleaned using preprocessing techniques:

(1) Binning,
(2) clustering,
(3) regression, and
(4) partitioning.

After the data set is cleansed, the attributes required for the analysis are selected using a select attribute constructor.

Implementation

The consolidated data is received from the National Oceanographic Data Center (NODC) which is a part of National Oceanographic and Atmospheric Administration (NOAA) which is an administrative organization under the United States government for the study of the atmosphere and the ocean.

Once the data set is downloaded, it is then added to RapidMiner as a data set, stored in the local repository. The loading is done by using the retrieve constructor from RapidMiner. Then a preliminary analysis will be done on it. This preliminary analysis refers to the cluster analysis that is conducted to find the best suitable clustering algorithm. The clustering algorithm will be chosen on the basis of the performance parameters. This is done because clustering algorithms need to be tailored to the data set they are used on. Once preliminary analysis is done, the data set will be cleansed, by rebuilding damaged data items, standardizing data, and normalizing it. Once this is done, then analysis of the data set begins with first an analysis of an unprocessed data set. Once that is done, the data set undergoes preprocessing which includes (Img. 2)

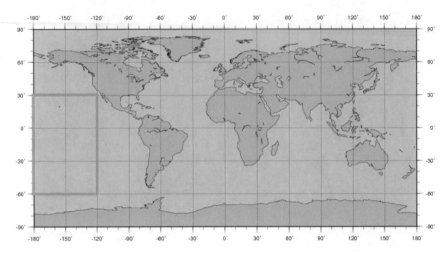

Img. 2 Area of sample collection

(1) data cleaning,
(2) normalization,
(3) selection of attributes,
(4) replacing missing attributes, and
(5) setting attribute roles.

Here, we see the processes of

(1) data retrieval,
(2) data cleaning,
(3) normalization,
(4) selection of attributes,
(5) replacing missing attributes, and
(6) setting attribute roles.

These take place through the use of constructors in RapidMiner which are mostly called in all exercises.

Post this, correlations with the nutrient and temperature, pH alkalinity, and dissolved oxygen take place. Furthermore, outliers are detected and various operations such as FP growth trees, decision trees, naïve Byes, and rule induction take place with the use of various constructors available on the RapidMiner. To find out and predict, future outliers in the data set. Outlier detection is necessary such that we can know under which conditions anomalous behavior occurs.

Here, we see post data preprocessing that the data set is run through a constructor which detects outliers by running the label and ID values against the normal behavior of the test data set. If the pattern of their behavior changes or is different from the normal, then they are classified as outliers. Most of the data passes through a constructor that uses information gain and entropy, and gain ratio as metrics to create a tree that gives us conditions that cause outliners to occur. Another predictive analysis model used here is in combination with decision tree naïve Byes and rule induction. It is important to note that the decision tree, rule induction, and the naïve Byes algorithm get the same preprocessed data set and not the output from the previous functions and constructors. The parameters for the decision tree include

(1) gain ratio,
(2) information gain,
(3) gini index,
(4) accuracy, and
(5) least square.

4 Results

The cluster analysis is done to ensure which clustering algorithm is best suited for the data set being analyzed.

There are multiple parameters that are used to ensure to judge the performance of the algorithm; these are within centroid distance and cluster density. The performance of each algorithm is then monitored and analyzed for the output.

In Fig. 1, we see that the data set used here has not been cleaned, missing values have not been replaced, and the data has not been normalized. Since the data here has not been preprocessed, therefore, multiple errors still exist in the data set. Therefore, no patterns can be generated or any analysis is done. The graph here attempts to find a correlation between the depth of the sample and the concentration of phosphate, silicate, and the temperature.

In Fig. 2, the graph looks at a correlation between the depth of the sample, the concentration of phosphate silicate, and the temperature of the samples to be ana-

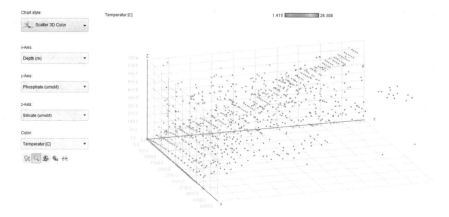

Fig. 1 Scatter graph of non pre-processed data

lyzed. Unlike the previous sample, the data set here has been cleaned, data has been preprocessed, and normalization has been done. In the correlation between depth phosphate levels and temperature, we see an explosive decrease in temperature when the phosphate levels increase from 0.8 (μmol/kg) and the depth exceeds more 500 m. The sudden reduction in temperature can cause long-lasting effects to marine life and has implications on the ecosystem as a whole (Table 1).

Figure 3 shows us a different aspect of the graph where we see that when the concentration of silicate increases beyond 20 μmol/l the temperature of the samples drops drastically. Similarly, when the concentration of phosphates increases beyond 0.9, we see the temperatures drastically reduce.

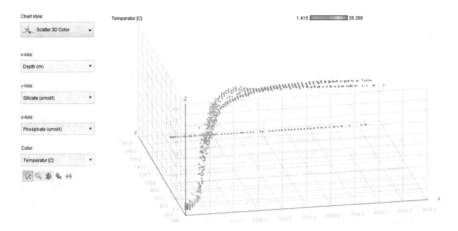

Fig. 2 Scatter graph of pre-processed data

Table 1 Outlier conditions

Condition	Outlier
Salinity > 36.52 and silicate < 1.51	Yes
Salinity > 36.483	Yes
Salinity > 36.57 and temperature = 23.074	Yes
Salinity > 36.57 and temperature > 27.074 and pressure > 9.45	Yes
Salinity > 36.57 and temperature > 26.453 and pressure > 49.250	Yes

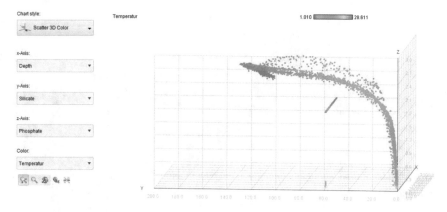

Fig. 3 Phosphate, silicate, depth 3D scatter graph

In Fig. 4, the graph shows how the temperature varies on the basis of the nitrate content of the samples in the data set. We see an initial slight increase in temperature with an increase in nitrate levels from 32.5 to 40 μmol/kg. Post this slight increase in temperature, we see a constant drop in temperature with a drop in concentration of nitrates in the samples from 40 to 2.5 μmol/kg. In Fig. 5, the main factors that are being considered are silicate and phosphate concentration levels with respect to temperature of the samples. When the concentration of silicate increases beyond 20 μmol/l, the temperature of the samples drops drastically. Similarly, when the concentration of phosphates increases beyond 0.9, we see the temperatures drastically reduce. Figure 6 shows us the correlation between silicate and phosphate concentration levels, depth, and alkalinity. The relation is not linear in nature and varies greatly with the ratio of depth and concentration of phosphate and silicate, but the sudden increase and gradual decrease with depth are based on the ratio of phosphate to depth. Figures 7 and 8 show us the correlation between silicate, phosphate concentration, nitrate levels, and alkalinity.

Outlier analysis seeks to detect anomalous behavior of the data, wherein each element is first taken and tested for its behavior against test data and normal behavior. The normal behavior is defined as the behavior with certain leeway followed by the majority, in this case, 90% of the data set. We see a decision tree used to predict outliers in the data set on the basis of the attributes that have been selected. The decision tree is shown in Fig. 9. The graph in Fig. 10 indicates under which pH level and silicate levels to potentially dangerous levels of phosphate occur. The phosphate danger level is classified based on the tendency for seawater to undergo eutrophication if the phosphate level increases any further. Figure 10 shows us the same correlation between eutrophication danger levels under silicate and pH levels (Fig. 11).

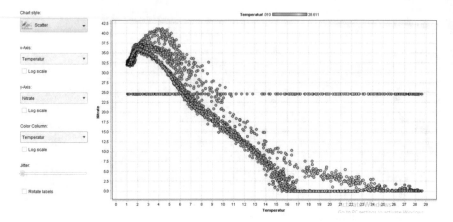

Fig. 4 Nitrate, temprature scatter graph

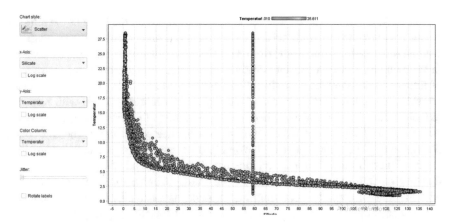

Fig. 5 Silicate temprature scater graph

5   Conclusion

There are quite a number of distinct conclusions that can be drawn from the research that has been conducted.

- The most apparent one is the importance of data preprocessing wherein nothing can be derived or observed when there have not been any preprocessing methods applied to the data. Whereas when the data undergoes preprocessing, we find that missing and incorrect values are corrected and normalized such that it helps with analysis.
- The most alarming conclusions of this research are that there is a direct correlation between high concentrations of phosphate and other nutrients and temperature of the ocean.

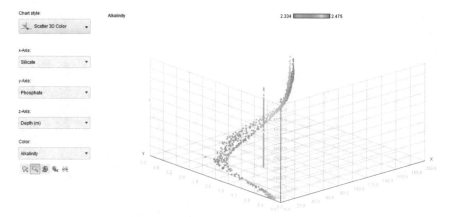

Fig. 6 Silicate, phosphate, depth scatter graph

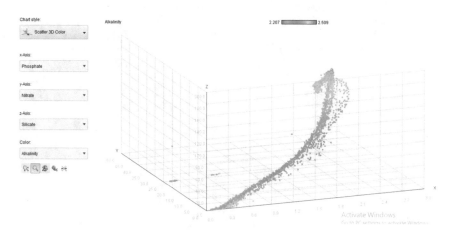

Fig. 7 Phosphate nitrate alkalinity scatter graph

- From previous studies conducted and cited here, we see that there exists a direct correlation of excess nutrients and nutrient pollution with the explosive growth of algal blossoms.
- We also see direct correlation between silicate, pH, oxygen levels, and dangerous phosphate levels that can lead to eutrophication.
- We see the concentrations of the nutrients like nitrates, phosphate, and silicates; there exist direct correlations between increasing temperatures and increasing nutrient concentrations.

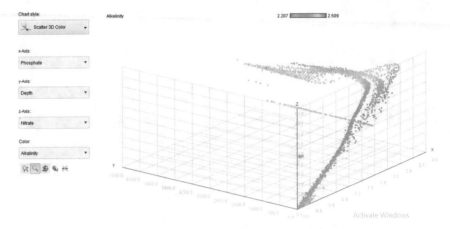

Fig. 8 Phosphate nitrate depth scatter graph

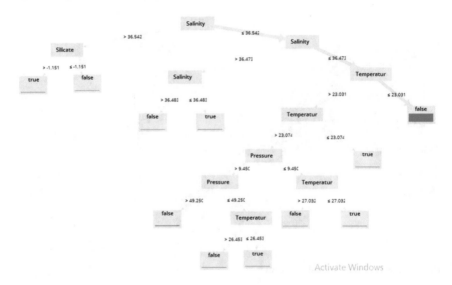

Fig. 9 Decision tree

- Another observation made is that depth and nutrients also play a role in the level of alkalinity of the water present and increases with the depth of the sample taken with varying concentrations.

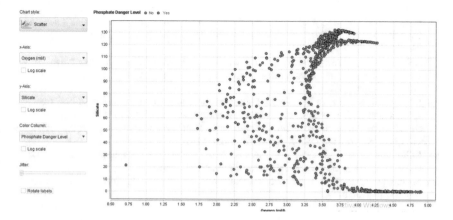

Fig. 10 Phosphate danger level, silicate, oxygen scatter graph

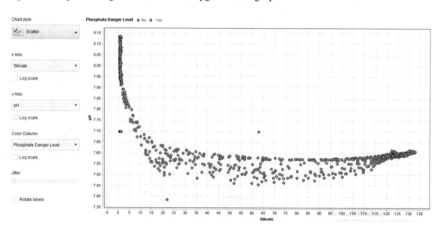

Fig. 11 Phosphate danger level, silicate, pH scatter graph

References

1. J.K. Hart, K. Martinez, Environmental sensor networks: a revolution in the earth system science. Earth Sci. Rev. **78**, 177–191 (2006)
2. D.M. Anderson, A.D. Cembella, G.M. Hallegraeff, Progress in understanding harmful algal blooms: paradigm shifts and new technologies for research, monitoring, and management. Ann. Rev. Mar. Sci. **4**, 143–176 (2011)
3. D.M. Nelson, M.A. Brzezinski, Diatom growth and productivity in an oligo-trophic midocean gyre: A 3-yr record from the Sargasso Sea near Bermuda. Limnol. Oceanogr. **42**(3):473–486 (1997)
4. Victor Smetacek, Diatoms and the ocean carbon cycle. Protist **150**(1):25–32 (1999)
5. E. Bucciarelli, W.G. Sunda, Influence of CO_2, nitrate, phosphate, and silicate limitation on intracellular dimethylsulfoniopropionate in batch cultures of the coastal diatom Thalassiosira pseudonana (Association for the Sciences of Limnology and Oceanography, 14 November 2003)

6. R.W. Howarth, A. Sharpley, D. Walker, Sources of nutrient pollution to coastal waters in the United States: implications for achieving coastal water quality goals. J. Coast. Estuar. Res. Fed. **25**, 656–676 (2002)

7. J.N. Galloway, W.H. Schlesinger, H. Levy, A. Michaels, J.L. Schnoor, Nitrogen fixation: Anthropogenic enhancement-environmental response. Global Biogeochem. Cycles **9**(2):235–252 (1995)

8. J.T. Sims, R.R. Simard, B.C. Joern, Phosphorus loss in agricultural drainage: historical perspective and current research eutrophication. J. Environ. Qual. **27**, 277–293 (1997)

9. S.S. Rathore, P. Chandravanshi, A. Chandravanshi, K. Jaiswal, Impacts of excess nutrient inputs on aquatic ecosystem. IOSR J. Agric. Vet. Sci. (2016)

10. M.E. Conkright, W.W. Gregg, S. Levitus, Seasonal cycle of phosphate in the open ocean (Laboratory for Hydrospheric Processes, 15 December 1998)

11. S. Gopal Krishna Patro, K.K. Sahu, Normalization: a preprocessing stage (Cornell University Library, 19 March 2015)

Study of Symmetric-Key Cryptosystems and Implementing a Secure Cryptosystem with DES

**H. Santhi, P. Gayathri, Sanskar Katiyar, G. Gopichand
and Sagarika Shreevastava**

Abstract The advancements in communication networks and computation over the years have led to a massive volume of data flowing from one place to another. A lot of cryptosystems have been proposed and implemented to keep the channel of communication secure. But with increasingly powerful computers rolling out every day, more and more complicated systems are being built with moderate data overheads to ensure maximum security over the channel. So, in this paper, we briefly discuss the strengths and weaknesses of the three most popular symmetric-key cryptosystems: AES, DES, and 3DES. It has been supported by the various organizations that it with the computers of today, DES can be broken using brute-force attacks in a reasonable amount of time. Given its simple nature of implementation and small key size, we propose a modified implementation of DES to enhance its security with a minimum computational trade-off.

Keywords Cryptography · Transposition · Symmetric key

1 Introduction

Network security has always been an area of extensive research. A lot of mechanisms have been proposed and adopted to ensure that no unwanted third-party can access the data other than the two nodes that have established the communication channel. Various mechanisms that involve routing control, authentication, etc., serve as decent mechanisms which try and block any unauthenticated party from having access to the data.

H. Santhi (✉) · P. Gayathri · S. Katiyar · G. Gopichand · S. Shreevastava
School of Computer Science and Engineering, Vellore Institute of Technology, Vellore, India
e-mail: hsanthi@vit.ac.in

© Springer Nature Singapore Pte Ltd. 2019
S. C. Satapathy et al. (eds.), *Information Systems Design and Intelligent Applications*,
Advances in Intelligent Systems and Computing 862,
https://doi.org/10.1007/978-981-13-3329-3_28

Cryptography is a branch of mathematics that, in simple terms, deals with the transformation of data, and back. This transformation helps in rendering data, referred to as plaintext, into something utterly meaningless and garbled such that it is protected even in the case where the third-party gains access to the data on the channel. This process is called encryption. The meaningless garbled data, known as the ciphertext, can be transformed back to the original data by using a set of operations called decryption, at the other end. The set of encryption and decryption processes constitutes a cryptosystem.

The field of cryptography has had a long history, right from the old Egyptian hieroglyphics, to ensure the security of messages by deviating from the rules of the language. The deviation is what interests most researchers today. The strength of a good cryptosystem lies in the fact that even when one has constant access to the ciphertext on a communication channel, one cannot analyze and decode that without an additional information, called the key.

Any Cryptosystem should be very secure since the main aim of cryptography is that no unauthorized person can access the data. It provides a number of security goals to ensure the safety of the data. Following are the main goals of cryptography:

Confidentiality—No unauthorized person can read the data other than the receiver.
Authentication—The system checks the sender's identity before processing the data.
Integrity—Only authorized party can make changes in the data.
Non-Repudiation—Both the sender and the receiver should not deny the communicated message.
Reliability—The quality of service provided to the users remains the same irrespective of the attacks on the system.

In further subsections of this section, we have discussed types of cryptosystems—symmetric and asymmetric, and the popular implementations of the same. In Sect. 2, we mention and acknowledge the work and people whose work inspired us to write this paper. Further in Sect. 2, we do a comparative study between AES, DES, and 3DES on certain parameters that will help identify the distinctions between these algorithms. In Sect. 3, we propose our implementation of a cryptosystem that involves the DES as a functional module and strengthens it by implementing an additional layer of transposition. Section 4 documents the results of our proposed system and a basic analysis of the working of it, using an example. The paper is concluded in Sect. 5.

1.1 Type of Cryptosystems

There are multiple ways to classify data encryption. Most commonly these are categorized into two types—asymmetric and symmetric.

Asymmetric-key encryption has different keys for the encryption and decryption process. One of them is public and another is private. It is also called public key encryption. If the encryption key is published, then the system enables private communication from the public to the user having the decryption key. If the decryption key is published, then the system serves as a signature verifier of the encrypted data. Public key methods can be used to transmit encrypted data or keys when both the users could not agree on one secret key. It improves the security of the data being transmitted. Algorithms like RSA and Diffie–Hellman are few of the examples of asymmetric encryption.

Symmetric-key encryption has one single secret key. It is one of the oldest methods used in cryptography. Single key is used to encrypt and decrypt the data. Any user having the key can decrypt the message and so the security of this key is very important. The symmetric-key system is an apt option when the application does not want external involvement of the user. Some of the examples are DES, 3DES, and AES. The paper's main focus is on these three algorithms.

Symmetric-key Cryptosystems

There are a lot of Symmetric-key algorithms, for example—AES, DES, 3DES, Blowfish, Twofish, etc. Out of these the most commonly used are DES, 3DES, and AES.

Data Encryption Standard (DES)

Data Encryption Standard was developed by IBM and adopted in the year 1977. It is an improved version of an older system called LUCIFER. Up until a decade ago, it was used as the main standard for encrypting data for many decades.

DES is based on Feistel block cipher. It has a 64-bit cipher block, which means it encrypts 64 bits at once. Two inputs are expected in this algorithm—plaintext and the secret key. The key size is 56 bits, even though 64 bits are input for the key. The least significant bit of each byte is either arbitrary or used as parity bit. The plaintext message is arranged into 64-bit blocks. If the input bits (plaintext) are not divisible by 64, then the last block is padded. Permutations and substitutions based on the key are incorporated throughout in order to make it more complex for the unauthorized users. 16 rounds of permutation and substitution are done to make the encryption more secure. Then the left and right halves are swapped for one final permutation. This round is the inverse of the initial permutation, and the output is the 64-bit ciphertext.

Triple DES (3DES)

Triple DES is nothing but DES with three different keys. It was introduced by IBM to overcome the shortcomings of DES. It is easy to implement and is more secure. 3DES has two forms—one requiring three completely different keys and another with only two completely different keys. It is very important to avoid having the same key for all the steps, or else it will just be a slower version of DES.

The method which uses three different keys uses the first encryption key K1, followed by K2, and finally, encryption is carried out with K3. This method provides high-level security but requires a very long key. 168 (56 × 3) bits are required for the key, which is difficult in practical situations. And hence, the method using two keys was introduced.

In the method having only two different keys, first encryption is done using key K1. The output of this step is then decrypted using key K2. And finally, the decrypted data is again encrypted using K1 as the key. This method is also referred to as Encrypt–Decrypt–Encrypt (EDE).

Even though 3DES is very secure, it is not unbreakable. It has a lot of advantages over previous algorithms but it is still vulnerable to certain attacks. It is also not very feasible to use 3DES when there is a large amount of data. When compared to modern algorithms, it is considered slow and outdated.

Advanced Encryption Standard (AES)

AES was developed by two cryptographers—Joan Daemen and Vincent Rijmen and published by National Institute of Standards and Technology (NIST). It was developed for a contest held by the NIST in the year 1997, and it was publicly disclosed by the year 2000. The algorithm used in this system is called Rijndael. It is considered the most secure symmetric cryptosystem currently. The data block processed is of 128 bits. The key size can be 128 bits, 192 bits, or 256 bits depending on the number of rounds. The number of rounds can be 10, 12, or 14, respectively. Each processing round involves four steps. These steps are as follows:

i. **Byte Substitution**: substitute byte-by-byte using an S-box.
ii. **Shift Rows**: simple permutation.
iii. **Mix Columns**: Substitution method where each column's data from the shift row is multiplied by the algorithm's matrix.
iv. **Add Round Key**: the processing round's key is XOR-ed with the data.

The first three functions are only to prevent cryptanalysis by confusion and diffusion. The fourth function is the one actually responsible for encryption. These four operations are performed in each round.

AES provides with more security as it is strongly resistant to different types of attacks. Since it has a very large key size, it makes it harder for even computers to consider all possible keys.

2 Literature Survey

In the paper [1] proposed by R. Tripathi and S. Agrawal, a comparative study between symmetric and symmetric cryptography techniques on the basis of keys, throughput, speed, etc., is given. Few cryptosystems from each are also compared, such as AES, DES, 3DES, and Blowfish from symmetric-key cryptosystem, and RSA and Diffie–Hellman from asymmetric-key cryptosystem.

Pancholi and Patel [2], Gordon et al. [3], Agnew et al. [4], Warjri [5] are all comparative study among different types of symmetric algorithms. From development to its effectiveness, most of the points are covered. Through these comparisons, one can understand how cryptosystems evolved over time and how the methods used today rectifies the issues caused in earlier methods.

The paper [6] by Noura Aleisa is also a comparative study between AES and DES for encrypting images, but also with theoretical and real experimental analysis. Many image encryption schemes are also proposed.

M. Mathur and Kesarwani [7] shows how DES is better at providing security than 3DES even though 3DES is still beyond capability of many attackers today. However, powerful 3DES is against regular attacks; it still is vulnerable to brute-force attacks using differential and linear cryptanalysis.

In [8, 9], a visual representation of DES and AES is provided, respectively. It was very helpful to understand how these cryptosystems work with the help of these papers.

In Table 1, all the papers with inspiring implementations have been discussed briefly across different parameters.

2.1 Comparison of Existing System

In Table 2, we have discussed three of the most popular symmetric-key cryptosystems. The table shows a comparison between the three on various parameters.

3 Proposed System Design

As discussed earlier how the advancements in computer architecture and parallel processing have made it practically possible to break the Data Encryption Standard. Various techniques have been successful in breaking DES with each varying in processing power and time. DES owes this unfortunate credit to its small key-space. Any 56-bit key comes from a possible key-space of 256 keys—however, some keys are too trivial to be used, which further reduces the key-space even more. Even a brute-force algorithm operating on today's budget laptop can break a DES cipher in about a few months—a supercomputer can do it within an hour.

While our proposed mechanism has been illustrated with DES in the paper, it can be implemented with 3DES or AES with certain adjustments.

We are proposing a system design (Fig. 1) which involves using a simple multilayer columnar transposition of the plaintext, followed by DES on the resulting text. So even though a system can break the DES, the resulting plaintext must go through a series of columnar transpositions that the sender and receiver know about before the actual plaintext is obtained.

Table 1 Survey of related work

Paper	Aim	Techniques	Outcomes	Merits	Demerits
Agnew [4]	To develop a high-speed public key cryptosystem	Proposes a hardware device design based on observations of structure of multiplication	Cryptosystem with low complexity and high speed	Can be implemented in VLSI since its architecture is regular	Extensive number of registers and gate arrays used
Bala [10]	To modify AES to make it more secure	Concealing messages by cryptic steganography	Security was improved by using this technique	High degree of security attained and time complexity reduced	Good processing power required to embed and extract, decode data
Banerjee [11]	To provide authenticated querying for wireless sensor network (WSN)	Symmetric-key-based scheme to enable authenticated querying	Queries from unauthorized users could also be authenticated	Improvised underlying pairwise key distribution technique	Does introduce extra cryptographic overhead
Gupta and Hussain [12]	To share the session key of symmetric-key system securely	Asymmetric mechanism for sharing the key	Enhanced security of the message transfer	Man-in-the-middle attack not possible	Complex for a symmetric-key system
Warjri [5]	Proposing a new symmetric-key algorithm	New key generation method using modulo 69	Large amount of data transferred with better security	Two keys and inverse function increases security	Limited symbol set; does not work for other symbols than mentioned
Mohit [13]	To make DES-RSA hybrid secure even after two rounds	Converting DES to asymmetric-DES using RSA algorithm	Improved and secure version of DES	Greater security to existing structure	Only 64-bit message encrypted
Tsouri and Wilczewski [14]	To design a symmetric encryption key tailored to body area networks	Using RSSI measurement from existing packets going back and forth	Secure multiple on-body communication links	Generates long and highly random bit sequences	Keys weakly matches with eavesdropper
Opritoiu [15]	To modify AES for Altera Cyclone II EP2C35 FPGA	Concurrently processing all the bytes of a data block and computes each round key on-the-fly	Verified using the Modelsim simulator	Exhibits parallelism and 805.24Mbps throughput	Modified for a specific device
Ghnaim [16]	Cryptanalysis of four-round DES	Through a fitness function called N-gram statistics	Proves inadequate security of DES	Efficient in finding missing key bit of DES	Only applicable for four-round DES

Table 2 Comparison of classical symmetric-key algorithms

Criteria	DES	3DES	AES
Developed in	1977	1978	2000
Developed by	IBM	IBM	Joan Daemen and Vincent Rijmen
Principle	Balanced Feistel network cipher structure	Feistel network cipher structure	Substitution and permutation
Derived from	Lucifer	DES	Square
Cipher type	Symmetric block cipher	Symmetric block cipher	Symmetric block cipher
Encryption mode	Single encryption	Encrypt–decrypt–encrypt	Single encryption
Plaintext size	64 bits	64 bits	128, 192 or 256 bits
Data block	Divided into two	Divided into two	Entire data block as single matrix
Key size	56 bits	168 bits (K1, K2, K3) 112 bits (when K1==K3)	128,192 or 256 bits
Possible keys	2^{56}	2^{168} or 2^{112}	2^{128}, 2^{192} or 2^{256}
Time to check all possible keys (at 50 billion keys/s)	400 days	800 days (112 bits)	$5 * 10^{21}$ years (128 bits)
No. of rounds	16	$16 * 3 = 48$	10 (128 bits), 12 (192 bits), 14 (156 bits)
Processing	Takes more processing power than AES	Takes most processing power	Takes less processing power
Encryption time (kbps)	0.3	0.8	0.3
Decryption time (kbps)	0.2	0.7	0.3
Speed	Slow	Slowest	Fastest
Security	Less secure, proven inadequate	Secure	More secure
Cryptanalysis resistance	Vulnerable to differential and linear cryptanalysis, weak substitution tables	Vulnerable to brute-force attacks using differential and linear cryptanalysis	Strong against differential, truncated differential, linear, interpolation, and square attacks

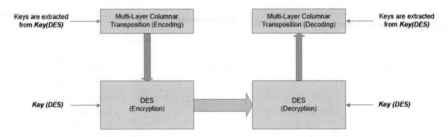

Fig. 1 Proposed system design flow

3.1 Salient Features of Proposed Work

The following are the salient features of our proposed work:

- Same relative frequency of symbols as the plaintext pre-DES, but the text is not readable. Transposition techniques do not affect the relative frequency symbols, thus confusing the reader.
- A simple columnar transposition is fairly intuitive and can be broken by anagramming. However, a layered column transposition cipher is very difficult to reroute, due to additional dependencies of rows and columns.
- The resulting text from the columnar transposition is the same size as the plaintext.
- No additional block required. An extra of 8 bits can be built into the DES key. This multiplies the security while keeping the same key size, i.e., 56 bits.
- Can completely utilize the 64-bit block size of DES, if required, to store additional values for the key that will be used in the columnar transposition.
- The method of choosing the key can be as sophisticated as required and completely depends on the user. It determines the number of columns that will render into something random. We recommend more number of columns for large text and less for shorter text.
- The number of layers of column transposition also depends on the user's liking.
- Given parallel processing and architectural features, it is fast to break when the user has the key.

3.2 Method of Layer and Key Selection

One may choose a different set of parameters for the multilayer columnar transposition depending upon the application. We can have any number of rounds, theoretically. We recommend having 3–8 rounds to be practical.

We can also decide how the columns are ordered, by using some criteria. One such criterion is to use the key of the DES system that we are going to use further. Another one is the timestamp.

Fig. 2 Key bits represented
by numbers

	Bit [1]	Bit [2]	Bit [3]	Bit [4]	Bit [5]	Bit [6]	Bit [7]	Bit [8]
K[1]	1	2	3	4	5	6	7	8
K[2]	9	10	11	12	13	14	15	16
K[3]	17	18	19	20	21	22	23	24
K[4]	25	26	27	28	29	30	31	32
K[5]	33	34	35	36	37	38	39	40
K[6]	41	42	43	44	45	46	47	48
K[7]	49	50	51	52	53	54	55	56
T	57	58	59	60	61	62	63	64

We suggest using a four-layered columnar transposition with number of columns as 8, 4, 8, 4, respectively. However, to simplify the explanation, we will not be using it in the example in Sect. 4.

We will now discuss how you can generate the column orders from the DES key. Note that this is only one of the possible variations. The reader may choose to devise their own criteria.

Let us assume that we are working on a 56-bit key K for DES and represent it in a binary form in a matrix, with an 8×7 representation (Fig. 2)—(1–56). We can append another byte T at the end such that it covers the complete block of 64 bits and works to our advantage, since cryptanalysts usually ignore the rest 8 bits in the block while trying to break the DES. And since the key for DES can be broken easily, it does not make sense to simply stick to the DES completely.

In Round 1, we choose a column number of 8, and to have each column number defined in a range 0–256, we need 8 bits per column. Thus, 64 bits are used for defining the entire set. This broad range is somehow not really a necessity. Hence, the permutations are going to be 8! at the best way, anyway. So, for the first round, we propose taking the 64 bits in the manner listed in Fig. 3. Take alternative rows and concatenating the bits in the order mentioned.

For every column slice a byte (8 bits) from the concatenated string. Convert this value to base 10 integer and order the columns in ascending order, and then write the values in the columns from top to bottom.

In round 2, we need 4 bytes for 4 columns (32 bits). The key bit selection has been numbered in Fig. 4. Again, every byte in the obtained bit-sequence is converted to a base 10 number and the columns are ordered accordingly.

Fig. 3 Sample key selection
for round 1

1	2	3	4	5	6	7	8
9	10	11	12	13	14	15	16
17	18	19	20	21	22	23	24
25	26	27	28	29	30	31	32

33		41		49		57	
34		42		50		58	
35		43		51		59	
36		44		52		60	
37		45		53		61	
38		46		54		62	
39		47		55		63	
40		48		56		64	

Now, alternatively for round 3, we propose using an 8-column sequence and that again requires 8 bytes and this time we can select the key in the manner done in Fig. 5.

To top the process off, we will use a 4-column sequence in round 4, the key selection of which can be done as shown in Fig. 6.

Having done all the ordering we can finally obtain a text that is more likely to be secure than the *simple columnar transposition* where the text is read in from top to down in columns and then printed row-wise to produce a transposed text.

Fig. 4 Sample key selection
for round 2

1		2		3		4	
5		6		7		8	
9		10		11		12	
13		14		15		16	
17		18		19		20	
21		22		23		24	
25		26		27		28	
26		30		31		32	

4   Result and Analysis

Now, we will illustrate using an example to better understand the working of the
proposed system. Note that our focus will be on the columnar transposition and not
exactly on DES. We know about DES' strengths and weaknesses from the discussion
Sect. 2.1 (Fig. 7).

In this section, an example is proposed. Further, a simple implementation of the
same is proposed in Python. The simulation results can be seen along with the DES
codes. Below is an example for better understanding.

VIT-is-in-Vellore-Tamilnadu	
Round 1	VSERMAIILEIDTNLTLUIVOANx
Round 2	VMLTLOSAENUAEIILINRIDTVx
Round 3	VLEEIDMONINTLSUIRVTAALIx
Round 4	VINLRALDISVLEMNUTIEOTIAx

Now the string obtained in Round 4 is passed through the DES encryption (EBC
Mode). A random key, $Key_{(DES)}$: **11101805** is used for this example. Below is a
screenshot of the same. The DES decryption process is followed being taken as the
one done on the receiver side.[1]

---

[1]Python has some issues printing a hexadecimal string; thus, a garbled message is seen.

Fig. 5 Sample key selection for round 3

1	2	3	4	5	6	7	8
9	10	11	12	13	14	15	16
17	18	19	20	21	22	23	24
25	26	27	28	29	30	31	32

	33		41		49		57
	34		42		50		58
	35		43		51		59
	36		44		52		60
	37		45		53		61
	38		46		54		62
	39		47		55		63
	40		48		56		64

Fig. 6 Sample key selection for round 4

	1		2		3		4
	5		6		7		8
	9		10		11		12
	13		14		15		16
	14		18		19		20
	17		22		23		24
	21		26		27		28
	25		30		31		32

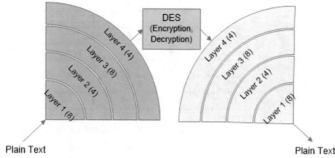

Fig. 7 Layered design of proposed framework

Table 3 Plaintext represented in matrix

V	I	T	I
S	I	N	V
E	L	L	O
R	E	T	A
M	I	L	N
A	D	E	X

Table 4 After round 1

V	S	E	R
M	A	I	I
L	E	I	D
T	N	L	T
L	U	I	V
O	A	N	X

Table 5 After round 2

V	M	L	T
L	O	S	A
E	N	U	A
E	I	I	L
I	N	R	I
D	T	V	X

The decrypted string is further fed into the columnar decryption transposition matrix, and after 4 rounds we will obtain the original plaintext. The sequence that follows is from Tables 3, 4, 5, 6 and 7. A screenshot of the same has been given for reference (Figs. 8 and 9).

Table 6 After round 3

V	L	E	E
I	D	M	O
N	I	N	T
L	S	U	I
R	V	T	A
A	L	I	X

Table 7 After round 4

V	I	N	L
R	A	L	D
I	S	V	L
E	M	N	U
T	I	E	O
T	I	A	X

Fig. 8 Sample code for Round 4

```
>>> #encrypting
>>> des = DES.new('11101805', DES.MODE_ECB)
>>> text = 'VINLRALDISVLEMNUTIEOTIAx'
>>> ciphertext = des.encrypt(text)
>>> ciphertext
'/\xcc+\xe7|\\m\x19\xe9\xa5BDq\xd5\x18\xcc\x
bb\x8bn\x84\x82p\x9d\xb7'
>>> print ciphertext
/İ+ç|\m┤éₐBDqÕↄↄ↕nↄ╜pↄ₀
>>>
>>>
>>> #decrypting
>>> plaintext = des.decrypt(ciphertext)
>>> plaintext
'VINLRALDISVLEMNUTIEOTIAx'
```

Fig. 9 Column transposition Example

5 Conclusion

In this paper, we conclude that DES has long been obsolete. With today's fast computers and algorithms, AES is the safest option of the three, out there among symmetric-key cryptosystems. We proposed a generic criterion of choosing different keys from the same key depending on the timestamp or any other criteria. We further proposed and implemented a cryptosystem involving the DES. And we showed how the cryptosystem adds another layer of security, which even if easy to identify but is difficult to reroute and crack.

References

1. R. Tripathi, S. Agrawal, Comparative study of symmetric and asymmetric cryptography techniques. Int. J. Adv. Found. Res. Comput. **1**(6), 68–76 (2014)
2. V.R. Pancholi, B.P. Patel, Cryptography: comparative studies of different symmetric algorithms. Int. J. Technol. Sci. **VI(I)**, 4–7 (2015)
3. D.S. Gordon et al., Complete fairness in secure two-party computation, in *Proceedings of the Fourtieth Annual ACM Symposium on Theory of Computing—STOC 08* (2008), p. 413
4. G.B. Agnew et al., An implementation for a fast public-key cryptosystem. J. Cryptol. **3**(2), 63–79 (1991)
5. J. Warjri, KED—a symmetric key algorithm for secured information exchange using modulo 69. Int. J. Comput. Netw. Inf. Secur. **10**(August), 37–43 (2013)
6. N. Aleisa, A comparison of the 3DES and AES encryption standards. Int. J. Secur. Appl. **9**(7), 241–246 (2015)
7. M. Mathur, A. Kesarwani, Comparison between DES, 3DES, in *Proceedings of the National Conference on Horizons IT-NCNHIT* (2013), pp. 143–148
8. Z.S. Stanisavljevic, Data encryption standard visual representation, in *2015 23rd Telecommunication Forum Telfor (TELFOR)*, vol. 7 (2015), pp. 946–953
9. Y.K. Apoorva, Comparative study of different symmetric key cryptography algorithms. Int. J. Appl. Innov. Eng. Manag. **2**(7), 204–206 (2013)
10. R. Bala, N.P. Gopalan, Security enhancement using a modified AES algorithm, in *Proceedings of the International Conference on Informatics and Analysis* (2016), p. 4
11. S. Banerjee, D. Mukhopadhyay, Symmetric key based authenticated querying in wireless sensor networks, in *Proceedings of the First International Conference Integrated Internet Ad Hoc and Sensor Networks* (2006), pp. 6–9
12. A. Gupta, M. Hussain, Secure session key sharing using public key (2015), pp. 573–576
13. P. Mohit et al., Modification of symmetric-key DES into efficient asymmetric-key DES using RSA (2016), pp. 1–5
14. G.R. Tsouri, J. Wilczewski, Reliable symmetric key generation for body area networks using wireless physical layer security in the presence of an on-body eavesdropper, in *Proceedings of the 4th International Symposium on Applied Sciences in Biomedical and Communication Technologies ISABEL'11* (2011), pp. 1–6
15. F. Opritoiu et al., A high-speed AES architecture implementation, in *Proceedings of the 7th ACM International Conference Computer and Front Matter* (2010), pp. 95–96
16. W.A. Ghnaim, Known-ciphertext cryptanalysis approach for, in *International Conference on Computer Information Systems and Industrial Applications* (2010), pp. 600–603

Big Data Analysis on Effective Communication Skills and Personal Grooming—A Key to Managers and Supervisors in Business Enterprises and Retail Outlets in Vellore and Katpadi

S. Horizan Prasanna Kumar and J. Karthikeyan

Abstract Communication skills and personal grooming are interwoven and considered to be predominant among other managerial skills. These skills add flavour to the in-depth business knowledge of an entrepreneur to a salesperson involved in any business enterprises. To a person in marketing, it is very much in need not only to please its customers but also to take the message of the company and products to the market. A survey to learn its importance directly from business outlets and enterprises will definitely help young learners to understand its nature and prominence from the root. This will mould them to be a better tycoon in near future. Their findings may help existing business to nurture by modifying their set back and to train their future employees with high standards that please the market.

Keywords Salesperson · Managerial skills · Survey · Communication
Personal grooming

1 Introduction

In the world of business, competition is inevitable. There is no shortcut to success in business other than distinctiveness in every aspect of sustaining the needs of the customer. In today's digital age, communication is effortlessly done in plentiful setups, but it is also important to realm the competency to verbally communicate as well. When communication lines are bare between businessmen and their customers, it can unswervingly distress the sales of the business. When a business man effectively

S. Horizan Prasanna Kumar (✉)
Department of English and Foreign Languages, SRM Institute of Science and Technology, Chennai, India
e-mail: horizan.s@ktr.srmuniv.ac.in

J. Karthikeyan
SSL, Vellore Institute of Technology, Vellore 632014, India

© Springer Nature Singapore Pte Ltd. 2019
S. C. Satapathy et al. (eds.), *Information Systems Design and Intelligent Applications*,
Advances in Intelligent Systems and Computing 862,
https://doi.org/10.1007/978-981-13-3329-3_29

Fig. 1 Research design

communicates the future trends, growth to their products and services and how it can be beneficial the listeners, it converts them into their clients. Good communication ultimately boosts the bottom line of a business. Clearly, the benefit of communication in a business is bountiful, and in fact, it is hard to visualize any kind of success at all short of it. No matter what kind of business you have, and whether it is big or small, one of the single most important factors of triumph is communication and how you impress customers [1] with your impressive style of appearance (Fig. 1).

2 Research Design

The research design [2] is formulated to ensure the quality of the work at every stage. It helps the researcher to go step by step, which leads to the success of the survey without any deviation. It also confirms whether the survey and documentation are in sequence as per the standards of any researches. A complete understanding of the business scenario with emphasize to communication skills is dealt with in the background of the study. More knowledge about the role of communication skills among managers and store in-charges will be understood by means of literary study. The target group is fixed to limit the focus of the study. Geographical location is fixed on the basis of feasibility to meet and collect data in person. It is believed that the outcome of this survey would be very useful for the local enterprises and business investors to think about enhancing effective communication skills among their managers who really run the business.

3   Background of the Study

Vellore and Katpadi are the twin towns. Vellore is the headquarters of Vellore District which is an upcoming business city. This district is known for its world-class finished leather goods industries. In addition, lots of fabrication industries that support huge machineries and boilers manufacturing plants are located across the district. These twin cities have India's top hospital, Christian Medical College and Hospital, Vellore Institute of Technology, Golden Temple, fort, biggest mosque, oldest church and many pilgrimage centres. These attractions, educational institute and hospitals attract millions of visitors every year to these towns. Increase in population flow paves way for India's renowned chain of stores, branded outlets, multinational chain of restaurants, hotels, spas and travel agencies. These business enterprises invest a huge sum to showcase their best products and services offered to their customers. They try unique marketing methods to reach and satisfy their customers. But the efforts taken by the investors sometimes miserably fail because of their staff, who are employed to run the business. With continuous monitoring of these employees, one can understand that the work commitment of these employees is positive. But when it comes to the way of their communication and approach, things go erroneous. Some huge invested businesses ended up miserably due to poor customer friendly relations the store employee had with their customers. This feedback is not purely unanimous but derived from entrepreneurs, sales managers and investors. A study of this kind will help young learners to understand the existing employee qualities, and the learning outcome of this study may be useful for them in their future business dreams. Moreover, suggestions can be carried to those business enterprises and outlets which need guidance.

4   Literary Survey

An informal study and analysis would be of great help to identify the existing problem among the business enterprises and outlets, and suggesting measures would be of great help for boosting business among these enterprises. So a clear understanding of earlier surveys and literature reviews would of great support to proceed further. Most of the studies conducted on the communication skill for the managers in various business organizations have been reviewed in the following section. The survey [3] which was conducted in Dhaka, Bangladesh among local private business concerns mainly focused on the need for communication skill among executives in various business organizations related to their profession. Questionnaire was used to collect data, and Likert scale method was adopted. Initially, 250 executives working in 10 private business organizations were chosen for this survey. Among which, 50 executives were selected from top level and 75 executives were from mid-level and the rest of 125 were selected from bottom level performance. The findings of this survey revealed that majority of the executives have a positive opinion about communication skill. This is because the majority of the organizations, where they work, train them intensively on communication skills. Some of the executives were dissatisfied with communication skill training offered to them. Finally, this study reveals how important is communication skills for their career and profession.

A study was conducted [4] with 225 employees who were graduates selected from the Universities in Turkey and worked at banks. It focused on the notion of leadership styles and communication competency of bank managers on their job satisfaction in Turkish banks. Data were collected using the questionnaires and analysis was made using SPSS tool. This study included 53% of female and 47% of men. Approximately, 83% of participants were below the age of 40. The result revealed that the employees had a good leadership style and competent enough to communicate with their colleagues and customers in the banks [5]. Another study revealed that strong communication skills not only helped to succeed at workplace but also to achieve in their career of education. One hundred and eighty human resource managers from various business organizations and 311 business school instructors from nine colleges in Minneapolis and St. Paul, Minnesota were involved in this survey. They were given 24 questions, and the data were collected through cross-sectional design. Random sampling method was used to analyse the data collected from 44 managers and 44 professors.

Rodzalan and Saat's [6] study conducted on students generic skills development in industrial training revealed that the students clearly understood that though integrated theory and practice is needed in a real workplace, having good communication skill is necessary for success. The researcher used both the quantitative and qualitative method, and the data were collected at the pre-study phase, which is before the commencement of the course, and the other post-study phase, which is towards the end of the industrial training. Data were collected from 30 engineering students who took that course. Moslehifar and Ibrahim's [7] studies reflect that speaking is the most important skill to the human resource development. Data from 136 HRD persons from different multinational companies in Malaysia were collected using questionnaire. This study recommended companies to focus on speaking skills and to HRD trainees in various organizations in Malaysia [8]. Research exposed the need for advanced level of English knowledge for recruiting manager and others top officials. The study also emphasized the need for written and oral skills. The success of a business organization is calculated by the effective communication skill used by employees working in that company.

Csapo and Featheringham's [9] study discloses the survey conducted for 5 years, 2000–2004, by a public University in Midwest which mainly focused on how communication skills are the most valuable for the executives in business organizations and graduate students. The questionnaire was sent through mail to 362 alumina students during the year 2005. The result of this study reveals that communication skills are very much in need at the workplace, and graduates without communication skill cannot succeed with just technical skill.

Mallett-Hamer's study [10] aimed at finding out communication gap between the leaders, the supervisors and the customer service representatives in business organizations. The data were collected quantitatively through the drop-off survey by asking multiple choice questions asked face-to-face to the supervisors, customer service representatives and organizations leader. The result shows that all the organizations were satisfied with communication between their employees except few roomers. A survey conducted by the University of Southern Maine [11] identified that listening skill is the most significant skill to every organizational staff and supervisors. Eight hundred and fifty-two employees of this University were chosen as samples to identify which skill is important at the workplaces. Both quantitative and qualitative analyses were followed and the questions were sent through email and survey monkey. The data were analysed through IBM and SPSS statistics tools.

5 Limitations of the Study

This analysis [12] was subjected to the managers and supervisors heading the business enterprises and stores. Five areas were concentrated that will give us an overall understanding of the person concerned. They are as follows:

1. Way of approach—Approach of the manager/supervisor towards the customers.
2. Communication skill—Effectiveness of verbal and non-verbal communication.
3. Information shared—Whether the manager/supervisor gives customized info about the product the customer ask for.
4. Appropriate answers for the questions—Whether the manager/supervisor gives relevant and useful info regarding the product the customer asks for.
5. Appearance (dress/personal grooming, etc.)—External appearance and personal grooming of the manager/supervisor.

6 Field Survey

Quantitative data gathering method [13] is adopted to gather data that include observing and recording well-defined events, obtaining relevant data from management information systems and administering surveys with closed-ended questions (Figs. 2, 3 and 4).

Soumadev Ghosh	Soumadev Ghosh	Ravi Teja
B. Chaithanya Kumar Reddy	M. Mahaboob Shariff	Dharm.Ch
Mithesh S R	Tanvi Mehta	Santhosh.J
K.S.Abdhur Raqueeb	Hitheshini.K	Susil Kumar
Bhawana Kodhati	Komal Jain J.C	Sindhu.S
Sanjitha.C.P	Bhargavi Mantha	Adhitya.Venkatesan
Ch. Keerthi Priya	A. Oviya	S. Sreenika Vidyavathi
Srinekethan	Vignesh Var	Gelbert Jaffery.G
T. Keerthi Margaret	S.Keerthiga	Vignesh.S
Mohammed Muzammil G	Gaurav Kumar.S	Krishnabala.K
S. Janani	Khushboo Bhagdani	Vignesh Var
Y. Amali Joy	Idris Nihad K	Janet Bafana
R.K. Abdullah Mujahid	Ajith Kumar M	Madhumidha M
Edrick Kevin Paul	Shankar.R	Duane Elsa Mathew

Fig. 2 List of students who are conducted the survey

Annai Pet Clinic	Mega Mart	TVS Motors
B&B Builders.	Hotel Rangalaya Royal	Poorvika
Coimbatore Home Appliances	Pantaloons	Tata Motors
Geetha Home Decors	Ibaco	Olive Kitchen
GRT Residency	Quality And Taste	Allmart
Indian Post	Gr8 Buys	Mr. Carl Edison
Louis Phillipe	Naturals	Hotel Lakshmi Bhavan
Poorvika Mobile World	Optico	Globus 2
Porvika Mobiles	Unlimited	Chandni Chowk
Top Gear Fashion	Flo Café	Naturals Saloon
TVS Showroom	Louis Philip	Wrangler
Unlimited Store	Quality Taste	Cafe Coffee Day
Waves	Puma	SBI
Woodland	Max	Darling Food Court

Fig. 3 List of enterprises where the survey was conducted

Ahmed Basha	Raj Kumar	Motilal S
Giridharan	Praveen	Renugopal
Gnanasekaran.S	Maunand	Satish.K
Gowri R	A.Ramesh	Sauood Danish
Karuppaiah K	Martin	Poomari Muthu Kumar
Mahesh	K.Sathish	Gayathri Sankar
Manuk Das	Dileep Chungath	Sudeep Nair
Muwahd	J.Merlin	Selva Kumar.S
Prabu S	Saravanan	Badrinath
Prasanth	Amruta Ganeshan	Babu Kamath
Raju	Prasanth	Bala
S.Arockia Swamy	Nanda Kumar	Sudheer

Fig. 4 List of managers involved in this survey

Fig. 5 Way of approach

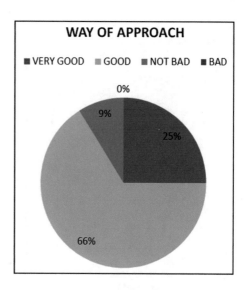

7 Findings and Analysis

The survey results reveal that the present status of managers in Vellore and Katpadi with reference to their way of approach, communication skills, information shared, appropriate answers for the questions and personal grooming. Each module is rated in the scale of four as very good, good, not bad and very bad.

Figure 5 shows that the majority of the managers and supervisors in Katpadi and Vellore have a good way of approach. 25% among them are rated very good and 19% are found not bad, which indicates that they need improvement by means of rigorous trainings. None of their approach is found very bad. Figure 6 indicates that 20% of managers have a very good communication skill both verbal and non-verbal, and 53% are found good. 25% of them need attention as they are in 'not bad' category. 2% of them are found bad; hence, special attention needs to be given to these employees. Overall result shows that their communication skill needs to be improved.

Figure 7 reveals that 60% of the managers and supervisors are found good for sharing information related to the products they sell. They have deep knowledge about their products and help buyers by giving sufficient inputs regarding it. 25% are found very good, and 14% fall under not bad category. Only 1% is found bad in sharing information as they were newly employed and with little education. Figure 8 shows that 48% of managers are good in giving appropriate answers to the questions raised by the customers, and 29% are found very good by impressing their clients by going an extra mile. 22% of them need serious attention as relevance lacks. 1% of managers gave totally inappropriate answers (Fig. 9).

It is well noticed that 43% of managers and supervisors in Vellore and Katpadi, in general, have a good sense of dressing and personal grooming. 41% of them are

Fig. 6 Communication
skills of the managers

Fig. 7 Information shared
by the managers

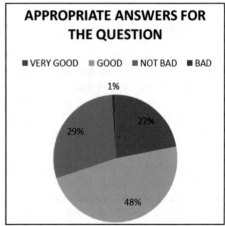

perfect match for their business as they are very good. Another 16% is found not bad, and they need special attention by means of training and development.

8   Recommendations

This survey is not only an eye opener for the budding graduates to understand the essential managerial skills. It also helped the investors and entrepreneurs to identify areas which need focus for their supervisors and managers. Training in these areas would help them to be better and outshine in their job. It is recommended to add business training programmes, soft skill training and communication development

Fig. 8 Appropriateness of the answers given by the managers

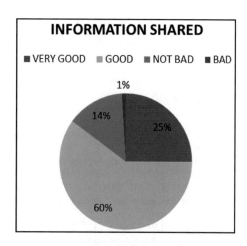

Fig. 9 Managers appearance

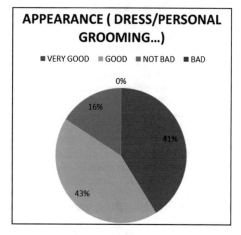

programmes at the earliest to enrich their customer relationship which directly leads to successful business. Effective use of social media is also a good way for informal development of spoken skills. The development of proper feedback system would also enable the managers to be perfect in that area.

9 Conclusion

The success of a business depends on the business skills the employee and investor have. To outshine in the global business, one has to update and incorporate the key skills required to satisfy clients. Continuous training and exposure to product updating will keep up the business in this tough competitive world. Students should be motivated to conduct studies and surveys for two reasons. One is to get awareness of current trends in business and another is to suggest the existing business enterprises the ups and downs. It will be a win-win strategy for both students and the investors.

References

1. J. Griffin, R.T. Herres, *Customer Loyalty: How to Earn It, How to Keep It* (Jossey-Bass, San Francisco, CA, 2002), p. 18
2. S. Lewis, Qualitative inquiry and research design: choosing among five approaches. Health Promot. Pract. **16**(4), 473–475 (2015)
3. N. Akhter, S.I. Khan, M.K. Hassan, Communication skill of the business executives: an empirical study on some local private business concerns in Bangladesh. Int. Bus. Res. **2**(4), 109 (2009)
4. M. Çetin, M.E. Karabay, M.N. Efe, The effects of leadership styles and the communication competency of bank managers on the employee's job satisfaction: the case of Turkish banks. Proc.-Soc. Behav. Sci. **58**, 227–235 (2012)
5. D. Conrad, R. Newberry, 24 business communication skills: attitudes of human resource managers versus business educators. Am. Commun. J. **13**(1), 4–23 (2010)
6. S.A. Rodzalan, M.M. Saat, The effects of industrial training on students' generic skills development. Proc.-Soc. Behav. Sci. **56**, 357–368 (2012)
7. M.A. Moslehifar, N.A. Ibrahim, English language oral communication needs at the workplace: feedback from human resource development (HRD) trainees. Proc.-Soc. Behav. Sci. **66**, 529–536 (2012)
8. H. Didiot-Cook, V. Gauthier, K. Scheirlinckx, *Language Needs in Business, A Survey of European Multinational Companies* (2000)
9. N. Csapo, R.D. Featheringham, Communication skills used by information systems graduates. Issues Inf. Syst. VI **1**, 311–317 (2005)
10. B. Mallett-Hamer, *Communication within the Workplace* (University of Wisconsin-Stout, 2005)
11. A.O. Sullivan, *The Importance of Effective Listening Skills: Implications for the Workplace and Dealing with Difficult People* (2011)
12. J. Dingle, Analysing the competence requirements of managers. Manage. Dev. Rev. **8**(2), 30–36 (1995)
13. B.L. Berg, H. Lune, H. Lune, *Qualitative Research Methods for the Social Sciences*, vol. 5 (Pearson, Boston, MA, 2004)

Technology-Enabled Digital Language Learning Using a Flipped Innovation Model

I. S. John Vijayakumar and J. Karthikeyan

Abstract The English Language has always played an important role in the teaching and learning process of academics and research across the globe. Hence, young learners of the twenty-first century are expected to possess a good level of proficiency in English. The aim of this study is to effectively use digital language learning platforms to upskill entry-level tertiary learners by adopting the newly proposed flipped innovation [FINN] model. The objective is to influence learners to get actively involved in the learning process, adopt outcome-based learning strategies and experience the realization of their goals by integrating modern tools in language learning. This learner-centric technology-enabled skills enhancement initiative will provide learners with a flexible, collaborative learning environment that will also lead them to record new learning strategies and find plausible solutions that will guide them to overcome issues related to the skills development process.

Keywords ELT · Digital language learning · Listening skills · Flipped innovation

1 Introduction

The rapid advancements in science and technology have induced phenomenal changes in the work style of the modern world. With scientific advancement comes rapid changes, and these changes have been happening at a faster pace than ever before. This has necessitated professionals globally to equip themselves with better skill sets to perform well in the intellectually demanding and performance intensive workplace. The complex work environment demands a systemic functional approach in discharging daily duties and expects professionals to possess adequate competencies to handle both regular and challenging situations. Roux and Du Toit [1] put for-

I. S. John Vijayakumar (✉) · J. Karthikeyan
SSL, Vellore Institute of Technology, VIT University, Vellore 632014, India
e-mail: isjohn.vijayakumar2015@vit.ac.in

J. Karthikeyan
e-mail: jkarthikeyan@vit.ac.in

© Springer Nature Singapore Pte Ltd. 2019
S. C. Satapathy et al. (eds.), *Information Systems Design and Intelligent Applications*,
Advances in Intelligent Systems and Computing 862,
https://doi.org/10.1007/978-981-13-3329-3_30

ward a provocative argument that the best way to manage the growing complexities of the society in the twenty-first century is through developing a 'systems thinking' capability. And one of the very important, desired attributes of a professional, especially an engineer, is good communication skills with the desired level of English proficiency.

With the emergence of English as the global language of business, science and technology, and its ever- increasing role in the teaching and learning process in the world of academics and research, young learners of the twenty-first century are expected to be proficient in English. With the objective of satisfying this demand, English language teachers across the world have used innovative strategies for training young learners. Computer-assisted language learning laboratories and technology-enabled digital language learning tools and platforms have played a vital role in this process. Although the outcome of such initiatives has yielded good results, the level of success achieved is questionable. Hence, ELT practitioners across the globe have started to realize the need for a paradigm shift in the role of technology in the process of language acquisition, the methods used in the teaching–learning process, the content and composition of the materials used and the objective and outcome of the activities, especially those available on digital learning platforms.

This article proposes a hybrid language learning model that can be made available to language learners, especially entry-level tertiary learners, on a technology-enabled digital language learning platform. Certainly, the effective implementation of this model will help bridge the gap in upskilling learners.

2 The Need for a New Eclectic Method

Although various approaches and methods have been formulated and widely practised in the world of ELT, the need for a new twenty-first-century model is vital to provide a robust process-based framework to enhance language proficiency of learners to the expected level within a given time frame-work. Riemer [2] has stressed the importance of language and communication courses in engineering curriculum that will contribute to the process of lifelong learning. For example, learners who are trained using a lot of authentic listening materials on digital language learning systems (DLLS) are found to acquire better language proficiency. The use of technology-enabled learning, blended with conventional classroom teaching methods has proved to be useful and has helped achieve tangible and measurable outcomes. Stepp-Greany [3] has explained the benefits of technology-enhanced language learning and its positive influence in the development of linguistic skills of language learners.

Most of the learning and testing materials on a technology-enabled learning platform have been product oriented and not process oriented. A new hybrid model has been proposed. The proposed hybrid language learning model tries to make use of the learners' interest and knowledge for better language proficiency development. The plan is to promote higher levels of learner engagement to aid better understanding and learning. A learner-centric, process oriented, technology-enabled model will influence students to get actively involved in the learning process, adopt outcome-based learning strategies and experience the realization of their goals by intelligently integrating modern tools in language learning. This learning-centric, skills enhancement initiative offered on a technology supported platform will also provide learners with a flexible learning environment. The ultimate objective is to promote learner autonomy that will provide individuals with the possibility of identifying the nature of content to be used for learning purposes. Learners will also be guided to record new learning strategies and find plausible solutions to overcome various issues related to the skills development process. This hybrid language learning model envisages to flip the role hitherto played by the teacher and in turn, encourage the learner to get involved in the design of materials, identifying the outcomes, chalk out strategies to achieve the desired results and map the learning experience and record it to create a repository that will serve as a guide for future learners.

3 Flipped Innovation Model (FINN Model)

The flipped innovation (FINN) model aims to adopt a two-fold approach. The approach is termed "flipped" because the learner takes a lead role, instead of the teacher, in planning the activity and in augmenting the use of the technology-enabled interface for developing language proficiency, both in the classroom or outside the classroom. The first aspect is related to the flipping of roles, as described earlier, which will also promote a learner-centric approach in the language acquisition process. The other aspect is the diffusion of innovation [4] in which not only the issues relating to skills development are taken for discussion and analysis, but also the underlying challenges faced while implementing the learning process are also investigated and practical solutions to redress those challenges are identified by the learners in real time and recommended to others, thereby making the whole experience rich and rewarding to all those involved it. From the teacher's perspective, the new model will provide an avenue to the teacher to allow the learners to get involved in designing the activity and in using it creatively. At the same time, with inputs from the students, the teacher will be able to fine-tune the pedagogic process in a better way. The very fact that the contribution to the innovative process is more from the learner than the teacher is another reason for terming it "flipped innovation".

4 Implementing the FINN Model in Listening Skills Development

Although an integrated learning of all the four skills, namely, listening, speaking, writing and reading, is usually recommended as it results in better linguistic skills development, this research article limits its scope in implementing the FINN model to listening skills only with the rationale that this procedure can be later applied to the other three aspects of language learning.

4.1   L2 Listening

The world of work will be a community of meaning by listening; produce what is desired and being a long-term partner to their customer(s) [5, 6]. Therefore in order to be successful in a competitive workplace, a professional must be ready to connect and be willing to listen. Listening comprehension is defined as 'an active process in which listeners select and interpret information that comes from auditory and visual clues' [7]. Listening comprehension is not an activity that is passive in nature, but is a complex process. It involves a lot of analysis and support [8]. For example, the effective use of listening comprehension activities to develop the proficiency skills of learners has been practised widely, and a lot of research has been carried out. Renandya and Farrell [9] have proposed that teachers must train learners by helping them to understand the method to process language when spoken and expose them to good listening comprehension exercises. The next generation will effectively use the Internet to develop language proficiency because of their confidence and ease with which they use technology [10]. As Warschauer [11] puts it, more and more non-native speakers will use the English language in daily communication, and the emergence of regional and local varieties of English will influence English language and learning in the twenty-first century.

4.2   Developing New Listening Activities Using a Technology-Enabled Interface

One of the important ways that can effectively develop listening comprehension skills is the time-tested method of exposing the learners to a lot of listening materials, preferably authentic materials, in a Computer-Assisted Language Learning (CALL) laboratory. Generally, pre-designed fixed content present in digital language platforms on technology-enabled interface is used both in training and testing across the world. L2 learners with support from language teachers complete the tasks by recording their responses as answers in the given worksheet or on the system interface itself. The learning is basically measured in terms of the correctness of the response and the marks obtained, making it more product based and testing oriented. The answer usually expected in a computer-based test, aimed at testing listening skills, is either providing a word or selecting a choice from the given options. Although pre-recorded situational dialogues and audio and video clips are used in training, learner's choice-based interactive activities, that will eventually promote higher order thinking skills and an effective language acquisition interface are absent in a conventional training platform. It is to overcome this short coming, an attempt was made to design an activity based on the FINN model.

Sample activities, like the one mentioned below, will enable the next-generation learning environments to provide next-generation learners with new ways to acquire English proficiency. The objective is to integrate the best of resources and expertise to promote an intelligent framework to empower learners in their pursuit of acquiring meta-competencies vital for academic excellence and professional development.

4.3 Sample Activity Using FINN Model

The learning happens in four phases:

a. Planning Phase: Learners will initiate activity creation by completing specific tasks-based proficiency development activities on the chosen material and make it available on a computer-assisted or a mobile-assisted language learning interface accessible to the others, both the teachers and group members.
b. Procedural Phase: The facilitator/teacher will then complement/support learner participation by outlining the things to be learnt by the learners.
c. Practice Phase: Each such created activity will be available on the digital platform to support collaborative learning and stimulate learners to realize their full potential and get involved in the activity to achieve tangible, desired outcomes.
d. Presentation Phase: Each learner will present details of the learning outcomes, strategies learnt and employed, issues faced and possible solutions if any and any new creative idea to move the whole exercise to the next level.

Sample Activity using FINN Model: Steps involved

Step 1:	The students in the class are divided into groups and briefed about the activity, the learning method and outcome
Step 2:	Each group will identify a short video or audio clip from any authentic source on the web or other available sources
Step 3:	Members of each group will complete specific tasks based on the short video or audio clip
	(a standard template may be prepared in advance by the teacher to record observations)
Step 4:	Each group will upload the completed task to the DLLS
Step 5:	The teacher will then complement by including more learning points to the activity based on the materials submitted by each of the groups and reload them to the DLLS
Step 6:	Each group will then identify new experiences, learning outcomes and problems and issues faced during the process
Step 7:	Each group will then share their experiences, observations and findings and map learning outcomes

4.4   Activity Based on Flipped Innovation (FINN) Model

An activity based on the FINN model created for use by Electronics and Communication Engineering students presents the innovative aspect expected as the outcome from learners. One of the responsibilities of an Air Traffic Controller, in which the use of language and communication skills are important, is chosen for designing the ELT activity. The job of air traffic controllers (ATC) is to regulate air traffic and they depend on radar images and voice interaction with pilots and other ATC personnel. A typical scenario, in which an aircraft in distress was identified from the resources available on the web and the interaction [voice communication] between the Pilot and ATC personnel, was used for listening skills development.

Step 1:	Each person in a group gets to know the other members in the group and together they discuss the activity on hand. The activity given is to identify a situational dialogue that belongs to the real world. They need to understand various language-related aspects of getting involved in a conversation
Step 2:	The group selects a video source from the world of aviation. Audio files, transcript [12] and video files [13, 14] of the communication between the air traffic controller (ATC) and the pilot of US Airways Flight 1549 are selected by a group from the YouTube and all the materials are uploaded to the DLLS
Step 3:	Members of the group listen to the audio in the video files and identify language aspects like linkers, discourse markers, terminology [words] associated with the aviation industry and the language used for communication
Step 4:	The group then uploads the newly completed work to the DLLS. They create new dialogues using the same scenario but with different endings. The proposed endings: 1. The pilot agrees to land in designated runway and makes a safe landing 2. The pilot requests to land in a nearby airstrip and makes a safe landing 3. The pilot sends a Mayday signal and finally the aircraft crashes
Step 5:	The teacher complements by suggesting necessary changes to the task carried out by the group and reloads it to the DLLS
Step 6:	Each group then does a role play with the newly proposed ending. Challenges faced while doing the activity are recorded and analysed. The teacher suggests improvements after providing suitable explanations
Step 7:	Members of each group then share their experiences, observations and findings and map learning outcomes, finally uploading them to the DLLS

A record of a few challenges faced by learners:

One of the challenges faced by students while doing the above-mentioned activity was the new experience the students came across listening to the English spoken by native speakers and that too in a professional scenario. Furthermore, the voices over the wireless communication channel with a lot of disturbances made the listening activity challenging. However, students felt that it certainly will make learners take extra effort to listen carefully and thereby get more involved in the process. Students also expressed their views by stating that the understanding of the challenges faced by them while doing the activity helped them to perform better. This at the same time provided valuable insights to the teacher about various intricacies of the language acquisition process.

5 Conclusion

The purpose of creating a new model, the FINN model, is to explore new possibilities of making the whole process of language acquisition meaningful and effective. The idea is to create the right environment for learners to perform better and to learn better. The focus is not on testing their skill sets only, but to first create a pervasive platform for learners to explore and experience the new learning methodology. The aim is also to empower teachers to facilitate and guide the learners in their efforts. The teacher also needs to record the new learning experiences of the learners and the challenges faced by them. The data thus collected can be effectively used to make the model better and be used in further research.

With emphasis not only on the experience of learners during the listening task but also in making learners get actively involved in the process, the diagnostic approach will certainly enhance the level of engagement, thereby increasing the level of learning. Fundamentally, the FINN model will help learners to innovate standing on the teacher's pedestal and also get involved in the process of learning. The aim of this model is to provide learners with more freedom to innovate, get involved in collaborative learning and record learning strategies and solutions to overcome issues related to skills development and create a repository using on a digital platform to be sued by others learners. The FINN model can also be implemented in any area of study or branch of learning, and not just in the field of ELT.

References

1. A. Roux, J. Du Toit, Thinking about the future and strategic transformation, in *Business Futures* (Institute for Futures Research, Stellenbosch, 2003)
2. M.J. Riemer, Intercultural communication considerations in engineering education. Glob. J. Eng. Edu. **11**(2), 197–206 (2007)
3. J. Stepp-Greany, Student perceptions on language learning in a technological environment: implications for the new millennium (2002)
4. J. McGovern, Changing paradigms. The Project Approach (1995)
5. G.W. Fairholm, Spiritual leadership: fulfilling whole-self needs at work. Leadersh. Organ. Dev. J. **17**(5), 11–17 (1996)
6. K.R. Hey, P.D. Moore, *The Caterpillar Doesn't Know: How Personal Change Is Creating Organizational Change* (Free Press, New York, 1998)
7. J. Rubin, An overview to a guide for the teaching of second language listening. A guide for the teaching of second language listening, pp. 7–11 (1995)
8. L. Vandergrift, Facilitating second language listening comprehension: acquiring successful strategies (1999)

9. W.A. Renandya, T.S. Farrell, 'Teacher, the tape is too fast!' Extensive listening in ELT. ELT J. **65**(1), 52–59 (2010)
10. K.W. Lee, English teachers' barriers to the use of computer-assisted language learning. Internet TESL J. **6**(12), 1–8 (2000)
11. M. Warschauer, The changing global economy and the future of English teaching. Tesol Q. **34**(3), 511–535 (2000)
12. Federal Aviation Administration (Internet) (U.S. Department of Transportation, Washington, DC) (updated 2009 Mar 02; cited 2016 Jan 20). Available from: https://www.faa.gov/data_research/accident_incident/1549/
13. https://youtu.be/jZPvVwvX_Nc
14. https://youtu.be/E8itHvXd0oM?list=PLuifHZ5dsQaCbvmAG2vQed5lf9dmXa-am

Factors Affecting the Adoption of Cloud Computing Among SMEs in Mauritius

V. Domun and H. Bheemul

Abstract In order to sustain, SMEs need to make use of technology to ease their work thus being more productive. As per the government of Mauritius through the "Achieving the second Economic Miracle and vision 2030", Cloud Computing is what we need. Cloud Computing allows many SMEs to expand and to be on the move in Mauritius, thus having operational expenditure rather than Capital expenditure. SMEDA propose its services for SMEs in Mauritius as well the NPCC. This research paper talks about the current situation concerning the adoption of cloud computing and the challenges we are facing to get SMEs to start using this technology. The most top rank issues are privacy and security. This research also proposes a number of recommendations to support SMEs in their journey of cloud adoption in Mauritius.

Keywords Cloud computing · SMEs · Cloud adoption

1 Introduction

IT allows organisations to sustain within a complex environment. Cloud computing provides services such as storage, platform and software to customer via the web from a data center.

SMEs contribute to the economic development of Mauritius. However, due to their size coupled with constraints, SMEs in Mauritius are facing many challenges and adoption of the ICT forms part it. Cloud Computing elucidates the logistics side of SMEs with the ICT adaption since hardware, software and maintenance costs are being relieved.

V. Domun (✉)
Université des Mascareignes, Rose-Hill, Mauritius
e-mail: cvdomun@gmail.com

H. Bheemul
Train 2 Gain Group, Kingston Upon Thames, London, UK
e-mail: director@t2g-group.com

© Springer Nature Singapore Pte Ltd. 2019
S. C. Satapathy et al. (eds.), *Information Systems Design and Intelligent Applications*,
Advances in Intelligent Systems and Computing 862,
https://doi.org/10.1007/978-981-13-3329-3_31

1.1 Statement of the Problem

According to Kim, cloud computing adoption differs as per the size of the organisation. They feel that bandwidth affects the cloud service availability mostly into developing countries. Security of data is often the main issues of SMEs towards their Cloud Service provider.

In the Mauritius Vision 2030, Mauritius is seen as a country whereby Cloud Computing is seen as a business enable tool among SMEs in Mauritius. Despite the efforts of the government and associated bodies like SMEDA towards a rise in competitiveness of SMES using ICT, such as a decrease in the cost of Internet and providing loan and lowering cost, the adoption and usage of Cloud computing has been very low and slow, though it brings operational expenditure [1–6].

1.2 Research Objective

The objective of this research is the following:

1. To find the factors and conditions that have impacts on the adoption of cloud computing among SMEs in Mauritius
2. To find the challenges that SMEs are facing for the adoption of cloud computing in Mauritius.

2 Methods

This research is based on a sectional survey design. As per collected data , descriptive research is made up of surveys and various kind of finding inquiries. The study was based on a sample of 40 SMEs and using mean and Standard deviation, the data has been analysed (Table 1).

This shows that the firms adopted cloud computing due to the availability and services provided by the cloud service provider. Management has been supportive for the adoption of this technology (Table 2).

This shows that SMEs are not using cloud computing or are reluctant due to the trust factor of the cloud service such as control, privacy, auditing, hacking, changing of providers.

Table 1 Factors that have an impact on the adoption of cloud computing

Factors	Mean	Std
The benefit of adopting cloud computing outweighs its cost	3.34	0.822
Cloud computing services offered are relevant to the company's business	4.55	0.567
Unreliability of in-sourced IT services drove us into the cloud	4.00	0.803
The need to have data backup necessitated the move to the cloud	4.37	0.678
It is easy to use cloud computing	4.56	0.496
Top management supports cloud computing adoption	4.45	0.587
The human resource is well trained to handle cloud computing (IT)	4.34	0.537
Our customers expect us to use technology such as cloud	4.35	0.721
Competition in the market has made adoption of cloud computing mandatory	4.39	0.680
Availability of cloud computing services has enabled us to adopt	4.56	0.496

Table 2 Challenges facing cloud adoption

Challenge	Mean	Std
No reason to throw away the existing system	2.87	0.973
Unaware of computing services or available options	2.27	1.549
Limited resources of cash to operate cloud	2.87	1.334
Firms that have adopted cloud services report problems with	3.39	1:008
Discouraged by the IT provider	2.93	0.940
Cloud computing can expose the firm to data security risks and minimize information privacy	3.30	1.271
I don't trust the effectiveness of cloud service usage and	3.92	1.029
With cloud computing, my firm can get bogged down by IT infrastructure inefficiencies	3.83	1.084
Difficulty in assessing the costs involved due to the on-demand nature	3.90	1.009
Difficulty in migrating in and out of the cloud and switching	3.74	0.927
The firm would spend more for the bandwidth than it would on hardware and in-house software	3.12	1.324
Cloud providers still lack round-the-clock service resulting in frequent	3.54	1.084

3 Conclusions

Cloud computing is a simple system and SMEs can easily learn how to use it to bring productivity and efficiency at work. The National productivity council in Mauritius, always train SMEs and more focus should be put towards the adoption of cloud computing.

With cloud computing, SMEs no longer need to have capital expenditure and no server room. As a result, this brings Green IT, to the motto of the Mauritian Government, 'Maurice Ile D'urable'. Concerning security, data centers are certified

ISO 27001, registered as Data Controller. IT audits are also being able to carry out nowadays on cloud services and this gives the assurance of the data being protected.

References

1. S.A. Brown, V. Venkatesh, Model of adoption of technology in households: a baseline model test and extension incorporating household life cycle. MIS Q. **29**(3), 399–426 (2005)
2. A.N. Mishra, R. Agarwal, Technological frames, organizational capabilities and post-adoption IT use: an empirical investigation of electronic procurement. Inf. Syst. Res. **21**(2), 249–270 (2010)
3. R. Saleem, *Cloud Computing's Effect On Enterprises: Cost and Security*. Master's Thesis (Lund University, 2011)
4. R. Srinivasan, G.L. Lilien, A. Rangaswamy, The role of technological opportunism in radical technology adoption: an application to e-business. J. Mark. **66**(3), 47–60 (2002)
5. D. Throng, How cloud computing enhances competitive advantages: a research model for small businesses. Bus. Rev. pp. 59–65 (2010)
6. Government of Mauritius "Achieving the Second Economic Miracle and Vision 2030"

An Investigation of the TCP Meltdown Problem and Proposing Raptor Codes as a Novel to Decrease TCP Retransmissions in VPN Systems

Irfaan Coonjah, Pierre Clarel Catherine and K. M. S. Soyjaudah

Abstract When TCP was designed, the protocol designers at this time did not cater for the problem of running TCP within itself and the TCP dilemma was not originally addressed. The protocol is meant to be reliable and uses adaptive timeouts to decide when a resend should occur. This design can fail when stacking TCP connections though, and this type of network slowdown is known as a "TCP meltdown problem." This happens when a slower outer connection causes the upper layer to queue up more retransmissions than the lower layer is able to process. Some computer scientists designed a Virtual Private Networking product (OpenVPN) to accommodate problems that may occur when tunneling TCP within TCP. They designed the VPN to use UDP as the base for communication to increase the performance. But the problem with UDP is said to be unreliable and not all VPN systems support UDP tunneling. This paper seeks to provide systems with low-latency primitives for reliable communication that are fundamentally scalable and robust. The focus of the authors is on proposing raptor codes to solve the TCP meltdown problems in VPN systems and decrease delays and overheads. The authors of this paper will simulate the TCP meltdown problem inside a VPN tunnel.

Keywords Tunneling · VPN · TCP · Raptor codes

I. Coonjah (✉) · K. M. S. Soyjaudah
Faculty of Engineering, University of Mauritius, Réduit, Moka, Mauritius
e-mail: irfaan.coonjah@umail.uom.ac.mu

K. M. S. Soyjaudah
e-mail: ssoyjaudah@uom.ac.mu

P. C. Catherine
School of Innovative Technologies and Engineering, University of Technology, La Tour Koenig
Pointes aux Sables, Port Louis, Mauritius
e-mail: ccatherine@umail.utm.ac.mu

© Springer Nature Singapore Pte Ltd. 2019
S. C. Satapathy et al. (eds.), *Information Systems Design and Intelligent Applications*,
Advances in Intelligent Systems and Computing 862,
https://doi.org/10.1007/978-981-13-3329-3_32

1 Introduction

What's wrong with TCP? TCP is a connection-oriented transport protocol which means that it establishes sessions between the sender and the receiver. TCP makes use of sequence numbers, checksums, acknowledgments, timeouts, and retransmissions to ensure reliability [1]. TCP waits for the receiving end to send an acknowledgement after successfully received packets. The sender retransmits packets whenever it does not receive an acknowledgement within a timeout period. When TCP resends unacknowledged packet, it slows down the flow of data packets to lower the throughput to ease the congestion caused. As a result, file transfer between two distant points over a Wide Area Network (WAN) using TCP does not make an efficient use of the available bandwidth, it therefore creates a lengthy transmission times. The mechanisms for TCP's flow control and congestion avoidance record the time from the transmitter to the receiver, assuming that long round-trip times are due to network congestion rather than actual the transit times [2]. Olaf Titz, in 2004, initiated the Crypto IP Encapsulation (CIPE) project. CIPE has been developed as open source and is believed to be the most secure approach in crypto protocol development. It aims at creating a lightweight VPN solution using UDP as base connection. It works by tunneling IP packets in encrypted UDP packets. The CIPE project was meant to overcome the problem of TCP's retransmission algorithm and the stacking problem. Olaf Titz showed that TCP retransmission increases congestion and produce an effect known as "meltdown". The CIPE project concluded to have slightly better performance than IPSEC and was designed to overcome the TCP over TCP problem. The CIPE project was one attempt to cater for the TCP meltdown problem [3]. When PPP over SSH solution was used, the performance was fairly unusable; even an optical link suffered frequent packet loss, 10–20% over an extended period of time. Users still face the problem of reliability when transmitting traffic into a UDP tunnel. The TCP transport layer protocol was only replaced by UDP in CIPE project.

In this proposed study, the authors will simulate TCP meltdown over a WAN environment using OpenVPN, OpenSSH, and CISCO VPN systems. Different packets were sent inside the TCP tunnel and the number of retransmission packets was recorded (overheads). Different VPN systems are expected to have different values. Three graphs will be plotted, retransmission packets versus packet size. The outcome of the experiment will determine whether raptor codes can be proposed as a solution to the TCP meltdown problem.

This document is organized into the following sections:

- Section 2 covers retransmissions algorithm of TCP in details.
- Section 3 covers TCP stacking.
- Section 4 details the experimental framework.
- Section 5 details the performance measurements.
- Section 6 describes raptor coding.
- Section 6 proposes raptor codes to cater for TCP meltdown.
- We then conclude by giving the results and observations in Sect. 7.

2 TCP Retransmission Algorithm

TCP breaks down the data stream into segments and sent as individual IP datagram [4]. The segments contain a sequence number which numbers the bytes in the stream, and an acknowledge number which tells the other side the last received sequence number. Sequence numbers are used to reassemble datagram that may be lost, duplicated or reordered [5]. When an acknowledgment number for a recently sent segment does not arrive in a certain amount of time, the sender marked the packet as lost and proceed with resends the whole segment again. Over the internet lines, factors like bandwidth, loss rate and delay and are very distinct from one connection to another [6]. On a gigabit LAN, a small timeout is considered to be inappropriate and also inappropriate on an international link which is congested. The increase in congestion will lead to an effect known as "meltdown" [7]. Whenever a segment timeouts, the following timeouts increases exponentially to avoid the meltdown effect [8].

Figure 1 shows the TCP retransmission policy. Three duplicate acknowledgments are received:

- For every packet received, the recipient returns an ACK
- Recipient sends duplicate ACKs if a packet is lost
- Sender retransmits lost packets

TCP uses a retransmission timer to ensure data delivery in the absence of any feedback from the remote data receiver. The duration of this timer is referred to as RTO (retransmission timeout). Research suggests that a large minimum RTO is needed to keep TCP conservative and avoid spurious retransmissions [9].

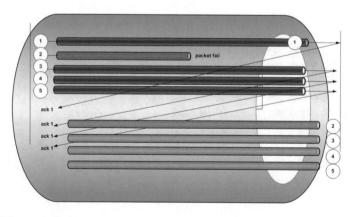

Fig. 1 TCP retransmission

3 TCP Stacking

TCP makes effort not to break connections; the timeout can rise up to the range of several minutes. For unattended bulk data transfer, the connection will slow down and are of course undesirable and will most likely be terminated by the user. The enhancement for reliability for stop working when TCP traffic is sent on top of another, and this issue was not cared by when designing TCP protocol. But it happens when running PPP over SSH or another TCP-based protocol because the PPP-encapsulated IP datagrams likely carry TCP-based payload, like Fig. 2.

The TCP timers for the lower layer and the upper layer have are different [10]. When an upper layer connection is fast, the connection timers is fast as well. The lower connection however can have timers which are slower, making it an unreliable base connection therefore undergoing packet loss. The timeouts increase when the TCP lower layer retransmission queues up. The fact that the connection is jammed for a certain amount of time, the upper layer TCP fails to get the ACK on time, and this eventually increases the retransmission queue. The upper layer connection becomes slow and the retransmission creates a meltdown effect and makes TCP unreliable in this situation [11].

4 Experimental Framework Description

The authors of this paper setup this framework which is in a WAN environment. The topology is displayed in Fig. 3 and the details are as listed below:

- private network is 10.1.2.0/24
- eth1 on the server has public IP 202.123.2.11
- eth0 on the server has private IP 10.1.2.13
- tun0 on the server has private IP 10.0.0.200/32
- tun0 on the client has private IP 10.0.0.100/32
- tun1 on the server has private IP 10.1.0.1/32
- tun0 on the client has private IP 10.1.0.2/32

Fig. 2 TCP stacking

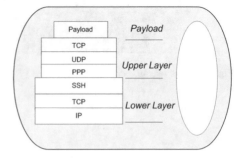

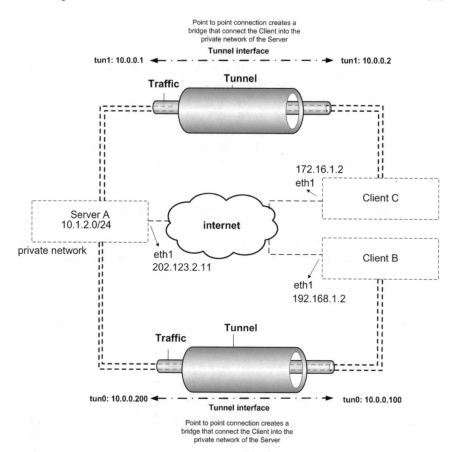

Fig. 3 Logical topology

The logical topology is made up of boxes representing servers and routers as displayed; the logical topology of a network is therefore not necessarily the same as its physical topology. The diagram represents three machines, one server and two clients. The server is connected to a private network 10.1.2.0/24 via Ethernet, and the two clients are connected to network 192.168.1.0/24 and 172.16.1.0/24, respectively. The server's IP address on the private network is 10.1.2.13 and the client IP addresses are 192.168.1.2 and 172.16.1.2/24. Each network has an Internet NAT gateway to allow for internet connectivity. In this topology, the client A is connected to the server via an SSH tunnel interface, tun0 and client B are connected to the server through the tunnel interface tun1. Both clients must be able to access the network 10.1.2.0/24 through their tunnel interface addresses. During the tests, traffic will be sent from the Client to the Server. The packet sizes will range from 1 to 100 Mbps and the retransmissions of packets will be recorded.

5   Performance Measurements

During the test, the authors used iperf to send data and record the percentage drops from the server to the client. Iperf allows the user to set various parameters that can be used for testing a network, or alternatively for optimizing or tuning a network. Iperf has a client and server functionality, and can measure the throughput between the two ends, either unidirectionally or bidirectionally [12] (Fig. 4).

It is important to note that there is no flag or unique identifier associated with a TCP retransmission. Wireshark calculates TCP retransmissions based on SEQ/ACK number, IP ID, source and destination IP address, TCP Port, and the time the frame was received.

It compares the sequence numbers to what it has determined to be the next expected sequence number from the last packet of the conversation into the same direction, by packet order (not by timestamps). It is very easy for Wireshark to count a duplicate packet as a retransmission. The authors of this paper however ensured that the same frame was not captured twice. This is very common in data center capture architectures [13] (Fig. 5).

The experimental testing procedure consists of three main parts, the values for the percentage TCP retransmission were recorded on the server using on iperf. The packet

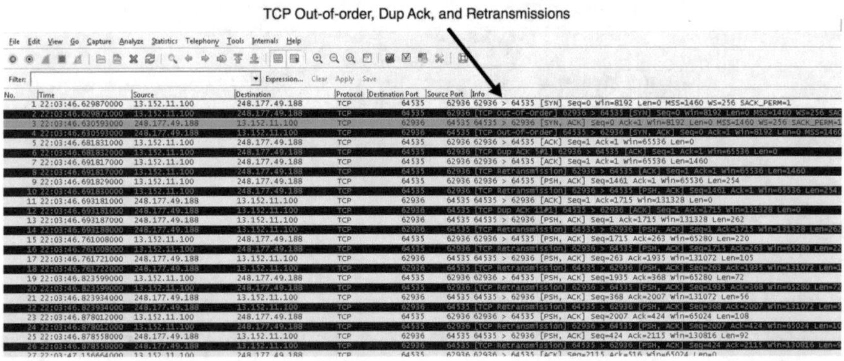

Fig. 4 Wireshark capture

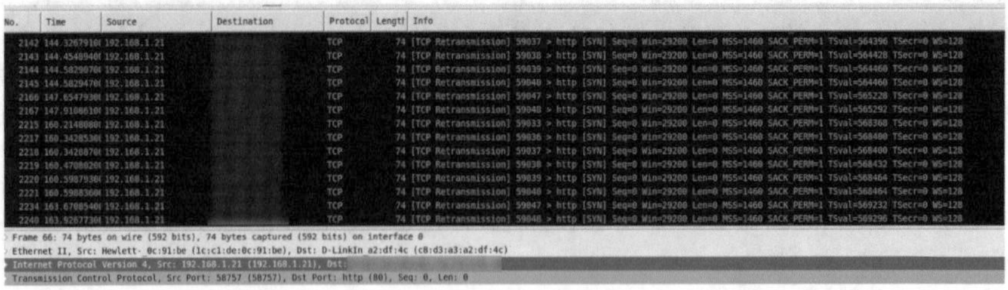

Fig. 5 TCP retransmission

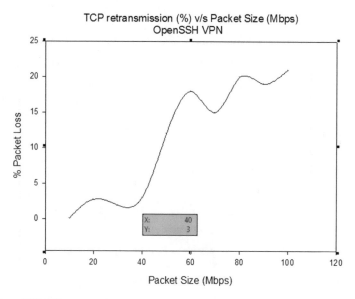

Fig. 6 OpenSSH TCP retransmission

sizes ranges from 10 to 100 Mbs. Three popular VPNs were set up with the logical diagram in Fig. 3. The tests were performed in a real-life scenario, therefore the authors could not cater for congested network and broadcast in the LAN environment. The tests were performed three times and three graphs were plotted using percentage packet loss v/s packet size. The different files ranging from 10 to 100 Mbps were sent from iperf using the TCP option, this ensures that we are sending TCP packets to the client in all three VPNs. It was observed that the packet loss increases drastically in all three scenarios when the sizes of the packet transferred are greater than 40 Mpbs. The value in Figs. 6 and 7 does not differ a lot when compared to the values in Fig. 8. Both OpenSSH and OpenVPN are open source whereas Cisco VPN is proprietary. The client VPN for Cisco was setup using an ASA version 5520. Based on the three graphs, we can say that TCP retransmission is negligible when the packet size is below 40 Mbs. TCP retransmissions are usually due to network congestion. Because we a dealing with TCP which is connection oriented, we can therefore say that TCP meltdown occurs when large packet size reaches 50 Mbps.

6 Proposing Raptor Codes for Less Overhead

Raptor codes were developed by Amin Shokrollahi in 2001 and published in 2004 [14]. Raptor codes are an advanced improvement over LT codes, which were the first practical class of fountain codes. Raptor codes encode a given message with of a number of symbols, k, into an unlimited sequence of encoding symbols. The

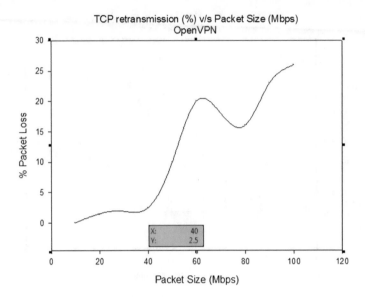

Fig. 7 OpenVPN TCP retransmission

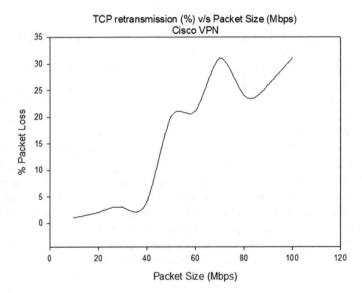

Fig. 8 Cisco VPN TCP retransmission

knowledge of any k or more encoding symbols allows the message to be retrieved with some nonzero probability. The probability that the message can be recovered depends on the number of symbols received. The probability increases when the number of symbols received above k becomes very close to 1 [15]. For example, with the latest

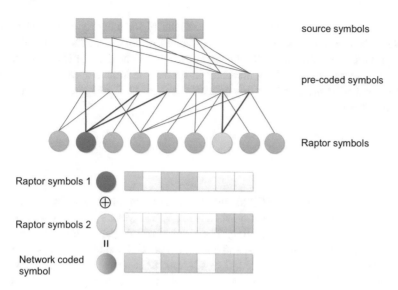

source symbols

pre-coded symbols

Raptor symbols

Raptor symbols 1

$\oplus$

Raptor symbols 2

$=$

Network coded
symbol

Fig. 9 Raptor codes

generation of Raptor codes, the RaptorQ codes, the chance of decoding failure when k symbols have been received is less than 1%, and the chance of decoding failure when $k+2$ symbols have been received is less than one in a million. A symbol can be any size, from a single byte to hundreds or thousands of bytes (Fig. 9).

Raptor codes may be systematic or non-systematic. In the systematic case, the symbols of the original message are included within the set of encoding symbols. TCP is most commonly used because it is a reliable mode of communication for file transfer. TCP, ensure reliability at the detriment of bandwidth efficiency [16]. Raptor codes ensure reliability through efficient available bandwidth. It therefore overcomes TCP's bandwidth inefficiencies caused by TCP meltdown by using an alternative way to transmit data over an IP network by fully using the available bandwidth even if there is network congestion, packet loss, or latency. To control the throughput, TCP uses TCP window size to define the amount of data in the queue—With 100% throughputs, the total number of unacknowledged data between the transmitter and the receiver equals to the round-trip time multiplied by the end-to-end data rate [17]. TCP continually takes into consideration this amount at the transmitter, with TCP window size to monitor how much data is transmitted before receiving an acknowledgement. When there is packet loss, the expected acknowledgment does not arrive and, the window size is therefore reduced. The window size is gradually enlarged to the maximum value whenever the communication is successful [18].

Reliability can still be maintained while using the maximum bandwidth If ever UDP rather than TCP is used to deliver packets, then the inefficiencies of TCP are not incurred [19]. UDP being connectionless does not need any handshaking to set up a link between the source and destination [20]. If the packets are encoded with

Raptor, the Raptor algorithm will ensure reliability for integrity and reordering of packets. Hence, the benefits of a connectionless protocol coupled with reliability will provide protection at the Raptor packet-level. The benefits of Raptor codes are for large file transfer over long distances in-between distinct service providers have large round-trip times, and, for any packets dropped, TCP will cater by limiting the transmission rate [21].

7 Conclusion

Tunneling TCP traffic into TCP tunnel creates a problem known as TCP meltdown and this problem increases latency when using VPNs. Most VPNs make use of TCP as the transport layer for tunneling. The authors of this paper showed when TCP retransmission occurs in a VPNs, and conclude that TCP meltdown increases drastically when the packet size is larger than 40Mbps. Some tweaks in the network layer can be down to reduce the retransmission problem by decreasing the MTU size, but the problem of retransmission remains. There is however the trade-off between reliability when using TCP tunnel and unreliability when using UDP tunnel. The authors introduced raptor codes as a solution to solve the TCP retransmission problem.

References

1. Y.-L. Chang, C.-C. Hsu, Connection-oriented routing in ad hoc networks based on dynamic group infrastructure, in *Fifth IEEE Symposium on Computers and Communications, 2000. Proceedings. ISCC 2000*, pp. 587–592 (2000)
2. R. Dasgupta, R. Mukherjee, A. Gupta, Congestion avoidance topology in wireless sensor network using Karnaugh map. Appl. Innov. Mob. Comput. (AIMoC) **2015**, 89–96 (2015)
3. I. Coonjah, P.C. Catherine, K. Soyjaudah, Experimental performance comparison between TCP vs UDP tunnel using OpenVPN, in *2015 International Conference on Computing, Communication and Security (ICCCS)*, pp. 1–5 (Dec 2015)
4. S. Nakazawa, H. Tamura, K. Kawahara, Y. Oie, Performance analysis of IP datagram transmission delay in MPLS: impact of both the number and the bandwidth of LSPS of layer 2, in *IEEE International Conference on Communications, 2001. ICC 2001*, vol 4, pp. 1006–1010 (2001)
5. J. Garcia-Luna-Aceves, H. Rangarajan, A new framework for loop-free on-demand routing using destination sequence numbers, in *2004 IEEE International Conference on Mobile Ad-hoc and Sensor Systems*, pp. 426–435 (Oct 2004)
6. J.-K. Choi, C. Un, On acknowledgement schemes of sliding window flow control. IEEE Trans. Commun. **37**, 1184–1191 (1989)
7. Z. Anwar, A. Malik, Can a DDoS attack meltdown my data center? A simulation study and defense strategies. IEEE Commun. Lett. **18**, 1175–1178 (2014)
8. A. Gurtov, R. Ludwig, Responding to spurious timeouts in TCP, in *INFOCOM 2003. Twenty-Second Annual Joint Conference of the IEEE Computer and Communications. IEEE Societies*, vol 3, pp. 2312–2322 (March 2003)
9. M. Haeri, A. Rad, TCP retransmission timer adjustment mechanism using model-based RTT predictor, in *Control Conference, 2004. 5th Asian*, vol. 1, pp. 686–693 (July 2004)

10. S. Seth, M. Venkatesulu, *TCP Timers*, pp. 323–375. (Wiley-IEEE Press, 2008)
11. M. Haeri, A. Rad, TCP retransmission timer adjustment mechanism using system identification, in *Proceedings of the 2004 American Control Conference, 2004*. vol. 3, pp. 2328–2332 (June 2004)
12. V. Barayuga, W. Yu, Packet level TCP performance of NAT44, NAT64 and IPv6 using iperf in the context of IPv6 migration, in *2015 5th International Conference on IT Convergence and Security (ICITCS)*, pp. 1–3 (Aug 2015)
13. S. Wang, D. Xu, S. Yan, Analysis and application of wireshark in TCP/IP protocol teaching, in *2010 International Conference on E-Health Networking, Digital Ecosystems and Technologies (EDT)*, vol. 2, pp. 269–272 (April 2010)
14. B. Sivasubramanian, H. Leib, Fixed-rate raptor codes over Rician fading channels. IEEE Trans. Veh. Technol. **57**, 3905–3911 (2008)
15. T.Y. Chen, K. Vakilinia, D. Divsalar, R.D. Wesel, Protograph-based raptor-like LDPC codes. IEEE Trans. Commun. **63**, 1522–1532 (2015)
16. P. Palanisamy, T.V.S. Sreedhar, Performance analysis of raptor codes in wi-max systems over fading channel, in *TENCON 2008 - 2008 IEEE Region 10 Conference*, pp. 1–5 (Nov 2008)
17. A. Shokrollahi, Raptor codes. IEEE Trans. Inf. Theory **52**, 2551–2567 (2006)
18. H. Zeineddine, L.M.A. Jalloul, M.M. Mansour, Hardware-oriented construction of a family of rate-compatible raptor codes. IEEE Commun. Lett. **18**, 1131–1134 (2014)
19. P. Cataldi, M. Grangetto, T. Tillo, E. Magli, G. Olmo, Sliding-window raptor codes for efficient scalable wireless video broadcasting with unequal loss protection. IEEE Trans. Image Process. **19**, 1491–1503 (2010)
20. M. Song, Y. Choi, C. Kim, Connection rerouting method for general application to connection-oriented mobile communication networks, in *EUROMICRO 96. Beyond 2000: Hardware and Software Design Strategies, Proceedings of the 22nd EUROMICRO Conference*, pp. 412–419 (Sep 1996)
21. B.J. Vickers, T. Suda, Connectionless service for public ATM networks. IEEE Commun. Mag. **32**, 34–43 (1994)

Big Data Hadoop MapReduce Job Scheduling: A Short Survey

N. Deshai, B. V. D. S. Sekhar, S. Venkataramana, K. Srinivas and G. P. S. Varma

Abstract A latest peta to zeta era occurs from various complex digital world information, continuously collecting from device to device, social sites, etc., expressed as large information (as big data). Because of that we are unable to store and process due to lack of scalable and efficient schedulers. A main reason that day by day data is twice over digital world is database's size changes to zeta from tera. An apache open source Hadoop is the latest and innovative marketing weapon to grip huge volume of information through its classical and flexible components that are Hadoop distributed file system and Reduce-map, to defeat efficiently, store and serve different services on immense magnitude of world digital text, image, audio, and video data. To build and select an innovative and well-organized scheduler is an important key factor for selecting nodes and optimize and achieve high performance in complex information. A latest and useful survey, examination and overview uses and lacks facilities on Hadoop scheduler algorithms that are recognized throughout paper.

Keywords Big data · Hadoop · HDFS · MapReduce · Scheduling

N. Deshai (✉) · B. V. D. S. Sekhar · S. Venkataramana · K. Srinivas · G. P. S. Varma
Department Information Technology, S.R.K.R Engineering College,
Bhimavaram, Andhra Pradesh, India
e-mail: desaij4@gmail.com

B. V. D. S. Sekhar
e-mail: bvdssekhar@gmail.com

S. Venkataramana
e-mail: vrsarella@gmail.com

K. Srinivas
e-mail: kasrinu71@gmail.com

G. P. S. Varma
e-mail: gpsvarma@gmail.com

© Springer Nature Singapore Pte Ltd. 2019
S. C. Satapathy et al. (eds.), *Information Systems Design and Intelligent Applications*,
Advances in Intelligent Systems and Computing 862,
https://doi.org/10.1007/978-981-13-3329-3_33

1 Introduction

We are working in the computerized advanced world information. In the real world moving to big data day by day, an enormous information is coming from different fields as Facebook, Yahoo, Google, YouTube, Amazon, Microsoft and Twitter, eBay, diverse sensor systems, Airlines record, Global Position System, RFID per users, PC logs, Closed Circuit cameras, Internet of Things. All these fields make complexity in dealing and using. Lesser and lesser the comprehension of information, greater and greater it would progress to becoming [1–9]. There are some planning viewpoints in huge information Hadoop and MapReduce. Schedulers are in charge of errand task in MapReduce where relegating the assignments to a specific information hub [6–20]. This paper offers a more widespread meaning of huge information that is Big Data that captures its exclusive and character. Handling with Big information look "V" challenges

a. **High Volume**: Now the world digital data moving day to day towards huge data called big data, because social n/w's, business field, astronomy field, health-care field, finance field, banking field are creating complex size of data sets. In every second, facebook has 2 billion users, YouTube 1 billion users, twitter 350 million users, and instagram 700 million users the insanely large amount—or Volume—of data.

b. **High Variability**: More importantly, the way the data is captured may vary from time to time or place to place. How to understand the distinctive words along with the various meaning or a dissimilar meaning.

c. **High Variety**: It refers to the types of information that can now be used, with certainty of 80% of the world's information is not structured (content, image, sound, video, etc.) with a huge information innovation, we can dissect and gather information of different types, for example, messages, discussions on online networks, photographs, sensor information, video or voice accounts, recordings, clicks, information on machines and sensors.

d. **High Visualization**: Parallel, cones and circular network diagrams. Combine the multitude of different variables that are the result of the variety, speed and the very complex interrelationships between them, that are not easy to develop meaningful visualization. Traditional images when we try to draw a trillion data points, we need different ways to display data, such as data grouping, tree maps, sunbeam,

e. **High Speed**: Many data require fast processing, so the speed is based on the velocity at which they are generated, created or updated and travels at high speed around the world. Sometimes 1 min is too late, so big data is time-sensitive. The Facebook data store stores over 300 bytes of data, but the speed will take at which new data is created.

f. **High vulnerability**: Huge information brings new security concerns. Insurance additionally implies building the notoriety and trust of the brand. Thorough security works on, including the utilization of cutting-edge diagnostic abilities to oversee protection and security issues, can separate organizations from the opposition and make solace and trust with the general population.

g. **Value**: The potential estimation of huge information is colossal. Esteem is the fundamental wellspring of enormous information since it is very important for organizations that the IT foundation framework stores a substantial number of qualities in the database. Numerous factors can influence the execution of the framework as the change in equipment assets and programming calculations, the system throughput, the rightness of information, area (locality), dealing with, investigation, stockpiling, the protection of information and job setting up. Each one of these issues makes challenges in huge data set as Big Data.

This paper structure is as per the following Sect. 2 discusses about associated working procedure, Sect. 3 examines Hadoop technologies, Sect. 4 shows the major requirements, and Sect. 5 examines various types of schedulers in Hadoop MapReduce. At last, Sect. 6 focuses conclusions.

2 Interrelated Works

This research paper considered some Apache Hadoop open source scheduler types and present a path on how to progress MapReduce job execution. The Hadoop schedulers focus on only homogeneous Hadoop clusters. The Hadoop, MapReduce schedulers are compared with some important features. A complete review focuses on processing large dataset, it is put based on the MapReduce program framework. It is also a set of recognized and designed systems covered, for to offer a good software interfaces on the Apache Hadoop MapReduce programming framework. The large data processing systems researched and also examined some valuable views associated to many types of applications scenarios of most popular MapReduce programming framework. In this survey, to improve the query execution performance as parallelwise using more popular MapReduce programming framework and presents the set of MapReduce software framework limitations, and few techniques to solve this. The job scheduling major issues and Apache Hadoop MapReduce classification of scheduling algorithms are presented in this paper.

3 Apache Hadoop

Apache Hadoop open source is a software framework for handling huge data set based on distributed wise storage and efficient process to clusters on commodity hardware [20–28]. Hadoop framework can handle hardware failure automatically; it used in several applications such as Twitter, Face book, Google, Yahoo, YouTube Amazon, etc.

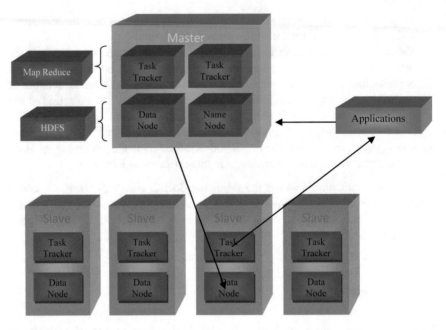

Fig. 1 Hadoop architecture

Apache open source Hadoop has more popular core components for efficient storing and processing large datasets on distributed file system, MapReduce framework in distributed and scalable the architecture of Hadoop is shown in Fig. 1. In Apache Hadoop splits the file into set of blocks (pieces) and each block length is 64 megabytes or 128 megabytes, these blocks are circulated across multiple nodes of cluster. To achieve efficient process, the large data in locality, open source Hadoop apply MapReduce processing to multiple nodes regarding parallel and distributes data processing. To manipulate huge number of nodes, handle large data, to allow more speed process of the large data sets inefficiently, scalability than conservative structural design with fast networking. Conventional architecture with fast. Hadoop Framework sequence of working is shown in Fig. 2.

3.1 Hadoop Distributed File System

It is implemented by Google as most popular distributed file system can handle the large dataset using commodity hardware clusters as more efficient and reliable. GFS is developed to produce tremendously high throughput, small latency and reduce each hardware failure in the server. Google file system guide by, Hadoop open source distributed file system use number of systems (machine) to keep many numbers of files. Through replicating factor on a number of servers we obtain good reliability. Large

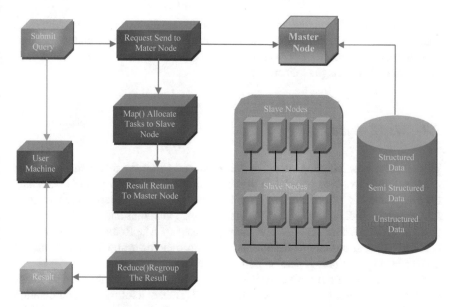

Fig. 2 Sequence in Hadoop framework

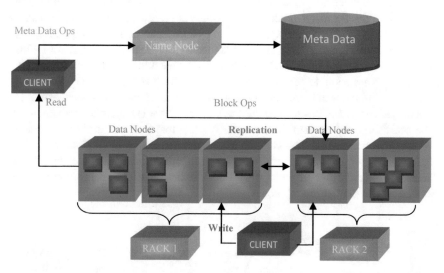

Fig. 3 HDFS architecture

data being stored on a number of multiple nodes, then we can gain fast computation and more reliable process. To place the large dataset through HTT Protocol, focus the client content accessing from web as shown in Fig. 3.

The Apache Hadoop open source distributed file system has two components are One and only One is Name Node uses about block of data map to Data Nodes and operate files like file rename, manage file directories, file open and file close, also Name Node maintains namespace of files, store file operations by clients. Various large number of Data Nodes are primary process is to create, delete number of data blocks, and also manage replication factor.

3.2 MapReduce

The Hadoop programming model is MapReduce is used to processing the large dataset in the cluster as parallel. It was originally developed by Google Company for a large number of web searching applications. The Hadoop MapReduce program framework is efficient can be used to data mining applications and machine learning tasks in different data centers. The main purposes of MapReduce are to give the permission programmers to design and develop their own applications from the drawbacks of partitioning, parallel process, and schedule the jobs.

Majorly all jobs are submitted to Hadoop MapReduce, it splits into small divisions and MapReduce design the tasks executed by Task Tracker it follows the computation in MapReduce phase. In MapReduce program model, the first phase is map it provides as key k1 and value v1 pairs. Second phase is reducing it follows sort and shuffle process then give the result. Each pair allows only one key per time then processes data per key k1 to yield key value (k1, v1) pairs. This programming model used different schedulers to assign jobs from queue designed by MapReduce Job Tracker to Task Tracker as shown in Fig. 4. Each scheduling will select the multiple nodes on what type of tasks are going to be executed. Also, reduce task carries summarized service, take care of all types of communication services and to transfer large data across the system and one more thing it can reduce redundancy, handle fault tolerance situations. The scalability and system fault tolerance are a major responsibility of Hadoop MapReduce framework. The Hadoop MapReduce distributing shuffle process can optimize the communication service cost. Through multithreaded implementation to handle the different datasets tasks as parallel. If reduced the total cost communication service on shuffle distributed process.

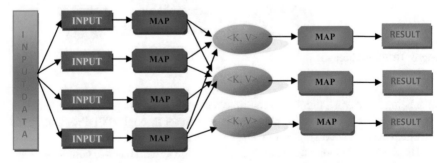

Fig. 4 MapReduce framework

First Step is Map: To execute the user (worker) nodes of local data by "map method" then the result (output) is written to temporary (private) storage.

Second Step is Shuffle: To reallocate the data of worker nodes regarding output keys, hence total data associated to few keys are placed on the same node.

Third Step is Reduce: each group output data is to process regarding key in parallel by a worker node.

4 Scheduling Considerations, Attributes and Implementations Proposals

To efficiently design set of schedule algorithms, to reach and progress some major requirements are high-quality service, fast response, data availability, efficient utilization of resources like communication, system hardwares, fairness, and good throughput. So it is critical to design such scheduling algorithms for fulfilling the requirement because of high complexity. So MapReduce scheduling algorithms are created to making the best required single attribute features. Also, it can be classified like adaptive means datasets, tasks, attributes, resources, and workloads and non-adaptive means fixed types.

4.1 Data Placement and Locality

The Hadoop MapReduce scheduler; the foremost decision is to arrange the compute methods close to the datasets. For the reason that it can decrease the input-output of clusters in disk, also finally it will raise the throughput. This paper wished for a k-means classical method of cluster analysis by putting the more relevant datasets inside the data center during the implementation and preparing stage. To analysis, the mutually dependent association between datasets and jobs follows the multi -task replication factor to decrease the size of middle data and send to data centers and also increase the job processing progress. To overcome stagger issue by using replication factor until to complete the total tasks in Hadoop MapReduce model.

4.2 Synchronization

To transfer Map function outputs (intermediate) submit to as Reducer input, also keys and values pairs are together as keys based on synchronization. This synchronization service is attained by MapReduce shuffle and sort methods, also it needs circulating sorting service and must involve every MapReduce iterations. The problem becoming a major, in different datasets due to various capabilities and configuration of every node, also influence various map and reduce finishing time. In MapReduce framework, the reduce process is unable to terminate unless each map function has completed, a general optimization to reduce function to drag the data from map function if completed.

4.3 Throughput

Throughput is number of Tasks completed in a unit of time regarding some characteristics and every MapReduce scheduler must produce good throughput along with high locality on data and more utilization of data.

4.4 Response Time

Time to interact to every request to process a task called response time sum of service and wait time. The service time varies for a specific request little if the work pressure increases. The response time is an important factor in QoS quality of service for every planner in MapReduce jobs and should be with some tolerance related to the amount of work sent.

4.5 Availability

It is a type of reliability metrics, it can be defined as the average time between failures and repetitions of user requests, or it is the time when the system supports and continues to execute to comply with future requests. All work must be completed completely and the planner must check it. This is a type of reliability measure. Repeat the client request between failures, this is also the length of the system that keeps the query and continues to work to meet the user's future demands. The total tasks must be completed after this verification via the programmer.

4.6 Resource Utilization

In many scientific areas, the resource major feature is utilization. Majorly, we maintain a maximum number of resources and efficient processing tools to handle massive data but in these is a lack in our process. The MapReduce scheduler must consume resources efficiently and particularly the scheduler should utilize their sources effectively, especially the nearest system hardware closes the communication employment. To optimize and manage resources to focus a number of tasks in clusters then achieving good results in our systems.

4.7 Fairness

It is a more significant feature which is through different types of Hadoop Schedulers. Fairness is to share the set of resources and task execution as poor fairly, uniformly, based on priority levels, size, and execution time between MapReduce Jobs.

5 Schedulers in Hadoop MapReduce

In this part, we will show the job made in the MapReduce scheduling regarding the fairness, priority, resource, time, utilization, and deadline of the system. If the scheduling process is static means offline since the allotments completed earlier than to start the execution or Dynamic means online since the allotments completed at the execution time regarding the lively atmosphere and desirable set of resources. Many times the scheduler is real-time computation system based on Storm distributed whenever manage the tasks regarding resource utilization.

5.1 FIFO Scheduler

It is the most defaulting MapReduce scheduling model working based on First-in first-out queue, when the jobs are schedule by the assigning order to Task Tracker nodes in Apache Hadoop open source. It can be used different systems, it cannot accept priority task due to first-come first-serve basis. If jobs are staying to turn as each job will use the total cluster. The major drawback is to decrease the data locality; also small queries will wait for results in unfair time. This paper target on task locality improves of FIFO scheduler by Apache Hadoop open source framework, presents the progress Task Locality scheduler. Tasks will be arranged and served too many job queues regarding the most feasible threshold value of job locality.

5.2 Fair Scheduler

It is developed by Facebook Company to obtain good fairness among all jobs, allocates every desire resources to each job through equal fairness sharing. This scheduler designs every job at minimum processing time to complete very quickly based on the equivalent distribution of resources. All users have pools along with at least desire sharing of resources and select the unused slot by assign the running tasks. This scheduler allocates the resources to every application on totally equivalent sharing. This scheduler working on memory only by default and organize the work based on memory and CPU. It used the total cluster if the only single application is working. Idle or free resource assigned to the tasks of applications, the cluster should be arranged between fixed numbers of users. This scheduler follows priority to assign weights for finding resources and allocate to the apps. It has a default queue is called as "default" it is shared among total users.

And create each username on each queue along with minimum sharing ability, minimum quota of slots, remaining slots means idle slots will be assigned to next pool. With fair scheduler, each application can be run by default, but the total running apps are limited regarding configuration file of each user queue. It advantage whenever user presents more apps at one time, also we can increase the scheduler performance by minimizing the intermediate data producing or switching. There are no chance applications failing because waiting for the scheduler queue up to complete the past users request.

5.3 Capacity Scheduler

It is presented by Yahoo, which is developed to handle complex clusters, also it can be applied to a number of applications when more number of clients are there, is designed especially to make efficient resource allotments of each computation between the number of clients and schedule the task allotment on domain resource. Also, this scheduler handles every job regarding user demands and memory utilization. Also, many queues are designed in the place of pools, to assign the tasks according to the queues of each client and to make MapReduce queue slot size. The remaining service of each queue is contributed among who queue needs. Once the task has been submitted then get a priority order on each to arrive task regarding time. The high priority level tasks will obtain the resources earlier than low priority tasks. Every cluster capacity is allocated among clients, not between tasks as like fair scheduler. We can manage the waiting time in capacity scheduler later than preemption applies to remain queues job whenever fair sharing is low.

5.4 Delay Scheduler

It is designed by Facebook, the delay occurs from an amount of time waiting to get the data on the local node to skip if in case not available then subsequent jobs. The purpose is to overcome the efficient share locality drawback. If your task is scheduled and will not launch the lack of job, lack of amount of time and reminding costs will launch its job instead, the intention of relational is to multiplex hafiz clusters based on statistical it has a minimum influence on fairness the gain more locality on data.

Finally, the designers follow delay scheduling because fairness is free meanwhile to increase locality on data. Tosca allotment has a free from job selection limitation.

5.5 LATE Scheduler

LATE scheduler (Longest surmised time to end) it is suitable to optimize the job performance and reduce job response time. It has a very fast response time during short jobs are running. Its purpose is more robust in heterogeneity but the drawback is not reliable.

5.6 Deadline Constraint Scheduler

It works under the user alloted time limitations, tries to meet the working target, and increases the system usages. This deadline follows MapReduce periods to come to an end task. The user must be noticed during ability scheduler experiment fails and go to new deadline and schedule the remaining jobs.

5.7 Resource-Aware Scheduler

Today's every system maintains multiple of nodes that follows varieties of services for allocating of nodes but many schedulers are not considered resources availability. This resource awaring scheduler would schedule the jobs are regarding resources availabilities. On each machine resource aware scheduler working depending on utilizability and availability and raise the utilization to reduce the resources using.

It follows major services

Free slots service: maximum slots for every node.
Free dynamic slot service: availability slots assign on every other node.

5.8 Node Selection Scheduling Algorithm

Its working and selecting on success rate regarding availability, ability, priority order. It follows nodes as queues to agree newly tasks along with their efficient node considering facility on node execution performance.

5.9 Round Robin Scheduler

It helps to decrease the responding time, also it helps to defeat the fifo scheduler drawbacks where jobs start working their arriving order, finally, it will introduce delay to new coming jobs. Scheduler will propose weightage round robin because Hadoop follows restrictions on weight improvements, for selecting various tasks to allocate a weight for every other queue, after scheduling the different tasks would be based on the mixed type of queue and sub-queue regarding weights are efficient.

5.10 Other Schedulers

More demands in open Hadoop source Apache schedule algorithms regarding progress and improvements: In [2] proposed a papular mining technique Bayes theorem classifier scheduler regarding learning their organizing execution state and gain self-classification.

In paper [3] is proposed to classical schedulers first high priority service in virtualization layers and the second service is batch processing to meet the deadline in Hadoop framework, for selecting best server to gain better performance to each tasks, also to make order jobs effectively by aware deadline scheduler. To combine these scheduler provide protections and guessing in resources in Hadoop environment.

In [4], authors proposed different computations for processing the massive information as stream oriented by a popular algorithm for intensive data workflow optimization, it provides limited latency period but more throughput. The main goal of this research paper is to study and analyse on different types of Hadoop and MapReduce algorithms and techniques which are useful to increase the Hadoop working performance in coming large application. These Hadoop and MapReduce features, properties and advantages and disadvantages and challenges and issues as shown in Tables 1 and 2.

Table 1 Hadoop schedulers features

Name of the scheduler	Speculative execution	Head of line problem	Sticky Slots	Locality problem	Job allocation	Implemented	Remarks
Default FIFO scheduler	NA	NA	NA	YES	Statically	YES	Homogenous system, static allocation, no
Fair scheduler	NA	YES	YES	YES for small jobs	Statically	YES	Homogenous static
Capacity scheduler	NA	NA	NA	YES	Statically	YES	Homogenous static, non primitive
LATE scheduler	YES	NA	NA	YES	Statically	YES	Homogenous Heterogeneous, static
Delay scheduler	NA	NA	NA	Improved compared to Fair scheduler	Statically	YES	Homogenous static
Dynamic priority scheduler	NA	NA	NA	NO	Dynamic	YES	Homogenous Heterogeneous, dynamic
Resource aware scheduler	NA	NA	NA	YES	Both	YES	Homogenous Heterogeneous, Dynamic
Deadline constraint scheduler	NA	NA	NA	YES for small jobs	Dynamic	Not in real time	Homogenous Static
Learning scheduler	ENABLED	NA	NA	YES	Both	YES	Homogenous Heterogeneous, dynamic
COSHH	NA	YES	YES	NO	Dynamic	Not in real time	Homogenous static
Aware scheduler	NA	NA	NA	NO	Dynamic	YES	Homogenous static

Table 2 Hadoop schedulers properties

Scheduler	Preemption	Priority in job queue	Fairness/fair sharing of resources	Better Working with	Homogeneous	Heterogeneous	Advantages	Disadvantages
Default FIFO scheduler	NO	No by default	NO	Small clusters	YES	NO	Easy design	Poor data availability and resource share, more waiting time on small jobs
Fair scheduler	YES	YES	Fair Share	Small clusters	YES	NO	Well resource sharing but poor response on shortest jobs	Poor balancing on workload, more settings needed
Capacity scheduler	NO	No by default	YES	Large clusters	YES	NO	Unwanted service apply on each job in queue	Implementing is very large
LATE scheduler	YES	YES	YES	Large clusters	YES	YES	applicable to heterogeneity	Lack of reliability
Delay scheduler	YES	YES	Less Fairness than Fair Scheduler	Small clusters	YES	NO	It is easy and simple to do large calculations	Difficult to work, lack of resource sharing
Dynamic priority scheduling	YES	YES	YES	Large clusters	YES	NO	Simple scheduler no affect of complex calculations	Lack on resource observation and fast network

(continued)

Table 2 (continued)

Scheduler	Preemption	Priority in job queue	Fairness/fair sharing of resources	Better Working with	Homogeneous	Heterogeneous	Advantages	Disadvantages
Resource aware scheduler	YES	YES	YES	Large clusters	YES	YES	Efficient resource sharing and executing	Lack on resource observation and fast network
Deadline constraint scheduler	YES	YES	YES	Large clusters	YES	YES	Excellent on optimization and timing	Poor predicting execution deadline
Learning scheduler	YES	YES	YES	Small clusters	YES	YES	Aware in timing and optimization	Lack on resource observation and fast network
COSHH	YES	YES	YES	Large clusters	YES	NO	Aware in timing and optimization	Lack on resource observation and fast network
Network aware scheduler	YES	YES	YES	Large clusters	YES	NO	Efficient resource sharing and executing	Need extra resource monitoring and high speed network

6 Conclusion

In today's market the Hadoop, MapReduce Algorithms are a big demand for big data. Every industry has a huge volume of data being stored but there is no tool to handle such complex data, Hadoop was introduced by Doug cutting to handle complex data at low cost and good speed, performance. We can see that most of these MapReduce schedulers are discussed in this entire paper and also addresses major problems. Throughout the paper, we studied, analyzed, and described the overview of various Hadoop and MapReduce algorithms for making efficient scheduler for speeding up our system and make the scheduler effective to gain fast processing.

References

1. A. Verma, et al., ARIA: Automatic Resource Inference and Allocation for MapReduce environments, in *8th Autonomic Computing ACM* (2011)
2. X. Yi, Research and improvement of job scheduling algorithms in Hadoop platform, Master Degree Dissertation. 45-51 (2010)
3. W. Zhang, et al., MIMP: deadline and interference aware scheduling of Hadoop virtual machines, in *14th IEEE/ACM International Symposium on Cluster, Cloud and Grid Computing* (2014), 978-1-4799-2784
4. S.G. Ahmad, et al., Data-intensive workflow optimization based on application task graph partitioning in heterogeneous computing systems, in *IEEE 4th BdCloud* (2014), 978-1-4799-6719
5. T.L. Casavant et al., A taxonomy of scheduling in general-purpose distributed computing systems. IEEE Trans. Softw. Eng. **14**, 141–154 (1988)
6. D. Yoo, K.M. Sim, A comparative review of job scheduling for MapReduce, in *Cloud Computing and Intelligence Systems* (IEEE, 2011)
7. S. Sakr et al., The family of MapReduce and large-scale data processing systems. ACM Comput. Surv. (CSUR) **46**(1), 11 (2013)
8. C. Doulkeridi, K. NØrvag, A survey of large-scale analytical query processing in MapReduce. VLDB J. **23**(3), 355–380 (2014)
9. N. Tiwari et al., Classification framework of MapReduce scheduling algorithms. ACM Comput. Surv. **47**(3), 49 (2015)
10. https://en.wikipedia.org/wiki/Apache_Hadoop
11. M. Malak, Data Locality: HPC vs. Hadoop vs. Spark. datascienceassn.org. Data Science Association (2014)
12. G. Sanjay, G. Howard, S.T. Leung, The Google file system, in *19th Symposium on Operating Systems Principles*, New York (2003), pp. 29–43
13. D. Yuan et al., A data placement strategy in scientific cloud workflows. Futur. Gener. Comput. Syst. **26**(8), 1200–1214 (2010)
14. W. Cirne, et al., UFCG/DSC Technical Report 07, 2005(4):225–246
15. M. Zaharia, et al., Delay scheduling: a simple technique for achieving locality and fairness in cluster scheduling, in *EuroSys'10*, pp. 265–278
16. P.S. Jun, et al., Optimization and research of Hadoop platform based on FIFO scheduler, in *7th ICMTMA*, https://doi.org/10.1109/.2015
17. http://hadoop.apache.org/docs/current/, 2.7.2 (2016)
18. J. Chen et al., A task scheduling algorithm for Hadoop platform. J. Comput. **8**(4), 929–936 (2013)

19. C. He, et al., Matchmaking: a new MapReduce scheduling technique, in *2011 IEEE 3rd International Conference on Cloud Computing Technology and Science (CloudCom)* (IEEE, 2011)
20. M. Zaharia, et al., Improving MapReduce performance in heterogeneous environments, in *OSDI'08: 8th USENIX Symposium*, Oct 2008
21. J. Geetha, et al., Hadoop scheduler with deadline constraint. IJCCSA **4**(5) (2014)
22. M. Yong, N. Garegrat, S. Mohan, Towards a resource aware scheduler in Hadoop, in *Proceedings of ICWS* (2009), pp. 102–109
23. G.K. Archana, V.D. Chakravarthy, HPCA: a node selection and scheduling method for Hadoop MapReduce, in *ICCCT'15* (2015), 978-1-4799-7623
24. Y. Wang, et al., A round robin with multiple feedback job scheduler in Hadoop, in *Progress in Informatics and Computing (PIC) International Conference*, Shanghai (IEEE, 2014) 978-1-4799-2030-3
25. J. Chen et al., A task scheduling algorithm for Hadoop platform. J. Comput. **8**(4), 929–936 (2013). https://doi.org/10.4304/jcp.8.4.929-936
26. S. Bardhan, D.A. Menasce, Queuing network models to predict the completion time of the map phase of map reduce jobs, in *ICMG* (2012)
27. J.V. Gautam, et al., A survey on job scheduling algorithms in Big data processing, in *ICECCT* (IEEE, 2015), 978-1-4799-6084-2
28. Hadoop, http://hadoop.apache.org/. Retrieved 29 Feb 2016

A Study on IOT Tools, Protocols, Applications, Opportunities and Challenges

N. Deshai, S. Venkataramana, B. V. D. S. Sekhar, K. Srinivas
and G. P. S. Varma

Abstract Today's latest marketing weapon is also new extremely urbanized innovative equipment to combined devices jointly to accomplish efficient computerization and hurriedly take accomplishment to things together and jointing each on with smoothly exchange communicate control observe the things compute like automation in current planet as IOT. It presents more immediate, instant, and fast responding facility to gadget concerning the things globe and anticipated reason that things direct to innovate, advanced, instantaneous, smart servicing, raise in potency and smoothness. Always an IOT preserve and ready with unmistakably and faultlessly oversized diversity of different systems, whereas it provided that open access services on required hardware which makes more interactive surroundings. Designing and constructing IOT worldwide which is really advanced task, its internal computation of gain strategies and connection layer knowledge, IOT would be real and advanced gadget-to-gadget (g2g) connecting model that likes very nearest to future, during which service of existence be going away to serve through microcontrollers; transceivers are used about digital things and suitable procedure and corresponds to everything also the total strategy support and efficient services for people's living manner. But, an important challenge still hinders its progress, particularly protection, power and speed problems. This entire survey paper is a center of attraction with an absolute study on (Internet of things) IOT important architectures, applications, platforms, designing issues, and protection.

N. Deshai (✉) · S. Venkataramana · B. V. D. S. Sekhar · K. Srinivas · G. P. S. Varma
Department Information Technology, S.R.K.R Engineering College,
Bhimavaram, Andhra Pradesh, India
e-mail: desaij4@gmail.com

S. Venkataramana
e-mail: vrsarella@gmail.com

B. V. D. S. Sekhar
e-mail: bvdssekhar@gmail.com

K. Srinivas
e-mail: kasrinu71@gmail.com

G. P. S. Varma
e-mail: gpsvarma@gmail.com

© Springer Nature Singapore Pte Ltd. 2019
S. C. Satapathy et al. (eds.), *Information Systems Design and Intelligent Applications*,
Advances in Intelligent Systems and Computing 862,
https://doi.org/10.1007/978-981-13-3329-3_34

Keywords IoT · Protocol · Gadget-to-Gadget · Application · Hardware
Architecture

1 Introduction

Today latest technology, IoT will become current marketing weapon as shown in
Fig. 2, that has touched every side angle in individual living and enters close to each
corner and affects in customers living place. In 1999, IOT was invented by Ashton
Kevin [1–3]. The key challenge is more interacting to the physical objects by Internet
service, the major foundation components on IOT: things, connections, and Internet
services as described in Fig. 1. Vermesan expresses an IoT like very advanced and
innovatively connected between the environment objects to the digital planet.

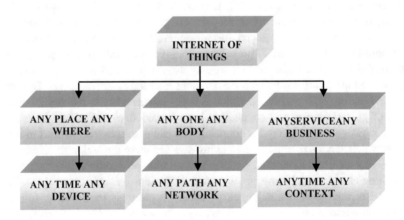

Fig. 1 IOT joins various things from edge to corner of universe

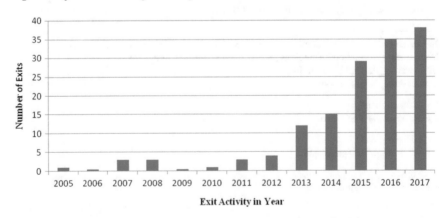

Fig. 2 IOT exit in history

The major intention is avoiding the length among universe gadgets and universe data performance machine objects personally strengthen the organism and growing individual system execution. An electronic environment assisted from the gadgets of the world use many sensors also actuators. Peña-López et al. express the IoT much like a paradigm in which workout, and also net potential are fixed in any sort of feasible things. In universal idiom, the advanced Internet-things means a latest, creative and modern of the globe wherever exactly every gadget and objects utilizes and joins with Internet [4]. Nowadays things on the Internet can connect houses, cities with the vehicle also roads to gadgets for performing and observe the things in valid time. Finally, enterprises must arrange fastidiously and perceive services and issues through IOT design. For that reason, during this survey paper, an examination and listing the IOT operational services that hold by enterprises and what target must concentrate on present decade business fields to widely follow such prepared IOT as shown in Figs. 3 and 4.

Fig. 3 IOT data on machine-to-machine

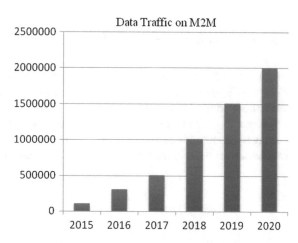

Fig. 4 Size of World IOT data

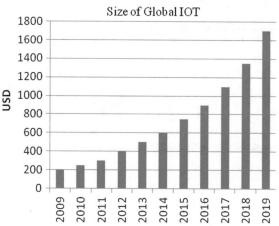

2 IOT Operational Models

IOT strategies are a unit interconnected and operated classified into.

2.1 IOT Gadget-to-Gadget Communication Model

It supplies a direct contact with many devices via shared n/w with independent manner without central service, serves together and collects, spread the information to the across the globe [5, 6]. Functions on gadget-to-gadget model generally in all industries and environments maintain many sensors utilize to detect and indicate different functioning values from force inside, additional effects [7, 8] as described in Fig. 5.

2.2 IOT Gadget-to-Cloud Model

In gadget-to-cloud model, various sensors and devices are directly connected to cloud [9]. To collect travel data from cities using IOT approach used for efficiently controlling traffic [10]. It follows efficient and popular communication tools are Ethernet along with wifi for arrangement connections to cloud [10]. The IP networks eventually communicate the cloud from the cloud resource as shown in Fig. 6. It handles actual humankind things in further distributed and dynamic oriented [11].

Fig. 5 Gadget-to-gadget connecting model

Fig. 6 Gadget-to-Cloud communication model

2.3 IOT Gadget-to-Gateway Model

As shown in Fig. 7, in gadget to the cloud model, various devices are directly connected to the cloud. To collect traveling information from all fields by IOT system intended for efficiently controlling traffic. It follows advanced and innovative communication apparatus are Ethernet beside with wifi to arrange the effective connection to cloud [5]. Wide area n/k eventually communicates the cloud service from the cloud resource. It handles actual humankind things in further distributed and dynamically oriented.

2.4 IOT Back-End-Sharing Knowledge Model

It enlarges one device to cloud (fog) connecting model. Because sensor and gadgets information will consume by real users behind it consumers would export and follow an examination of gadget information on the cloud.

It enlarges one gadget to cloud (fog) connecting model [5]. Because real users behind it would utilize sensor and gadgets data, consumers would export and do analysis on device data from the cloud as shown in Fig. 8.

Fig. 7 Gadget-to-Gateway communication mode

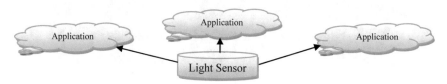

Fig. 8 Back-End-Sharing knowledge model

3 IOT Functions

From those apparatus and mechanization utilize and benefited by business and environment [1]. This part gives analysis on every tool their purposes and uniqueness bring from this model as shown in Fig. 9.

3.1 Control and Examine on Transport and Traffic

Internet-things are capable to react automobile for efficiently communicate and interact as the tidy-oriented manner and presents actual moment non-delay atmosphere also to forecast belongings on vehicles [12]. IOT can employ many things lying on traffic cars, buses, trains and road illumination be assembled by sensors plus RFID to gather traffics information, it can support to decrease the traffic time [13].

3.2 Manufacturing

IOT functions commonly follow in manufacturing and factories for getting automation to organize and handle designing things. IOT can embedded with manufacturing things to supply sequence and intend faultless dataflow during customer requirements. IOT can make free to use manufacture things by using things remotely [14].

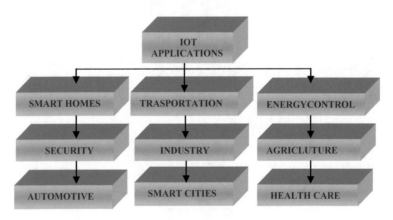

Fig. 9 IOT functions

3.3 Smart (Innovative) Houses

IOT organized its services to manage the house things similar to on/off lights, open/close gates, doors, area heat, etc., by schedule smartly from a remote place [1].

3.4 Public Safety (Protection)

IOT would helpful to direct accident situations, crimes, normal disaster and threats [1].

3.5 Digital (Electronic)-Health Responsibility

In hospital environment Internet-things services would be controlled, examine patients decrease, gather patient physical also inner condition information to compute analysis, investigation and take intelligent decisions will guide, and help for quality treatment by doctors [1]. These Internet-things apparatus installed directly patient body to observe and control their health disorders present different operating systems as shown in Table 1.

4 IOT Architecture

IoT policy includes many levels. It begins at application starting level(stage), it is primarily useful service to clients which deals with apparatus also collect information under smart tools, also this information distribute between sensors, actuator (mover) through their dissimilar network protocols, and lastly the physical level, maintains sensors, mover, controller make linking between things by interconnected service like gateway as shown in Table 2.

4.1 IOT Application Layer

It has accountability to offer efficient service also characterize messages which is transient by this level. The usability through simple graphic-interface (GI) for all customers, like mobile (cellular) phone application which makes IOT application easy [15]. It again follows the main service which responsibility about information spreading and safe.

Table 1 Comparison between different open sources operating systems

	Contiki	RIOT	Tiny OS	Uc Linux	Mbed	Free RTOS
Architecture	Monolithic	Microkernel RTOS	Monolithic	Monolithic	Monolithic	Mixrokernel RTOS
Scheduler	Cooperative	Preemptive Tickless	Cooperative	Preemptive	Preemptive	Preemptive
Model	Event driven Proto thread	More threading	Event driven	More threading	Event-driven Single thread	More threading
Class	Class 0, 1	Class 1, 2	Class 0	>Class2	Class 1, 2	Class 1, 2
Support MCU Families or Vendors	AVR MSO430^{TM} ARM Cortex-M PIC 32,6502	AVR MSO430^{TM} ARM Cortex-M PIC ×86	AVR MSO430^{TM} Px27ax	AVR ARM Cortex-M	AVR ARM Cortex-M	AVR MSO430^{TM} ARM ×86 8052, Reness
Programming languages	C	C, C++	nesC	C	C, C++	C
License	BSD	LGPLV2	BSD	GPLV2	Apache Liense2.0	Modifies GPL

Table 2 Internet-Things architecture

Application layer	Application layer		Application	Business layer
	Middleware layer		Service composition	Application layer
	Coordination layer		Service management	Service management
Network layer	Backbone N/w layer		Object abstraction	Object abstraction
Perception layer	Existing alone application system	Access layer	Objects	Objects
		Edge technology		

4.1.1 Smart (Innovative) Homes

In many houses includes many devices (apparatus) controlling is made by an individual application policy, as mobile (smart) phone application. Smart residences are added with additional financial feasibility and friendly customer interface relatively compare to performance [16].

4.1.2 Automatic

Its latest development in automatic manufacturing can be intended for own driving auto finetunes equipment which will control from remote area. It necessary fast interacts, high security, durability (sturdiness) and trustworthy services [1].

4.1.3 Traffic Administration

Always vehicles size is radically increasing also road jamming will rise in all the globe cities. A advanced smart automatic device needs to diminish accidents on roads. The main intention, which is, control traffic and touch density, cautious actions and urgent situation [17].

4.1.4 Security

IoT apparatus (devices) demands a quick reaction, extremely secure communication, financial achievability and trustworthy services [18].

4.1.5 Microchips

Microchips are extremely ample generally uses various fields as organic given that self-maintaining applications, but this is familiar, size is small for the entire system [19].

4.2 IOT Communication Layer

Which is a more needed channel among number of events in internet-things. Mainly it distributes information to devices. It wishes to intend a useful connection linked to mixed kind of objects (devices) by using protocols. Regarding distance, it follows short, medium and long [19].

4.2.1 Short Distance Communication

Many OSI and TCP/IP suite protocols organize data shared between objects (devices) by WiFi and without wired n/w service on signals to dispense and collect the gadget information and RFID regularly used service to do making and industry things [19].

Table 3 Key technologies in IoT

Key technologies	Performance specification
RFID technology	Noticeable and recognized objects. useful for investing things
Sensor technology	It gives very sensitive data from nature
Embedded technology	It creates things in the IOT encompass and achieve high intelligence devices
Nanotechnology	To respond and interact very quickly to things

4.2.2   Middle Space Communication

Linking several gateway needs communicated protocol on average distance. IBM Desktop and laptop n/w is a model as results of which manages connections linking the related gateways throughout system [19], as shown in Table 3.

4.2.3   Extended-Distance Communication

Connect a terribly sizable quantity of native gateways in an associate extremely heterogeneous environment through multi-model technologies and fully totally different light-weight protocols like 5G [19].

LAN (Medium)

Normally follows LAN man Wan services which are created in 1980 and primary version in 1983 which called it IEEE 802.3 it manages more bitrates also increase distance linking.

Bluetooth (Short)

Like communicating without material contact service for send and receive data on limited distances from various objects (devices) it can help to design PAN implemented by Dutch in 1994 called as Rs 232 IEEE 802.15.

Wireless Fidelity (Short)

Needs for wireless LAN (Local area network) on IEEE 802.11 introducing 1998 IT. The vary of WiFi reaches a hundred meters.

Zig Bee (Short)

This is capable to design a PAN by radio signals for automation, as wireless Network likes IEEE 802.15.4. This made zigbee Alliance 1998 up to 10 to 20-m range.

Frequency Identification Tags (RFID) (Short)

Finds and tracks tags automatically each tag has stored information mostly used for getting automatically recognition also information capture, also it keeps 2000 KB of data working at low frequency, high frequency, and ultra high frequency first made in 1940 commonly used in 1970.

Fifth Generation (5G) (Long)

It brings more throughput, mobility, density, small latency support up to 20 GB per second.

Satellites Networks

Satellites have a high responsibility to build IoT like automation. To offer IoT's through classical facilities those could not take from wireless fidelity-wifi, Bluetooth like wireless networks. Satellite networks handle and focus a complex area, then IoT's are to serve many devices across globe mainly in isolated regions.

4.3 IoT Physical Layer

This is the superfluous depth of generalization in Internet-things. It mostly includes sensors acquires the information to system and mover perform several events also execute commands on devices.

4.3.1 Sensor

Sensation like little object to assist actions on detailed objective amount measuring. Plenty of IoT apparatus (device) support numerous sensors. Those are extremely very essential responsibility in each piece of dealings of individual people life. A massive aspect on Internet-things is not regarding sensible items, though regarding sensors. These small modernizations linked the complete thing from food cups to cement in bridges and in adding to gather and drive knowledge to the cloud. This may possibly permit businesses to collect several precise feedbacks on however product or instrumentations are used, after the break, and still what users would possibly need within the opportunities.

4.3.2 Actuators

Accountable for sending and check a gadget or object. An actuator is an apparatus (devices) which organizes system and take actions in IOT atmosphere.

4.3.3 Controllers

The organizer is a smart innovative apparatus which gathered signals from efficient sensors and make some calculations and perform some computations on, finally, submit instructions to the actuator.

5 IOT Challenges

Understand an (it) is a major goal while this is a straightforward task to issues because it needs concentration on problem likes availableness, responsibility, adjustability, action, scalable, ability, secure, manage, believe [14]. To beat issues allow suppliers, utilize services expeditiously safety also confidentiality play major thing in the market.

5.1 Security

IOT apparatus are needed huge protection to carry on that various apparatus, i.e., devices, sensors linked by Internet so this is additionally complicated to care for protection and confidentiality on Internet-things functions. While designing time security has enormous, challenge to eliminate malicious things [1].

5.2 Privacy

To design extraordinary devices and sensors directly for gathering valuable information for well-organized automation in the actual world, there are various houses and metro cities collecting the data and governed popular aspects, finally secrecy (confidentiality) has to more important significant issue, which is a particular consideration that wants to construct safe data of individuals from coverage in the Internet-things at surroundings in which everything can preserve an individual identifier, now ready to unite autonomously over the n/w [20].

5.3 Interoperability

This is completely interoperable surroundings also, IOT tools and sensors should be communicated and distributed the raw information. This interoperability is enormously long because IOT gadgets (apparatus) are joined to various layers with communicated protocols between devices [5].

6 Conclusion

Internet-things have the latest vision but which is still improved there. The thought to attach the entire thing and everything and anywhere and always is attractive. IOT apparatus accomplish tremendously high notice, for the cause that which touch each place of life. More investigations, analysis, study, and developments are mandatory because technologies and functions are still at starting place in all region. The full paper focuses and covers the outline of features and issues.

References

1. A. Bassam, S. Omar, Internet of Things: an exploration study of opportunities and challenges, in *IEEE 2017 International Conference on Engineering & MIS (ICEMIS)* (2017), pp. 1–4
2. P. Charith, Z. Arkady, C. Peter, G. Dimitrios, Context aware computing for the Internet of Things: a survey. IEEE Commun. Surv. Tutor. **16**, 414–454 (2014)
3. Y. Lobna, K. Ayman, D. Ashraf, Hybrid security techniques for Internet of Things healthcare applications. Adv. Internet Things **5**, 414–454 (2015)
4. Y. Madoka, K. Takayuki, Sensor-cloud infrastructure, in *IEEE 2010 13th International Conference on Network Based Information Systems* (2010), pp. 1–8
5. Y. Rose, S. Eldridge, L. Chapin, The Internet of Things: an overview understanding the issues and challenges of a more connected world, in *The Internet Society* (2015)
6. H. Stephan, M. Carsten, The real-time enterprise: IoT-enabled business processes, in *IETF IAB Workshop on Interconnecting Smart Objects with the Internet* (2011), pp. 1–3
7. O. Bello, S. Zeadally, Intelligent device-to-device communication in the Internet of Things. IEEE Syst. J. **1**, 1–11 (2014)
8. D. Bandyopadhyay, J. Sen, Internet of Things: applications and challenges in technology and standardization. Wireless Pers. Commun. **58**(1), 49–69 (2011)
9. W. He, G. Yan, L. Xu, Developing vehicular data cloud services in the IoT environment. IEEE Trans. Ind. Inf. **10**(2), 1587–1595 (2014)
10. A. Botta, W. de Donato, V. Persico, A. Pescape, Integration of cloud computing and Internet of Things: a survey. Futur. Gener. Comput. Syst. (2015)

11. C. Doukas, I. Maglogiannis, Managing wearable sensor data through cloud computing, in *IEEE Third International Conference on Cloud Computing Technology and Science (CloudCom)* (IEEE, 2011), pp. 440–445

12. S. Narasimha, N. Shantharam, M. Vijayalakshmi, Analysis on IoT challenges, opportunities, applications and communication models. IJAEMS 2(4), 75–97 (2016)

13. T. Zachariah, N. Klugman, B. Campbell, J. Adkins, N. Jackson, P. Dutta, The Internet of Things has a gateway problem. In: *Proceedings of the HotMobile* (2015)

14. Y.-K. Chen, Challenges and opportunities of Internet of Things, in *17th Asia and South Pacific Design Automation Conference* (IEEE, 2012), pp. 383–388

15. B. Ramachandran, IOE-IoT anything connected, in Connectedtechnbiz.wordpress.com (2017)

16. X. Li, R. Lu, X. Liang, X. Shen, J. Chen, Smart community: an Internet of Things application, in *IEEE Communications* (IEEE, 2011), pp. 68–75

17. K. Rob, The real-time city Big data and smart urbanism. GeoJournal **79**, 1–14 (2014)

18. Y. Mehmood, C. Görg, M. Muehleisen, et al., Mobile M2M communication architectures, upcoming challenges, applications, and future directions. EURASIP J. Wirel. Commun. Netw. 243–251 (2015)

19. B. Eldin El-Shweky, K. El-Kholy, M. Abdelghany, M. Salah, M. Wael, O. Alsherbini, Y. Ismail, K. Salah, Internet of Things: a comparative study, in *2018 IEEE 8th Annual Computing and Communication Workshop and Conference (CCWC)* (IEEE, 2018), pp. 622–637

20. J. Nurse, A. Erola, I. Agrafiotis, M. Goldsmith, S. Creee, Smart insiders: exploring the threat from insiders using the Internet-of-Things, in *International Workshop on Secure Internet of Things* (2015), pp. 5–14

Land Use Land Cover Classification Using a Novel Decision Tree Algorithm and Satellite Data Sets

K. V. Ramana Rao and P. Rajesh Kumar

Abstract Since many changes are taking place on the surface of the earth for various reasons such as human activities, natural calamities, and environmental changes, the up-to-date regional land information of an area has got lot of importance for local authorities. Though this information can be collected using traditional land surveying methods it suffers from certain issues such as lot of man power involvement, more time consumption, etc. With the advent of technology the satellite data images available from various optical and polarimetric SAR (Synthetic Aperture Radar) sensors can be used for this land use land cover classification purpose. Since the polarimetric SAR (polSAR) data can provide more useful information than optical data it is more preferable for the land use land cover (LULC) classification. PolSAR data available from various airborne sensors and space-borne sensors which are launched into the space during the recent past can be best utilized for LULC classification of any area irrespective of seasonal effects. Though the available classification algorithms are efficient and produce good classification accuracy results they have their own drawbacks for some reasons. The decision tree algorithm developed for LULC classification in the present study is based on Gumbel distribution of nonlinear regression model [1]. Because of the flexibility in the design and other advantages this decision tree algorithm produced consistent classification accuracy results for all the data sets used in the present study [2].

Keywords Polarimetric SAR data · Classification · Decision tree algorithm
Accuracy

K. V. Ramana Rao (✉)
Department of ECE, VIEW, Visakhapatnam 530046, Andhra Pradesh, India
e-mail: ramanadeepu@yahoo.co.in

P. Rajesh Kumar
Department of ECE, A.U.C.E, Visakhapatnam 530003, Andhra Pradesh, India
e-mail: rajeshauce@gmail.com

© Springer Nature Singapore Pte Ltd. 2019
S. C. Satapathy et al. (eds.), *Information Systems Design and Intelligent Applications*,
Advances in Intelligent Systems and Computing 862,
https://doi.org/10.1007/978-981-13-3329-3_35

1 Introduction

The availability of satellite data sets paved the way for more research work and uti-lizing them for various applications. LULC classification is one such application in which this polSAR data can be used ensuring better results compared to the tradi-tional land surveying methods. The polarimetric SAR data is available in the form of phase data or amplitude data depending on the type of sensor design. Accordingly, suitable methodologies have been developed for processing the available data from the respective sensors. In the present study, both phase data and amplitude data have been considered for the analysis of LULC classification. In addition to the polSAR data optical data also has been considered as a reference for the proposed work. The proposed decision tree classification (DTC) algorithm has been implemented with the polSAR data sets along with the optical data. Co-polarization amplitude data (HH, VV) of ENVISAT-ASAR [3] and cross-polarization amplitude data (VV, VH) of SENTINEL-1 [4, 5] have been preprocessed using SNAP software and classifi-cation has been done using decision tree algorithm. Dual hybrid polarimetric phase data (RH, RV) of RISAT-1 has been decomposed into m-alpha, m-chi, and m-delta components by Raney-decomposition method, using polSARpro software and then classification has been done [6]. ENVI 5.2 version software has been used to perform the classification algorithm of all these data sets including optical data.

2 Visakhapatnam Location Map

In this paper, Visakhapatnam, one of the fastest developing cities in Asia, has been selected as the study area which exists on the sea coast of Bay of Bengal in India. In fact Visakhapatnam city is the head quarters of a district which is also called as Visakhapatnam in the state of Andhra Pradesh. The city is surrounded by Bay of Bengal on eastern side and by land and hills on the other boundaries. The city is basically an industrial area. Entire Visakhapatnam city including the outskirts is selected for classification purpose in the present study. Visakhapatnam city lies between $17°10'-17° 56'$ N latitudes and $83°08'-83° 40'$ E longitudes as shown in Fig. 1. Visakhapatnam is the biggest city in the state of Andhra Pradesh in India.

3 Data Used

In the present, three types of polSAR data sets from three different microwave sensors have been used along with one optical data imagery which was considered as a reference for the purpose of comparing results. Dual right circular polarization phase data with RH & RV of RISAT-1, dual cross-polarization amplitude data with VV & VH of SENTINEL-1 and dual co-polarization amplitude data with HH & VV

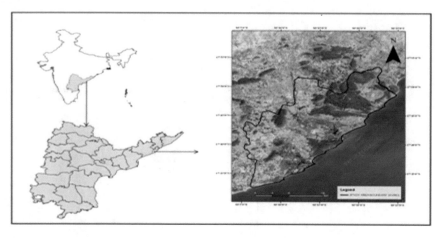

Fig. 1 Visakhapatnam location map

channels of ENVISAT-ASAR of the same study area have been used. Multispectral optical data of LANDSAT-8 is used as a reference data in the present study [7].

4 Decision Tree Classification (DTC) Algorithm

The DTC algorithm is proposed for the analysis of LULC classification based on Gumbel distribution statistical model. In this algorithm, proper information of different features along with their physical parameters is very much required. The objective of the proposed algorithm is to identify the pixels in the image according to the type and nature of various features. The flexibility in designing the decision tree algorithm allows selecting the variables appropriately so that the dimensionality of the data can be reduced for analyzing the data images. This algorithm is easy to understand, analyze, and robust in nature. Its large data handling capacity of both categorical and numerical data types are the added advantages [8]. ENVI software is used in the proposed work to distinguish various threshold values of the available data sets. These threshold values are then used for designing the decision tree with the help of statistical model to perform classification process [8].

5 Accuracy Results and Discussions

Initially, the DTC algorithm has been developed for all the individual data sets. Then the corresponding decision trees and decision tree classified images of all the polarimetric SAR data sets (ENVISAT-ASAR, RISAT-1 and SENTINEL-1) and the optical data (LANDSAT-8) have been generated and are shown along with the color

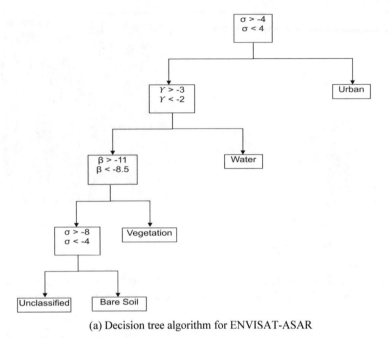

(a) Decision tree algorithm for ENVISAT-ASAR

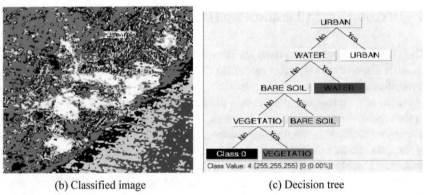

(b) Classified image   (c) Decision tree

Fig. 2 Decision tree classification using the polSAR data of ENVISAT-ASAR

code of different classes in the Figs. 2, 3, 4 and 5 respectively [9]. Classification accuracy results have been calculated with the help of confusion matrix and are mentioned in Table 1.

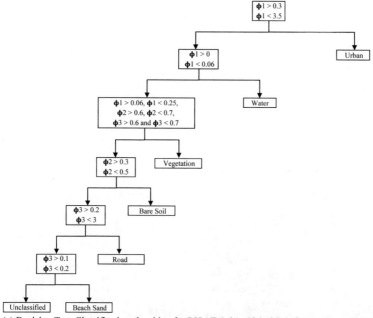

(a) Decision Tree Classification algorithm for RISAT-1 data (ϕ1, ϕ2 & ϕ3 are Even, Diffuse & Odd bounce respectively).

(b) Decision Tree classified Images of RISAT-1 21st October 2014.

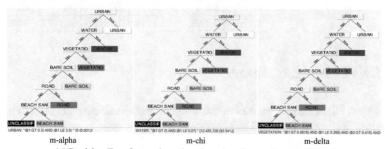

(c) Decision Tree feature branches (based on Raney decomposition)

Fig. 3 Decision tree classification done with RISAT-1data

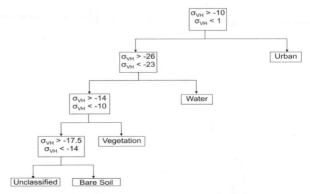

(a) Decision Tree Classification algorithm for the polSAR data of SENTINEL-1

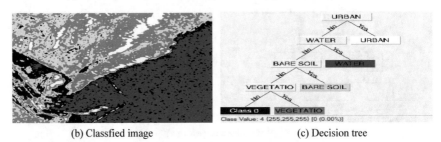

(b) Classfied image   (c) Decision tree

Fig. 4 Decision tree classification done with SENTINEL-1data

Table 1 Overall accuracy (OA) and Kappa coefficient (KC) of different data sets

RISAT-1					
m-Alpha		*m-Chi*		*m-Delta*	
OA	KC	OA	KC	OA	KC
99.743	0.997	96.873	0.962	99.2575	0.9909
ENVISAT-1		SENTINEL-1		LANDSAT-8	
OA	KC	OA	KC	OA	KC
99.729	0.9962	98.1104	0.9742	98.7717	0.983

5.1 ENVISAT-ASAR Sensor

Yellow, Red, Blue, and Green colors are assigned to the selected classes of Urban, Vegetation, Water and Bare soil as shown in Fig. 2b.

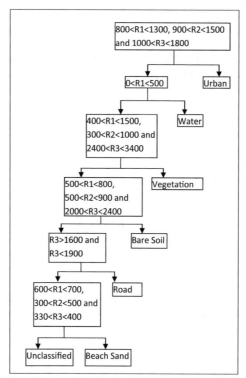

(a) Decision Tree Classification algorithm for LANDSAT-8 data. (Where R1, R2 and R3 are Green, Red and Near Infrared bands reflectance respectively)

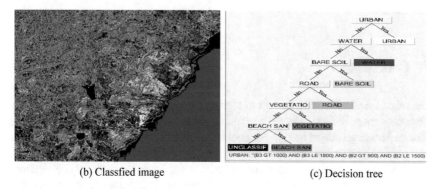

(b) Classfied image   (c) Decision tree

Fig. 5 Decision tree classification of LANDSAT-8 data

5.2 RISAT-1 Sensor

White, Red, Blue, Cyan, Yellow, and Magenta colors are assigned to the selected classes Urban, Vegetation, Water, Beach sand, Bare soil, and Road respectively as shown in Fig. 3b.

5.3 SENTINEL-1 Sensor

White, Red, Blue, and Yellow colors are assigned to the selected classes of Urban, Vegetation, Water, and Bare soil as shown in Fig. 4b.

5.4 LANDSAT-8 Sensor

White, Red, Blue, Cyan, Yellow, and Magenta colors are assigned to the selected classes Urban, Vegetation, Water, Beach sand, Bare soil, and Road respectively as shown in Fig. 5b.

6 Accuracy Assessment

See Table 1.

7 Conclusion

The response of radar with respect to various land cover feature types depends upon the polarization, incidence angle, and frequency of the electromagnetic waves. Hence various parameter values have been checked for deciding the boundary values in order to separate the selected classes. The proposed decision tree algorithm is applied on polSAR data sets of various types and an optical data. Table 1 shows that decision tree classification method is very consistent in giving good accuracy results irrespective of the data type.

References

1. D. Singh, P. Mishra, Y. Yamaguchi, Land cover classification of PALSAR images by knowledge based decision tree classifier and supervised classifiers based on SAR observables. Prog. Electromagnet. Res. B **30**, 47–70 (2011)
2. M.A. Friedl, C.E. Brodley, Decision tree classification of land cover from remotely sensed data. Remote Sens. Environ. **61**, 399–409 (1997)
3. D. Wang, H. Lin, J. Chen, Y. Zhang, Q. Zeng, Application of multi-temporal ENVISAT ASAR data to agricultural area mapping in the Pearl River Delta. Int. J. Remote Sens. **31**(6), 1555–1572 (2010)
4. S. Abdikan, F.B. Sanli, M. Ustuner, F. Calo, Land cover mappping using SENTINEL1-1 SAR data, in *The International Archives of the Photogrammetry, Remote Sensing and Spatial Information Sciences,* vol. XLI-B7 (2016)
5. T. Nagler, H. Rott, M. Hetzenecker, J. Wuite, P. Potin, The SENTINEL-1 mission; new opportunities for ice sheet observations. Remote Sens. **7**(7), 9371–9389 (2015)
6. R.K. Raney, J.T.S. Cahill, G.W. Patterson, D.B.J. Bussey, The m-chi decomposition of hybrid dual-polarimetric radar data with application to lunar craters: M-chi Decomposition of Lunar Craters. J. Geophys. Res. Planets 117(E12) (2012)
7. H.S. Srivastava, Y. Sharma, P. Patel, Feasibility of use of optical and SAR data for land cover mapping, in *Proceedings of ISRS-2008 National Symposium*, 18–20 Dec 2008
8. M. Pal, P.M. Mather, An assessment of the effectiveness of decision tree methods for land cover classification. Remote Sens. Environ. **86**(4), 554–565 (2003)
9. J.S. Lee, M.R. Grunes, E. Pottier, Quantitavie comparison of classification capability: fully polarimetric versus dual and single-polarization SAR. IEEE Trans. Geosci. Remote Sens. **39**(11), 2343–2351 (2001)

Image Denoising Using Wavelet Transform Based Flower Pollination Algorithm

B. V. D. S. Sekhar, S. Venkataramana, V. V. S. S. S. Chakravarthy,
P. S. R. Chowdary and G. P. S. Varma

Abstract Image Denoising is a consistent problem from long period of time and still a challenging task for researchers. There evolved many techniques for image denoising which involves filtering techniques in spatial domain, Transform techniques in transform domain (Sekhar et al. in IRECOS 10(10):1012–1017, 2015 [1]), and more recently evolutionary computing tools (ECT) and genetic algorithms proved more effective in denoising of images. There are many ECT available which can be applied for denoising problem (Sekhar et al. in JGIM 25(4) 2017, [2]). In this paper we made an attempt to Denoise both color and grayscale images by applying a new ECT which emerged out with more efficient results. Peak Signal to noise ratio (PSNR), Structural Similarity Index Metric (SSIM), Mean Structural Similarity Index Metric (MSSIM), etc., are considered in this paper as Image quality Assessment metrics. Comparison of proposed method is also compared with state-of-the-art techniques.

Keywords Image denoising · Evolutionary computing tools (ECT)
Flower pollination algorithm (FPA) · Optimization · Wavelet transforms
Peak signal to noise ratio (PSNR) · Structural similarity index metric (SSIM)
Mean structural similarity index metric (MSSIM)

B. V. D. S. Sekhar (✉) · S. Venkataramana · G. P. S. Varma
Department of Information Technology, S.R.K.R Engineering College, Bhimavaram, Andhra
Pradesh, India
e-mail: bvdssekhar@gmail.com

S. Venkataramana
e-mail: vrsarella@gmail.com

G. P. S. Varma
e-mail: gpsvarma@gmail.com

V. V. S. S. S. Chakravarthy · P. S. R. Chowdary
Department of Electronics and Communication Engineering, Raghu Institute of Technology,
Visakhapatnam, India
e-mail: sameervedula@ieee.org

P. S. R. Chowdary
e-mail: satishchowdary@ieee.org

© Springer Nature Singapore Pte Ltd. 2019

S. C. Satapathy et al. (eds.), *Information Systems Design and Intelligent Applications*,
Advances in Intelligent Systems and Computing 862,
https://doi.org/10.1007/978-981-13-3329-3_36

1　Introduction

Noise inducted during the image reception and transmission lead to disruption of information contained in it. Several conventional techniques are used to Denoise the image and restore the information quality. In the recent past several intelligent techniques based on meta-heuristics, evolutionary computing, soft computing, and nature-inspired methods are proposed to overcome the disadvantages due to limitation of conventional techniques. These techniques are employed in image compression, watermarking, denoising, and other image processing applications [3–13].

In this paper, such an attempt of combining the advantages of wavelet theory and intelligent decision-making on the wavelet parameters using evolutionary computing tool like Flower Pollination Algorithm(FPA) is demonstrated in denoising the image adulterated with Gaussian noise. The general implementation of the FPA using wavelet is incorporated in this methodology is also discussed. Figure 1 depicts a general flow diagram for proposed technique with proposed FPA combined with Discrete Wavelet Transform (DWT). An Objective function is defined to evaluate the obtained result.

Fig. 1 General flow for proposed technique

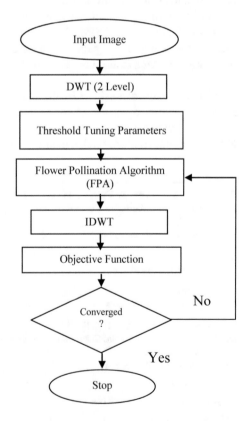

The paper is organized as follows. Description of the FPA algorithm is given in Sects. 2 and 3 discusses the proposed model based on the selected algorithm. This is followed by Sect. 4 with results and discussions. Finally conclusions of the work and possible future expansion are given in Sect. 5.

The experimentation is carried out on MATLAB platform. The image database is taken from standard images from image processing tool box available in MATLAB tool.

2 Flower Pollination Algorithm

The FPA is proposed by Yang [14]. It has been adapted to several engineering applications [15–17]. It typically mimics the pollination process in flowering plants. In optimization problems pollen is considered as a solution i, the corresponding solution vector is given as x_i and the solution space being the whole flower population considered. The two forms of pollination known as local and global pollination are used as local and global search mechanisms respectively. Successful pollination and reproduction yields the most fittest (g_*) known as the global pollination. For updating x_i the following mathematical relation of flower constancy is used:

$$x_i^{k+1} = x_i^k + L\left(x_i^k - g_*\right) \tag{1}$$

The levy distribution (L) is given as

$$L \sim \frac{\lambda \Gamma(\lambda) \sin(\pi \lambda/2)}{\pi} \frac{1}{s^{1+\lambda}} (s \gg s_0 > 0) \tag{2}$$

Similarly for local pollination the following expression is used for updating the x_i:

$$x_i^{t+1} = x_i^t + \varepsilon\left(x_j^t - x_k^t\right), \tag{3}$$

x_j and x_k are pollens from flowers of same plant which effectively represents the local random walk. ε random number with uniform distribution with the range [0, 1].

The fitness formulation uses the computed SSIM index of every individual image in the iteration and further tends to minimize the maximum value of the (1-SSIM) which is also known as dissimilarity index. This is given as follows:

$$f = \text{Min}(\text{Max}(1 - \text{SSIM}(X_N)))$$

Here 'I' refers to the iteration number and N refers to the number of individuals.

3 Proposed Model

The general procedure of optimization involve in performing Wavelet Transform on the image followed by appropriate thresholding. Later, after the wavelet coefficient mitigation using thresholding, the inverse transformation is performed to reproduce the enhanced image.

According to the literature survey it is evident that the threshold function has a major impact on the quality of the denoised or enhanced image. The conventional and fixed threshold cannot produce good results. In order to overcome this disadvantage several function formulations are suggested [18]. The approach in this paper inspired by [18, 19]. However, the algorithm and the corresponding fitness along with the upper and lower bounds of the adaptive threshold parameters are varied for better insight And tested only on available natural images. The technique involves in determining the threshold steering or tuning parameters that produce desired or best denoising coefficient. The adopted process can be explained using the following flowchart in Fig. 2. The evolutionary computing techniques mimic the natural behavior of any biological system or process in the natural world.

The implementation of flower pollination algorithm for the current problem can be demonstrated using the flowchart in Fig. 2. The initial population is a set of randomly generated threshold parameters within the range specified. The random number (r) is used to switch to either global search or local search when compared with the probability of switching. Further the population is updated according to the switched search mechanism. Later, each individual of the population is evaluated and the corresponding fitness is obtained. The process is terminated if problem is converged or maximum number of iterations or generations is reached.

4 Results and Discussions

Experimental results of different images are shown below. Figures 3, 4 and 5 shows input Test image 1 or original Test image 1, noisy image and enhanced or denoised image. Figures 6, 7 and 8 shows color images input or original image, Noisy image and denoised image.

After the implementation of the proposed technique, the analysis is carried out in terms of several parameters like PSNR, MSE, FSIM, and MSSIM. The magnitude of noise is estimated using the first two parameters while the SSIM is used to understand the optical illusion when compared with the noisy image. The performance of the FPA can be evaluated in terms of several algorithm-specific parameters and other statistical parameters like Gbest (global best), Mean of the best, and the variance of the best. The simulation-based experimentation is run for at least 10 times till the image metrics are guaranteed with better values. While evaluating the objective function, in this case the corresponding mean is 110.9729 and the respective variance is 43.6479. The optimal parameters so obtained for this example test image

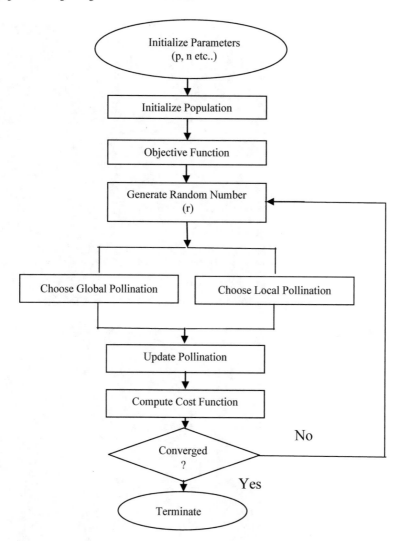

Fig. 2 Proposed method of Image denoising

as produced by FPA are 158.8513, 0.5597, 6.6807, and 2.6925. The simulation of denoising yielded a denoised image with PSNR of 31.4992 and the corresponding MSE is 46.0422. Similarly, the corresponding MSSIM is 0.99494 and FSIM is 0.8498. The respective input test image 1, noisy image with the corresponding denoised images are given in Figs. 3, 4 and 5. Similar Experimentation is carried out on color Lena image and analysis is carried out in terms of same image metrics. The optimal parameters so obtained for this example color image as produced by FPA are 31.5689, 0.7471, 9.9876, and 1.8848. The simulation of denoising yielded a denoised image with PSNR of 36.1682 and the corresponding MSE is 15.7129.

Fig. 3 Input image

Fig. 4 Noisy image

Similarly, the corresponding MSSIM is 0.9992 and FSIM is 0.9662. The respective input color test image, noisy image with the corresponding denoised images are given in Figs. 6, 7 and 8. Table 1 shows experimental results for the performance of Flower Pollination Algorithm on image denoising in terms of different image quality metrics. Experiment was carried out on test image and color images.

Fig. 5 Enhanced image

Fig. 6 Color input image

Fig. 7 Color noisy image

Fig. 8 Color enhanced image

Table 1 Optimal parameters and metrics

Input/metrics	Optimal parameters	Mean optimized	Variance optimized	MSE	PSNR	MSSIM	FSIM
Test image 1	158.8513, 0.5597, 6.6807, 2.6925	110.9729	43.6479	46.0422	31.4992	0.9949	0.8498
Color image	31.5689, 0.7471, 9.9876, 1.8848	110.0730	50.8420	15.7129	36.1682	0.9992	0.9662

5 Conclusion

Novel flower pollination algorithm is applied to denoising problem in image processing successfully with wavelet transformation. The corresponding image metrics like SSIM and MSIM were observed. Excellent magnitude to evaluate the performance of the algorithm. Further, the work can be extended to multi-objective optimization and blind denoising where the reference image is not available.

References

1. B.V.D.S. Sekhar, P.V.G.D. Prasad Reddy, G.P.S. Varma, Novel technique of image denoising using adaptive haar wavelet transformation, in *IRECOS*, 2015, vol 10, No 10, pp 1012–1017 ISSN 1828–6003
2. B.V.D.S. Sekhar, P.V.G.D. Prasad Reddy, G.P.S. Varma, Performance of secure and robust watermarking using evolutionary computing technique. JGIM **25**(4), Article 5 (October–December 2017) https://doi.org/10.4018/jgim.2017100105. Pages 61–79
3. F. Luisier, T. Blu, M. Unser, A new SURE approach to image denoising: interscale orthonormal wavelet thresholding. IEEE Trans. Image Process. **16**(3), 593 (2007). (Biomed. Imaging Group, Swiss Fed. Inst. of Technol., Lausanne)
4. B.C. Buades, J. Morel, On Image Denoising Methods, Technical Report 2004-15, CMLA 2004
5. B.C. Buades, J.M. Morel, A non-local algorithm for image denoising, in *IEEE Computer Society Conference on Computer Vision and Pattern Recognition*, 2005. CVPR 2005, vol 2, pp. 60–65 (2005)
6. N. Azzabou, N. Paragios, F. Guichard, Image denoising based on adapted dictionary computation, in *IEEE International Conference on Image Processing*, 2007. ICIP 2007. pp. III - 109-III -112 (2007)
7. M.R. Bonyadi, Z. Michalewicz, Particle swarm optimization for single objective continuous space problems: a review (2017)
8. A.P. Engelbrecht, *Computational Intelligence: An Introduction* (Wiley, New York, 2007)
9. J. Kennedy, Particle swarm optimization, in *Encyclopaedia of Machine Learning* (Springer, Berlin, 2011), pp. 760–766
10. Y. Shi et al., Particle swarm optimization: developments, applications and resources, in *Proceedings of the 2001 Congress on Evolutionary Computation*, 2001, vol 1. (IEEE, New York, 2001), pp. 81–86

11. Y. Shi, R. Eberhart, A modified particle swarm optimizer, in *The 1998 IEEE International Conference on Evolutionary Computation Proceedings*, 1998. *IEEE World Congress on Computational Intelligence* (IEEE, New York, 1998)

12. R. Eberhart, J. Kennedy, A new optimizer using particle warm theory, in *Proceedings of the Sixth International Symposium on Micro Machine and Human Science*, 1995. MHS '95, pp. 39–43 (1995)

13. R.C. Eberhart, Y. Shi, Comparing inertia weights and constriction factors in particle swarm optimization, in *Proceedings of the 2000 Congress on Evolutionary Computation*, 2000, vol 1, pp. 84–88 (2000)

14. X.-S. Yang, Flower pollination algorithm for global optimization, in ed. by J. Durand-Lose and N. Jonoska *Unconventional Computation and Natural Computation*. vol 7445 of Lecture Notes in Computer Science (Berlin, Springer, 2012), pp. 240–249

15. V. Vedula, S.R. Paladuga, M.R. Prithvi, Synthesis of circular array antenna for sidelobe level and aperture size control using flower pollination algorithm. Int. J. Antennas Propag. (2015)

16. V. Chakravarthy, P.S.R. Chowdary, G. Panda, J. Anguera, A. Andújar, B. Majhi, On the linear antenna array synthesis techniques for sum and difference patterns using flower pollination algorithm. Arab. J. Sci. Eng., 1–13

17. C.S.R. Paladuga, C.V. Vedula, J. Anguera, R.K. Mishra, A. Andújar, Performance of beamwidth constrained linear array synthesis techniques using novel evolutionary computing tools. Applied Computational Electromagnetics Society Journal. pp. 273–278 (ACES JOURNAL, Vol. 33, No. 3, March 2018)

18. A.K. Bhandari, D. Kumar, A. Kumar, G.K. Singh, Optimal sub-band adaptive thresholding based edge preserved satellite image denoising using adaptive differential evolution algorithm. Neurocomputing **174**, 698–721 (2016)

19. A.K. Bhandari, A. Kumar, G.K. Singh, V. Soni, Performance study of evolutionary algorithm for different wavelet filters for satellite image denoising using sub-band adaptive threshold. J. Exp. Theor. Artif. Intell. (2015)

Comparative Analysis of PSO-SGO Algorithms for Localization in Wireless Sensor Networks

Vyshnavi Nagireddy, Pritee Parwekar and Tusar Kanti Mishra

Abstract The Wireless sensor networks (WSN) combine autonomous wireless electronic devices which have abilities like sensing, processing, and communication. It is a self-organizing network constructed with immense number of sensors. Localization is about detecting a node at particular geographical position usually titled as range. Nodes in WSN can be installed uniformly, with formation of grid or randomly. When nodes are installed randomly it is important to determine the exact location of the node. But this approach is expensive and not always feasible using geographical positioning system (GPS). It will not provide definite location results in indoor surroundings. The challenging task of WSN includes improving accuracy in approximating position of a sensor node based on anchor nodes. They are incorporated in a network, such that their coordinates play an essential role in location estimation. A well-organized localization algorithm is capable of determining the accurate coordinates for position of nodes by making reference from sensor nodes. Optimization algorithms like Particle swarm optimization (PSO) and Social group optimization (SGO) are implemented with the fitness equation and the performance of both the algorithms are compared. This paper projects a fitness equation such that the results of PSO and SGO are validated by comparing error accumulation factor in both the algorithms.

Keywords Wireless sensor networks (WSN) · Localization · Anchor nodes
Global positioning system (GPS) · Optimization

V. Nagireddy · P. Parwekar (✉) · T. K. Mishra
Anil Neerukonda Institute of Engineering and Technology, Visakhapatnam, India
e-mail: pritee.cse@anits.edu.in

V. Nagireddy
e-mail: nvyshnavi6@gmail.com

T. K. Mishra
e-mail: tusar.cse@anits.edu.in

© Springer Nature Singapore Pte Ltd. 2019   401
S. C. Satapathy et al. (eds.), *Information Systems Design and Intelligent Applications*,
Advances in Intelligent Systems and Computing 862,
https://doi.org/10.1007/978-981-13-3329-3_37

1 Introduction

Wireless sensor networks are evolving technology in recent times. It is an ecosystem of various wireless apparatus that are positioned to evidence the changes in the network. It is expanding technology with enormous applications in it extending steadily.WSN applications constitute environmental monitoring, health care, household and industry, transport and communication applications [1]. Traditional applications of WSN are motivated to military services named defense advanced research projects age. Research challenges include hardware limitations which affects the overall design of the network. Resource constraints have limitations like low bandwidth, communication disruptions, limited energy, and storage. Specific location of sensor nodes leads to better consequences in the WSN applications but getting precise location of every node in the network is not an easy task. Modern networks are bi-directional. From this rapid development arise the new challenges to overcome as nodes have limited capabilities.

Localization is a phenomenon to evaluate the locality of nodes in a sensor network. Mainly the localization techniques are classified into range-free and range-based. Range-based can be further sorted into GPS, Angle of arrival (AOA), Received Signal Strength (RSS), Time of arrival (TOA). In range-free, there are activities related to connectivity patterns and distances based on hop counts [2].Current localization applications consist of tasks like tracking animals in their habitat, raise and fall of temperature or rainfall, military-based surveillance, smart houses, and smart health care. In wide range of applications related to surveillance and health, delay in detecting exact location of injured victim should be accurate at that particular span. For such exact locations, it is essential to deploy anchor nodes. Location estimations reduce the effort when the location of anchor nodes and their count is known [3]. Consequently a GPS is implemented as it is manageable in equipping it to the nodes due to the advantage of determining exact location. But this does not seem to be cost-effective for enormous nodes in a network. Limitations of using GPS are that some of them consume more battery due to the severity of the application in which they are used and the other limitation is when deployed in thick vegetation, GPS accuracy is affected [4]. Optimization is crucial part in WSN. Optimization is to consider best solution among the feasible solutions. Fitness function for optimization can be of single objective or multi-objective. In single-objective optimization scheme the optimizer either minimizes or maximizes one objective. In multi-objective optimization, as name suggests, multiple objectives are optimized simultaneously [5]. Nature-inspired algorithms can be categorized as if inspired from swarm, it can be termed as Particle swarm optimization (PSO), if inspired by a colony it is called as ant colony optimization. Similarly, algorithms like Genetic algorithm (GA), Social group optimization (SGO) are the other familiar algorithms that are in use. In this work, fitness equation which is proposed by authors in [6] is compared with SGO to check the performance of it. Figure 1 shows randomly deployed nodes in a region of interest and once the localization is done they can form a network for communication.

Fig. 1 Localization in
Region of Interest (ROI)

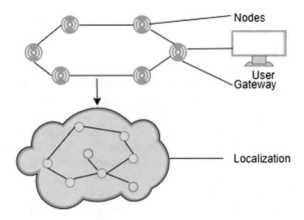

This paper has literature survey which is explained in Sects. 2 and 3 illustrates the two optimization methods PSO and SGO, Sect. 4 envisions simulation, results and conclusion is analyzed in Sect. 5.

2 Literature Survey

Majority of the smart environments today are established using wireless sensor networks. Enormous surveys are presented on localization in diverse applications of WSN. Recent developments in WSNs are portable sensors of low power, low cost. Due to the uncertainties in determining the exact location of sensors, localization is important. The primary goal in this paper is to have an efficient localization technique in finding a sensor at specific area and thereby to reduce the error in estimating the locations of sensors. Typically, to have the exact location of a node is tough for fields which make use of sensitive information through their applications [7]. So in this paper comparison is made and determines which method, either PSO or SGO, is best suited for the localization with the minimum error. In the previous research a work has been done in [6] where nodes are located using algorithms like PSO and Artificial bee colony (ABC). Multi stage localization is introduced in that work and is compared with single stage localization. A general distance equation is used such that distance between actual and anchor nodes are calculated. For localization problem, an objective function is developed with predefined count of anchor nodes.PSO and ABC algorithms are improved with respect to the current objective function. In multi-stage localization, an iterative process is performed such that all the unknown nodes are localized.

In [8] position estimation is achieved using PSO algorithm and Quasi-Newton approach for a better value ranging methods and localization model to find the error occurred in estimating the nodes but the accuracy in estimating location is not considerable due to the ranging errors. Initially PSO is used to select initial location and then after finite iterations Quasi-Newton gives the final result. By using this particle swarm optimization Quasi-Newton (PSO-QN) method, localization accuracy is improved.

In [9], three distinct algorithms are projected iDV-Hop1, iDV-Hop2 and Quad DV-Hop, as advancement for the existing Distance vector hop (DV-Hop) algorithm. Proposed algorithms have additional steps which are compared with traditional DV-Hop algorithms. Results were given for different scenarios iDV-Hop2 shows the finest performance, compared to DV-Hop and improved DV-Hop in entire scenarios due to its minimum error in localization. In the scenarios like uniform random and gridy, iDV-Hop2 is preferred compared with iDV-Hop1 and Quad DV-Hop. The iDV-Hop1 has its benefits in scenarios with uneven topologies; this algorithm minimizes localization error likely three times when compared to DV-Hop and improved DV-Hop. In all scenarios, Quad DV-Hop presents better results in comparison to existing DV-Hop algorithm.

In this work, PSO and SGO are compared to check their performance to enhance the location accuracy such that it helps to minimize the localization problem.

3 Optimization Methods

As this paper deals with the comparison of PSO and SGO algorithms, the idea behind the work is to check the new optimization technique that is SGO can it be used for localization efficiently in place of PSO. Sections 3.1 and 3.2 gives a brief introduction of algorithms.

3.1 Particle Swarm Optimization

PSO is nature-inspired computational method used for optimization. This works to improve the candidate solution and measured quality using iterative approach. Swarm can be defined as a group that moves in large number with similar behavior examples is flock of birds; school of fish. PSO technique in WSN artificially mimics the behavior of swarm. Design principles of such algorithms are inspired from complex biological approach [10]. PSO technique works efficient in optimizing the routing phenomenon in a network. In PSO at initial stage initialize the particles and determine their position and velocity. For each particle, the proposed fitness value is computed. The algorithm then tracks the personal best and global best that have to be noted for particular particle. Global best is set by considering the best among the local best of each candidate. Positions and velocities are updated accordingly. Fitness is again calculated with respect to the updated values and this process is repeated till the desired solution is achieved. Stop the procedure if the termination criterion is met. The cost of objective function represents the particle locations which are evaluated [11].

3.2 Social Group Optimization

SGO is population-based algorithm [12]. There exist many behavioral qualities in human beings which enables an individual to fix complex tasks. Not too many of them are able to solve their problems effectively and efficiently. When it comes to complex problems, which is solved by taking the reference of other individuals in a group, with this concept SGO is projected [12]. In SGO, the population is nothing but person (candidate solutions). This technique includes each candidate solution (person) with some sort of knowledge which results in solving a problem. SGO has mainly two phases improving and acquiring phase. The first phase concentrates on the each person knowledge and is improved with the impact of best person who has the high knowledge within a group. In the second phase every person strengthens their knowledge by interacting with another person in a group. Through this, knowledge levels of all individuals in a group increases [13]. The individual best is represented by considering the one which has best fitness.

3.3 PSO and SGO for Node Localization

Commercial applications of WSNs consider localization a predominant factor. There exist some applications where sensors are affixed to objects, animals or human beings which are moving. It is important to locate such sensors in applications related to public safety, health, security, and surveillance. Localization is used to provide the location information of the unknown nodes in particular network. The goal of this work is to make use of effective node localization algorithm which reduces the error accumulation and improve the accuracy in determining the location [14]. For the comparison of PSO and SGO the simulate environment has two categories like anchor nodes and unknown nodes. The location based calculation can be done in the following manner:

This paper has assumed m anchor nodes followed by n undetermined nodes in the network using the information of m-n anchor nodes position. The unknown nodes position is determined by usage of known coordinates. Assume the coordinates of anchor nodes is denoted by $l = [l_x, l_y]$; where $l_x = \{x_1, x_2, \ldots, x_n\}$, $l_y = \{y_1, y_2, \ldots, y_n\}$, are to be projected using the coordinates of anchor node $\{x_{n+1}, \ldots, x_n\}$ and $\{y_{n+1}, \ldots, y_m\}$. As these approximations are done, there exists error in estimating nodes. Sum of squared range errors is represented using d_k is given in Eq. (1), which is the difference between the nodes and the anchor nodes which are in range considered as objective function. kth anchor node coordinates are assumed as (x_k, y_k) and the coordinates of the unknown nodes are (x_m, y_m).

$$d_k = \sqrt{(x_k - x_m)^2 + (y_k - y_m)^2} \tag{1}$$

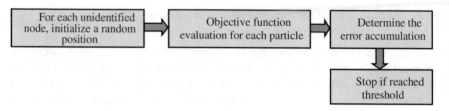

Fig. 2 Workflow diagram

The location estimation of specific coordinates of unknown nodes is an optimization problem, which involves minimizing an objective function representing error in locating the target nodes. The objective function [6] explained in Eq. (2), can be formulated as

$$f(x, y) = \frac{1}{M} \left(\sum_{i=1}^{m} d_k - \hat{d}_k \right) \tag{2}$$

where, $\boxed{\hat{d}_k}$ be the value of d_k obtained from the noisy range measurements. M is the count of anchor nodes that are fixed where $M \geq 3$. Localization error is calculated from the objective function where the difference of actual position and estimated position is considered to be the error. To obtain location accuracy the parameter considered in this work is the error obtained. The error value should be reduced such that the performance of the network can be improved. Figure 2 explains the workflow of the entire process.

4 Simulation and Results

In this paper, the criterion is to utilize the existing localization based objective function which is applied for algorithms like PSO and SGO to minimize the error and improve the network accuracy. Localization error is calculated by considering the difference between actual position and predicted position of the nodes in network. Through these aspects, WSN network lifetime is improved hence algorithm's performance is validated with MATLAB where error is considered as a vital parameter, in order reduce it algorithms like PSO and SGO are implemented and simulations are done using the objective function. Nodes are randomly deployed in 100 m × 100 area. Anchor node count is a predefined value which will be altered accordingly to have a detailed performance comparison of PSO and SGO. Error is determined for 500 nodes with 1000 iterations. In Fig. 3, the plot represents error occurred for 1000 iterations in SGO. In Fig. 4, the plot represents error for 1000 iterations in PSO. For an explicit contrast in projecting error, overall average is calculated and

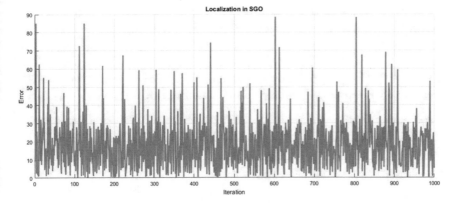

Fig. 3 Plot representing error in SGO

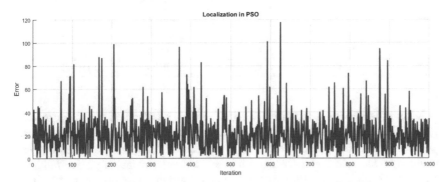

Fig. 4 Plot representing error in PSO

envisioned which is shown in Fig. 5. It is seen that PSO outperforms SGO. Considering optimization concept in localization, the objective function outcome should be minimum. To improve network performance, computational complexity should be avoided hence the error accumulated in locating a node must be minimal. From Fig. 5, it is clear that PSO performance is considerably best rather SGO.

From Table 1, it is observed that error is minimized with respect to the increase in percentage of anchor nodes in a specific area. On min max error comparison between PSO and SGO it is clear that PSO is best rather SGO. Analysis from Fig. 4 and Table 1, shows that PSO performs better than SGO.

5 Conclusion

In this work, location-ased error estimation is observed. Localization is one among the various factors that restrict the regular execution in WSN. Minimizing the error

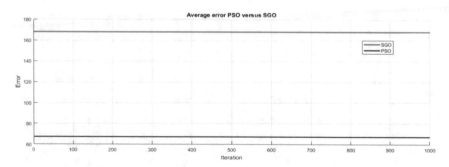

Fig. 5 Comparative plot PSO versus SGO

Table 1 Performance evaluation for different % of anchor nodes

Algorithm	Iterations	Anchor node %	Min error	Max error
PSO	1000	5	0.01330	112.6320
SGO	1000	5	0.0806	118.0425
PSO	2500	10	0.0068	77.7877
SGO	2500	10	0.0123	119.9541

in locating the nodes is the primary criteria in any localization based approaches. Initially PSO algorithm is applied then the objective function is evaluated for each particle until the termination criterion is met. Similar steps are followed for SGO then the performances of these algorithms are reviewed by making use of minimum number of anchor nodes. Effective choice of algorithm is also one of the vital tasks for obtaining location accuracy hence according to this work, optimized results are observed for PSO. Testing these algorithms using the objective function resulted in minimum value for PSO than SGO. Hence PSO is better compared to SGO in terms of error accumulation. PSO is preferred in terms of localization as the objective function outcome is observed to be minimum in PSO. Future work can be extended to propose only one optimization method which will be commonly applied for both localization and clustering of nodes to improve the network life with the minimum errors.

References

1. L. Cheng et al., A survey of localization in wireless sensor network. Int. J. Distrib. Sens. Netw. **8**(12), 962523 (2012)
2. V. Nagireddy, P. Parwekar, A survey on range-based and range-free localization techniques (2017)
3. P. Singh et al., A novel approach for localization of moving target nodes in wireless sensor networks. Int. J. Grid Distrib. Comput. **10**(10), 33–43 (2017)
4. J. Kuriakose et al., A review on localization in wireless sensor networks, in *Advances in Signal Processing and Intelligent Recognition Systems* (Springer, Cham, 2014), pp. 599–610

5. M. Iqbal et al., Wireless sensor network optimization: multi-objective paradigm. Sensors **15**(7), 17572–17620 (2015)
6. D. Lavanya, S. Udgata, Swarm intelligence based localization in wireless sensor networks, in *Multi-Disciplinary Trends in Artificial Intelligence*, pp. 317–328 (2011)
7. G. Han et al., Localization algorithms of wireless sensor networks: a survey. Telecommun Syst **52**(4), 2419–2436 (2013)
8. J. Cao, A localization algorithm based on particle swarm optimization and quasi-newton algorithm for wireless sensor networks. J. Commun. Comput. **12**, 85–90 (2015)
9. S. Tomic, I. Mezei, Improvements of DV-Hop localization algorithm for wireless sensor networks. Telecommun. Syst. **61**(1), 93–106 (2016)
10. A. Datta, Nandakumar S., A survey on bio inspired meta heuristic based clustering protocols for wireless sensor networks, in *IOP Conference Series: Materials Science and Engineering*, vol 263. No. 5. (IOP Publishing, 2017)
11. A. Gopakumar, L. Jacob, Localization in wireless sensor networks using particle swarm optimization, pp. 227–230 (2008)
12. S. Satapathy, A. Naik, Social group optimization (SGO): a new population evolutionary optimization technique. Complex Intell. Syst. **2**(3), 173–203 (2016)
13. A. Naik et al., Social group optimization for global optimization of multimodal functions and data clustering problems. Neural Comput. Appl. pp. 1–17 (2016)
14. C.-S. Shieh et al., Node localization in WSN using heuristic optimization approaches (2017)

Inclusive Use of Digital Marketing in Tourism Industry

Gunjan Gupta

Abstract With the advancement of Internet technologies and accessibility of the Internet, new marketing avenues have emerged thus facilitating the development of Internet marketing. It is evident that digital marketing is a high potential component of any marketing plan incorporated by successful businesses to increase their outreach and achieve desirable results. The customers have immediate access to all sorts of information like images and videos of destinations thereby highlighting the USP of that product through visual appeal and influencing the decision of customer toward the desired destination. The tourism companies are promoting their brand and reaching their potential customer by providing best travel deals through various digital technologies thus influencing the whole tourism industry. This paper discusses and analyzes the components and significance of digital marketing in tourism industry in contemporary times. It also explores the influence of digital marketing in tourism sector and indicates the further issues which need to be addressed.

Keywords Digital marketing · Social media · Tourism industry · Websites
E-mail marketing

1 Introduction

The rapid advancements in digital ecosystems, which include Internet technologies and social media platforms like Facebook, Twitter, etc. coupled with accessibility and affordability of electronic gadgets have transformed the way people communicate and share information in today's world. The Internet has emerged as a very powerful marketing tool which has revolutionized the way product and brand marketing is being undertaken by companies. Companies are now adopting advanced technological business models which help them to keep pace with current digital trends.

G. Gupta (✉)
Noida, India
e-mail: gunjangupta182@gmail.com

© Springer Nature Singapore Pte Ltd. 2019
S. C. Satapathy et al. (eds.), *Information Systems Design and Intelligent Applications*,
Advances in Intelligent Systems and Computing 862,
https://doi.org/10.1007/978-981-13-3329-3_38

411

Nowadays, most of the marketing strategies adopted by travel companies definitely involve some or other digital platforms.

Digital presence has become essential for any company in order to deliver the most significant information to their target audience, attract new visitors, and develop long-term relationships with their customers. Businesses and retailers are delivering information and promoting their products and services in an efficient and cost-effective manner to attract target audience, influencing their decisions thus increase their customer base. Digital marketing also helps organizations to track and analyze user behavior on their company website, social media, and other digital platforms in real time in terms of reach, engagement, and conversion. This tracked data provides opportunity to the organizations to analyze statistics on user behavior, their requirements and thus enabling the firms to comprehend the needs of the consumer which was unprecedented in earlier times. By analysis of customer data on user requirement or say, satisfaction surveys, travel companies modify their deals and packages and present new or improved offers which would meet the user requirement and expectations. Technology has further enabled automation in data analysis which provides the right triggers to the firm management in almost real time without human intervention.

The Internet has given a new dimension to the manner information that is being disseminated by companies and perceived by the people around the world. Irrespective of demography, region, location, any and every person can access, consume, and engage with entire gamut of information at their convenience due to Internet explosion and digital revolution. This availability of enormous information coupled with easy accessibility to wide-ranging products and services across various digital platforms has brought about a considerable change in the purchasing behavior of the consumer which invariably has affected the way the travel companies promote their products.

The Internet has become a remarkable channel for both marketers and customers to connect. Marketers use various marketing strategies to entice and attract customers by providing information about their best deals and offers. The customer is also able to interact with diverse brands, access massive information about various products offered by brands which impacts their buying behavior. The customer buying decision will be totally dependent on which brand is able to satisfy his needs and requirement and bestow him with maximum benefit in minimum cost. Thus, digital marketing has influenced the overall marketing strategies and contributed toward attaining the business goals by the brands and also giving the optimum opportunities to the consumer to take the ideal purchase decision.

2 Literature Survey

Sivasankaran [1] talks about how the Internet and other electronic commerce technologies have transformed the entire economy and new business models, revenue streams, customer bases, and supply chains have emerged in every industry. Travel industry and ticketing have seen a drastic change in the last decade.

Pitana and Pitanatri [2] say that customer behavior has drastically changed in tourism industry due to the advancement of Internet technology and digital marketing. Digital marketing is the latest trend and future of every industry, tourism industry has realized the benefit and incorporated digital marketing with their offline marketing which is still used. The online travel agents progress is incredibly high, both in terms of number and in terms of volume. There are a number of significant benefits in using digital marketing, since these characteristics cannot be found in offline marketing, i.e., more global, yet at the same time more personal.

Kaur [3] proposed that digital marketing is a significant channel for every business regardless of its category and area of expertise. It has completely transformed the way; businesses advertise their products and services to their visitors and potential customers. Digital marketing plays a crucial role in the success of every industry and the tourism industry has been influenced predominantly because of easy access to lot of information related to best deals and offers available to the consumer.

Batinić [4] says that businesses have realized the importance of online marketing and are implementing diverse marketing strategies to attain optimum results. Customers are increasingly getting the required information about various offers on holiday packages through the Internet and for marketing professionals; the Internet has become a very significant success factor in the tourism industry.

Avinash et al. [5] proposed that with the evolution of Internet and expansion in the number of Internet users, consumer is provided with huge amount of information in the virtual space. Upgradation in the social status of the people has resulted in the inclination toward travel and tourism. Travelers are consistently looking for latest information and updates on travel deals, and digital marketing has made the content easily searchable and accessible. Tourist industry and brands are consistently doing online marketing by providing lot of details to reach and capture consumer. The paper evaluates the effect of digital marketing on travel industry and provides suggestions to handle several issues and challenges faced by tourism service providers.

Bolek and Papińska-Kacperek [6] say that tourism is rapidly growing in European industry. Availability of the Internet has significantly increased the opportunities and options available to consumers for traveling and staying outside their usual environment. The paper aims to verify whether IT tools support an individual tourist in formulating his comprehensive travel projects in Poland by themselves. They also establish whether consumers browsing for travel services in Poland are aware of various IT tools and are able to utilize these tools to devise their travel plan.

Stone and Woodcock [7] discussed how social media can be used as a very effective channel during customer buying cycle. The authors suggest that a complete social media marketing strategy must be developed to exploit the capabilities of this channel effectively for promoting products or connecting with customers. The companies must also keep track of the effects of social media on their business, accordingly formulate a strategy and schedule every activity to achieve maximum benefits from social media.

As Hays et al. [8] put it, social media is expanding day by day and has attained popularity as one of the prominent elements of destination marketing organization (DMO) marketing strategy. DMO exploits social media to reach their audience globally within a limited budget. The aim is to study the DMOs of top 10 countries and explore how these social media channels are utilized by the tourist globally and analyze the impact of social media marketing strategies.

Bradbury [9]: The conference talks about the global shift of tourism industry toward social media and how social media has become the vital component in tourism marketing. Social media is the premier channel to interact with consumers, boost promotion, gain popularity, increase customers, their length of visits, and number of return visits.

3 Objectives of the Study

The paper has four main objectives as follows:

To recognize how digital marketing and its components is influencing the tourism industry.

To determine the impact of various components of digital marketing on consumer's traveling decision.

To understand and analyze data on how tourism sector has escalated due to digital marketing.

The factors on which further study needs to be done.

4 Digital Marketing and Tourism Industry

Digital marketing has substantially transformed the manner in which tourism industry is reaching out to customers, by providing plethora of information through different channels and reaching out to their potential customers. Nowadays, travel companies have realized the significance of digital marketing and enhancing their digital presence, i.e., website which is a very simple and cost-effective mode of dissemination of information about the availability of innumerable travel destinations for vacations or for conducting seminars, conferences, variety of holidays packages offered by them, varied customized options concerning restaurants with different food options, categories of hotels, and tour guides to their consumers. To attract potential customers and keep their customer base on their online platform, companies must consistently provide reliable and updated information and website navigation should be effortless to enable the customers to skim through the website pages seamlessly for a user-friendly experience.

Before the emergence of digital marketing, consumer had limited information at their disposal which was accessible via various offline marketing channels like print media newspaper, magazines, electronic media like television/radio advertising, and to a much smaller extent through outdoor media and word of mouth.

Internet and digital marketing have made worldwide information easily accessible to users around the world which enables consumer to search for locations easily and undertake overview of available options regarding exotic destinations which they had never heard of. The consumer can visit numerous travel websites, with a click of button to research; compare abundant options, select options as per their requirement and budget prior to taking decision related to holiday destinations and bookings. Customers can also compare and evaluate varied prices offered by different companies, which hotel to stay, significant places to visit, best restaurants, events happening in that location during their visit, and other relevant information about their planned destination. So, now customers are better informed through information available on the Internet and thus are not dependent on travel agencies for location information rather they are able to obtain excellent alternatives and options in their budget according to their requirements.

4.1   Social Media and Tourism

Marketing in tourism industry has been greatly influenced with the emergence and popularity of social media. Every company is analyzing and implementing varied social media marketing strategy to exploit the maximum benefit of this channel for brand visibility and growth.

Social media is one of the most significant and cost-effective platforms of digital marketing for tourism sector as this medium provides opportunity to companies to know the user profile of their customers and also help companies to understand the user requirements. Companies must treat their customers with respect by providing relevant and trustworthy content and be honest about every detail so that the customers remain loyal to them. Social media is not only about promoting content and offers but also having a conversation with their users so that whenever the customer is actually thinking of availing the services or product, you are the first to be considered by the customer. Brand recall plays significant role in attracting potential customer to the company website. The quality of the posted content should add value to visitor's lives thus increasing the engagement rate of the visitor with your content.

Today's travelers love to share their experiences and opinions. And social media has facilitated this by expanding people's ability to reach a wider audience than ever before with minimal effort and cost. Social media acts as the most prominent knowledge sharing platforms through which users are sharing their travel experience, valuable feedback, writing reviews, and sharing images which actually attract huge audience. This social sharing has the most profound effect on the tourism industry and worldwide travelers are using third-party review sites such as TripAdvisor as a source for planning holiday.

Digital marketing is a very powerful marketing tool for tourism industry to reach avid travelers who are eagerly searching for adventurous destinations to travel. Facebook and Twitter are most effective and useful social media platform for tourism industry and act as a source of bringing traffic to your website.

People determine their travel plans by referring to various websites, tour aggregators, blogs, comments, and social media channel to find best brand with a positive reputation and reviews. The customers buying decision solely depends on the brand identity and people reviews about that company. These crowd-sourced review sites display genuine feedback and ratings from consumers and industry experts, which actually inspires travelers to take informed decision.

4.2 E-mail Marketing and Tourism

E-mail marketing is the best performing channel for ROI and companies can leverage this channel for best returns. People start planning their trips in advance and when they are still in the process of taking decision, companies can use e-mail marketing to provide immense information to woo customers and by extending profitable deal and offers, they are actually tempting customers to go for vacations even if they are not planning to. Companies can encourage customers to go for tour by providing details of new destinations, variety of activities in which they can engage in, and even travel tips for perfect travel experience.

Companies can create an effective e-mail marketing strategy to target travelers by reaching the people at the right time with their best offer. E-mail marketing is a tool to establish a direct communication with prospects and customers, nurture their clients by delivering tailor-made proposals keeping in mind their convenience and satisfaction. The offer should be interesting and convincing enough that the customer should make a purchase without even comparing it with competitors. Segmentation and profiling of the prospects and customers should be done by tourism companies to target them systematically by providing personalized information to the specific person. Tourism companies should regularly send newsletters filled with various travel destination details, travel tips, flight alerts, travel reminders, vacation deals, high resolution, and attractive images and videos to establish themselves as an expert in the tourism industry.

5   Findings

These figures show that people use the Internet and various digital platforms before taking any decisions regarding their holiday destinations, mode of booking, expenses, hotels to stay, places to visit, etc.

Almost 40% of all tour and activity bookings are being made online (Rezdy data), and 60% of leisure and 41% of business travelers are making their own travel arrangements, generally via the Internet (Amadeus). Internet travel booking revenue has grown by more than 73% over the past 5 years and more than 148.3 million people use the Internet to make reservations for their accommodations, tours, and activities (Statistic Brain) [10].

Leisure travelers take at least 1–2 vacations in a year and 30% feel if they are getting good promotional offer would take a trip even if there were not planning. 95% of respondents read an average of 6–7 reviews and spend 30 min before booking (Tnooz). 70% of travelers look at up to 20 reviews in the planning phase (Tnooz) and 85% of consumers read up to 10 reviews before they feel that they can trust a business. The most important sources that influenced the decisions of global travelers were travel review websites (69%), online travel agencies (57%), and tour operator sites (56%) (Tourism Research Australia). 70% of global consumers say online consumer reviews are the second most trusted form of advertising (Edelman).TripAdvisor has 260 million unique monthly visitors and 125 million travel reviews and opinions from travelers around the world (TripAdvisor) [10]. 92% of consumers say that they trust earned media, such as social media, word of mouth, and recommendation from friends and family, above all forms of advertising (Webbed Feet) [11].

82% of consumers trust a company more if they are involved with social media (Forbes) [11]. 36% of online travelers visit social networking sites as main source of travel ideas and inspiration and 52% of travelers use social media to plan a trip (Eye for Travel) and were so influenced by social media that they even changed their original plan. 80% of travelers are more likely to book a trip from a friend liking a page rather than responding to a traditional Facebook ad (Eye for Travel) [10].

6 Analysis

Based on the available data, it is evident that travel industry has grown drastically in last 5 years due to online marketing. People feel that digital platforms are convenient way of finding detailed and updated information and are easily accessible. Most travelers refer to TripAdvisor and compare various travel websites before taking decisions. People read lot of reviews before planning any trips and their decisions get affected by their family and friends response. The word of mouth publicity, either through social media like Facebook or Instagram, contributes the maximum toward travel decisions. Increasing middle class with disposable income in the second and third world countries have also largely contributed toward boosting the travel industry. A media-aware traveler looks for deals and therefore promotional offers would have a better impact on their travel decisions.

The study of the statistics does not factor the demographic distribution of the travelers, their financial robustness, their age profile, and their professional affiliations. For example, a creative professional like photographer, artist, interior decorator, etc. would like to invest his income to travel to nonurban locales and acquire his inspirations from offbeat destinations. Similarly, younger travelers would prefer destinations that could be accessed in shoestring budgets as well as having theme parks and adventure sports. This study would form the part of the future project.

To summarize, the weightage factors for leisure travel (out of 10) could be as follows:

Accessibility to the destination: 4,
Cost of travel: 7,
Conventional destinations: 6,
Offbeat destinations: 4,
Age of the traveler: 8,
Destination safety/security: 9, and
Professional affiliation: 5.

7 Conclusion

By this study, we can establish that digital marketing is unavoidable for the progress and development of tourism industry and every firm should implement digital marketing strategies to achieve maximum returns. The tourism industry, to have a competitive edge over others should explore this growing technology and implement all the techniques to attract potential customers and convert them into long-lasting clients. Companies must enhance user experience on their website by delivering useful and relevant information and should keep user personal data secure and confidential to generate trust. Tourism industry should invest their efforts on social media as customer decision is solely dependent on other people's experiences, feedbacks, and reviews. E-mail marketing is also an effective channel for having a direct interaction with the customers and companies can track and analyze user behaviors on the website through analytics platform to recognize the precise requirements and modifications can be done accordingly. Initially, digital marketing can be nerve wracking but with consistent efforts, companies can acquire new customers and by providing good customer service, they become their leading online advertisers by giving excellent reviews and feedback.

References

1. S. Sivasankaran, Digital marketing and its impact on buying behavior of youth. Int. J. Res. Manage. Bus. Stud. **4**(3) (2017)
2. G. Pitana, P.D.S. Pitanatri, in *Digital Marketing in Tourism: The More Global, The More Personal* (2016)
3. G. Kaur, The importance of digital marketing in the tourism industry. Int. J. Res. GRANTHAALAYAH **5**(6) (2017)
4. I. Batinić, The role and importance of internet marketing in modern hotel industry. J. Process Manage. New Technol. Int. **3**(3), 34–38 (2015)
5. B.M. Avinash, S. Harish Babu, B. Megha, Digital marketing—its impact on travel and tourism industry in India. J. Tourism Hospitality Sports **21** (2016)

6. C. Bolek, J. Papińska-Kacperek, Modern IT tools supporting individual tourist projects. Eur. Sci. J. September, Special edition vol. 2. ISSN: 1857–7881 (2010)
7. M. Stone, N. Woodcock, Social intelligence in customer engagement. J. Strateg. Mark. **21**, 394–401 (2013)
8. S. Hays, S.J. Page, D. Buhalis, Social media as a destination marketing tool: its use by national tourism organisations. Curr. Issues Tourism **16**, 211–239 (2013)
9. K. Bradbury, The growing role of social media in tourism marketing. Retrieved from http://kelseybradbury.weebly.com
10. Travel Statistics for Tour Operators. Retrieved from www.redzy.com
11. J. Urlaub, Building trust through social media for sustainability is more critical than ever (2014). Retrieved from http://taigacompany.com

Design of Differential Amplifier Using Current Mirror Load in 90 nm CMOS Technology

Payali Das, Suraj Kumar Saw and Preetisudha Meher

Abstract In this article, a differential amplifier with a moderate gain of 40.56 dB is achieved. The UGF of 46.985 dB and phase margin of 84.15 degrees with a low power consumption of 61.084 uW. It is designed by using 90 nm CMOS technology in CADENCE VIRTUOSO platform by applying a supply voltage of 1.2 V with an ICMR of −296.098 mV–1.158 V. The post-layout simulations are also carried out and the results are being compared to the pre-layout results. Finally, a comparison is also drawn with some recent works.

Keywords Differential amplifier · UGF (unity gain Frequency) · Post-layout CMRR (Common mode rejection Ratio)

1 Introduction

Many electronic devices use differential amplifiers internally. It is used as a series negative feedback circuit by using op amplifier. Generally, we use differential amplifier that acts as a volume control circuit, also it can be used as an automatic gain control circuit. Differential signaling areas like driving or receiving signals over long cable lengths, driving a balanced ADC input, or buffering complementary output DACs having differential amplifiers as one of the main building blocks. Differential amps have innate common mode rejection properties, have excellent output gain and phase matching, and provide low harmonic distortion. In analog domain, broad application of diff-amps include products that are optimized for precision, high speed, and low

P. Das · S. K. Saw · P. Meher (✉)
Department of Electronics and Communication Engineering, National Institute of Technology, Arunachal Pradesh, Yupia 791112, India
e-mail: preetisudha1@gmail.com

P. Das
e-mail: payalidas26@gmail.com

S. K. Saw
e-mail: srjsaw98@gmail.com

power. In analog signal processing at high frequencies, CMOS differential amplifiers are very efficient in current days. They are comprised of current mirrors which are easy to design and also very flexible to implement. By substituting a current mirror (diode voltage source biasing a current source transistor), this amplifier has a stable biasing situation.

2 Literature Survey

Power and operating temperature are those parameters which are very important in circuit design, to maintain their correctness circuit designers must focus on power efficiency. A low power fully differential current-starving amplifier is designed in this paper to meet all these specifications. For energy efficiency purpose, the inverter-based differential amplifier is one of the best circuit topology [1]. Conventional CMOS differential amplifiers have lower DC gain than that of the diff-amp circuits with positive feedback. But there are two significant drawbacks which are higher noise and lower unity gain bandwidth [2]. The lower performance of a circuit is due to the present noise in it. Therefore, high-performance circuits require methodology to minimize noise in the circuit. In this paper, an HPSO-based differential amplifier has been designed to meet the low noise specification [3]. Op-amps, using gain-boosted circuit could enhance the whole op-amp's gain in case of unaffecting the frequency characteristics of main op-amp [4]. By using resistor and input-impedance boosting technique, a 1 V wide-input range and highly linear transconductor has been proposed in this paper. Afterwards, an op-amp and a DDA were implemented by using the transconductor [5]. Rail-to-rail operation of bootstrapped switches allows robust switched-capacitor designs in standard CMOS technologies with very low voltage. A total harmonic distortion of—24 dB is achieved [6]. A fully differential amplifier has been proposed by using fully balanced operational amplifier. These op-amps work with both differential mode and common mode signals having a high gain [7]. Three stage differential low noise amplifier has been proposed with a 10 dB differential mode gain [8]. A 21 GHz ultra wide band differential low noise amplifier has been presented in this paper. Only 20.76 mW power has been consumed by this amplifier. Noise frequency of the LNA is 4.4 dB at the center frequency [9]. Fully differential folded cascode op-amp is used for achieving large output swing. The disadvantage is that it uses common mode feedback circuit in its topology [10]. Darlington pair and internal circuit biasing technique are used to enhance the slew rate as well as gain and unity gain bandwidth. This circuit achieves a slew rate of 2791 V/μs, gain of 70 dB and unity gain bandwidth of 1.74 GHz [11].

Fig. 1 Proposed differential amplifier

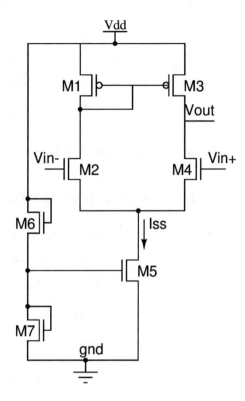

3 Proposed Design Methodology

3.1 Proposed Design

Here, in this design, we use current mirror load. To operate in a good stability operation all the transistors are placed in saturation region. Transistor M_1 and M_3 are sized such that they maintain perfect current division through this. We provided the input from transistor M_2 and M_4. The current I_{ss} is the summation of both the source currents of M_2 and M_4. The transistors M_6 and M_7 are used as the current reference circuit and give the proper biasing voltage to M_5 and a load capacitance of 1 pF is driven at the output to perform at better performance (Figs. 1 and 2).

We draw the small signal model only for the half differential amplifier circuit, as the circuit is properly symmetrical in both sizing and current point of view. Let us assume a point X from where the current I_{ss} is taken and assume it as a virtual ground.

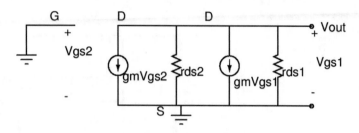

Fig. 2 Small signal model of proposed differential amplifier

3.2 Mathematical Expressions Used

$$\text{Output resistance, } R_{\text{out}} = r_{ds3}//r_{ds4}. \tag{1}$$

$$\text{Voltage gain, } A_v = g_m R_{\text{out}} = g_m(r_{ds3}//r_{ds4}). \tag{2}$$

$$\text{Slew rate, } \text{SR} = \frac{dV_{\text{out}}}{dt} = \frac{I_0}{C_l} \tag{3}$$

$$\text{Power dissipation, } P_{\text{diss}} = (V_{dd} + |V_{ss}|)I_0 \tag{4}$$

$$\text{Transconductance, } g_m = \frac{\delta I_d}{\delta V_{gs}} \tag{5}$$

$$\text{UGB} = \frac{g_m}{2\pi C_l} \tag{6}$$

Thermal noise of differential amplifier,

$$\bar{V}^2_{n,in=8\text{KT}} \frac{2}{3} \left[\frac{1}{2K'_2\left(\frac{W}{L}\right)_1 I_{D1}} + \frac{\sqrt{2K'_3\left(\frac{W}{L}\right)_3 I_{D3}}}{2K'_2\left(\frac{W}{L}\right)_1 I_{D1}} \right], \tag{7}$$

where K = Boltzman Constant = $1.3807 \times 10^{-23} \text{JK}^{-1}$
T= Temperature
Frequency response ($f_{-3\text{dB}}$), determination of R_{out},

$$f_{-3\text{dB}} = \frac{1}{R_{\text{out}} C_l} \tag{8}$$

4 Result and Discussions

4.1 Transient Response

Figure 3 represents the transient response at 1 GHz frequency and output is obtained. Figure 4 shows the gain phase plot in which 40.4695 dB and phase margin of 84.5165° and of UGB of 46.985 MHz are obtained, and Fig. 5 demonstrates the gain, phase and UGB plot at different corner is obtained. Figure 6 represents the output noise plot which is 400.43 nV/sq Hz.

Figure 7 represents the layout of the proposed circuits which has an area of 78.966 um^2 is obtained. Figure 8 shows the pre- and post-layout simulation of gain UGB and phase plot which has very least variation of 0.06 dB gain. Figures 9 and 10 show the power at pre-layout and Post-layout of 61.084 μ and 60.891 μ. Figure 11 shows the ICMR range of −296.098 mV to 1.1587 V (Tables 1, 2 and 3).

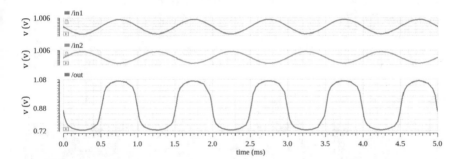

Fig. 3 Transient response

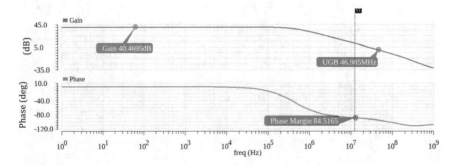

Fig. 4 Gain phase plot

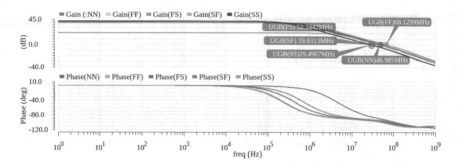

Fig. 5 Gain phase corner analysis

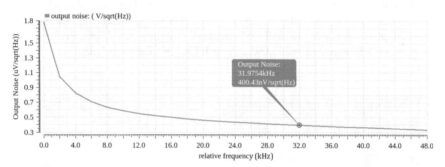

Fig. 6 Output noise plot

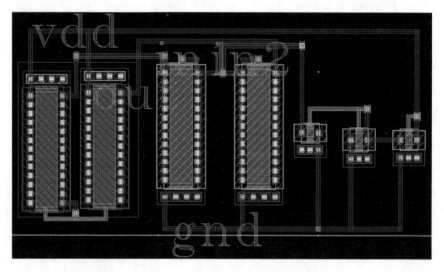

Fig. 7 Power plot of pre-layout

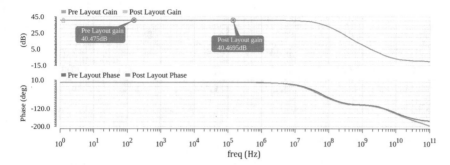

Fig. 8 Layout design of proposed circuits

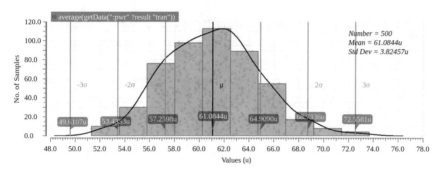

Fig. 9 Gain phase pre-post-layout

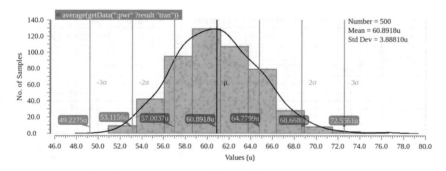

Fig. 10 Power plot of post-layout

5 Conclusion

The proposed circuit consumes a very low power and moderate gain, the circuit removes the current mirror topology by applying current mode technique by using perfect biasing the stability is being achieved. A wide-input common mode range is obtained by applying a low supply voltage. This proposed circuit has been used as reference circuits in the charge pump applications designed.

Table 1 Corner analysis of different parameter

Parameter	NN	FF	FS	SF	SS
Gain (dB)	40.47	39.3083	21.0	41.5	42.023
UGB (MHz)	46.985	68.1299	52.3442	29.8313	29.49
Phase margin (Deg)	84.516	84.0577	89.438	85.8149	84.87

Table 2 Performance summary

Parameter	This work
1. Technology (nm)	90
2. Power supply(V)	1.2 V
3. Gain (dB)	40.46
4. Phase margin (deg)	84.51
5. UGB (Hz)	46.985 MHz
6. C_l (pF)	1
7. Output noise	521.9 nV/$\sqrt{\text{Hz}}$
8. Slew Rate	67 V/μs
9. Power	60.88 μWatt
10. ICMR ($-$)	-1.2 V
11. ICMR (+)	$+1.2$ V
12. Layout area (μM)	78.966

Table 3 Comparison table of the proposed differential amplifier with other works

References	[8]	[13]	[2]	This work
Technology (nm)	180	90	NA	90
Power supply (V)	NA	1.2	1	1.2
Gain (dB)	10	11	50.8	40.46
Phase margin (deg)	NA	NA	111	84.51
UGB (MHz)	NA	NA	NA	46.985
Load cap (pF)	NA	NA	NA	1
Output noise (nV/$\sqrt{\text{Hz}}$)	NA	NA	223 μV	521.9
Slew rate (V/μs)	NA	NA	NA	67
Power consumption (Watt)	25.6 m	84 m	10.8μ	61.084μ
ICMR	NA	NA	NA	0.296–1.258 V
Layout area (m^2)	0.55 m	0.15 m	NA	78.966 μ

Fig. 11 ICMR plot of proposed circuits

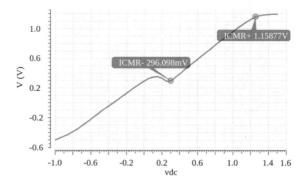

References

1. W. Wilson, T. Chen, R. Selby, in *A Current-Starved Inverter-Based Differential Amplifier Design for Ultra-Low Power Applications* (IEEE, 2013)
2. P.T. Tran, in *Operational Amplifier Design with Gain-Enhancement Differential Amplifier* (IEEE, 2012), pp. 6248–6253
3. C.L. Singh, A.J. Gogoi, C. Anandini, K.L. Baishnab, in *Low-Noise CMOS Differential-Amplifier Design Using Automated-Design Methodology* (IEEE, 2017), pp. 326–330
4. S.M.S. Rashid, A. Roy, S.N. Ali, A.B.M.H. Rashid, Design of A 21 GHz UWB differential low noise amplifier using .13 μm CMOS process, in *ISIC* (2009), pp. 538–541
5. E.K.F. Lee, Low-voltage op amp design and differential difference amplifier design using linear transconductor with resistor input. IEEE Trans. Circ. Syst. II Analog Digital Signal Proc. **47**(8), 776–778 (2000)
6. T.-S. Lee, H.-Y. Chung, S.-M. Cai, Design techniques for low-voltage fully differential CMOS switched-capacitor amplifiers, in *ISCAS* (2006), pp. 2825–2828
7. E.M. Spinelli, M.A. Mayosky, C.F. Christiansen, Dual-mode design of fully differential circuits using fully balanced operational amplifiers. IET Circ. Devices Syst. **2**(2), 243–248 (2008)
8. H.-K. Chiou, H.-Y. Liao, K.C. Liang, Compact and low power consumption K-band differential low-noise amplifier design using transformer feedback technique. IET Microwave Antenna Propag. **2**(8), 871–879 (2008)
9. A. Pandey, S. Chakraborty, V. Nath, Slew rate enhancing technique in Darlington pair based CMOS op-amp. ARPN J. Eng. Appl. Sci. **10**(9), 3972–3973 (2015)
10. L. Wang, Y.-S. Yin, X.-Z. Guan, in Design of a gain-boosted telescopic fully differential amplifier with CMFB circuit (IEEE, 2012), pp. 252–255
11. N. Bako, Ž. Butkovi, A. Bari, Design of fully differential folded cascode operational amplifier by the gm/ID methodology, in *MIPRO* (2010), pp. 89–94
12. D. Guckenberger, K.T. Kornegay, Design of a differential distributed amplifier and oscillator using close-packed interleaved transmission lines. IEEE J. Solid-State Circ. **40**(10), 1997–23007 (2005)
13. T. LaRocca, J.Y.-C. Liu, M.-C.F. Chang, 60 GHz CMOS amplifiers using transformer coupling and artificial dielectric differential transmission lines for compact design. IEEE J. Solid-State Circ. **44**(5), 1425–1435 (2009)

A Survey on On-Chip ESD Protection in CMOS Technology

Lukram Dhanachandra Singh and Preetisudha Meher

Abstract Nowadays, electrostatic discharge (ESD) protection is an important part of the fast I/O applications. In recent years, the number of ESD damages to ICs has considerably increased because of huge energy dissipation in an exceptionally small time in terms of the nanosecond. As the number of damages increase, the proposal of methods and designs for ESD protection by researchers has been increased subsequently. As this research area is undeveloped and still unexplored deeply, with this paper we intend to present a prearranged and broad outline of the study on ESD protection in Complementary Metal–Oxide–Semiconductor technology. This paper's surveys present a review on on-chip ESD protection designed in favor of the applications of high-speed I/O using CMOS technology. The solutions to overcome different issues caused due to ESD are also discussed for on-chip ESD protection in Complementary Metal–Oxide–Semiconductor technology.

Keywords Electrostatic discharge (ESD) · ESD protection circuits · CMOS
Silicon-controlled rectifier (SCR) · High-speed I/O

1 Introduction

With the enhancement of inexpensive and high integration ability, many ICs have been made-up using CMOS process, together with high-speed I/O interface circuits. And for CMOS ICs, electrostatic discharge (ESD) is one among many severe reliability concerns and it is defined as the transmission of electrostatic charges among the objects having dissimilar electric potentials. ESD protection design for CMOS radio frequency ICs [1] was proposed, which used stacked polysilicon diodes to over-

L. D. Singh
Department of ECE, NIT Manipur, Imphal, India

P. Meher (✉)
Department of Electronics and Communication Engineering, National Institute of Technology,
Arunachal Pradesh, Yupia 791112, India
e-mail: preetisudha1@gmail.com

© Springer Nature Singapore Pte Ltd. 2019
S. C. Satapathy et al. (eds.), *Information Systems Design and Intelligent Applications*,
Advances in Intelligent Systems and Computing 862,
https://doi.org/10.1007/978-981-13-3329-3_40

come some limitations such as, low parasitic capacitance, steady input capacitance and avoided the noise coupling from the common substrate. It is also included that a turn-on well-organized power-rail electrostatic discharge clamp circuit, designed using substrate-triggered technique significantly increased the overall ESD level of RF ICs.

An on-chip ESD protection device in CMOS IC [2] was also proposed to solve two convenient issues of latch-up induced by transient and high switching voltage, which is based on silicon-controlled rectifier (SCR). To decrease switching voltage of SCR-based device, a customized device structures using circuit techniques assisted by trigger was introduced. It also solved the latch-up issue by rising triggering current or holding voltage. For reliable technology scaling and high-frequency CMOS ICs, above concern is one of the problems to be resolved with main concern.

The remaining part of this paper is constructed with the sections, which give the descriptions of various ESD protection methods, solving different ESD problems proposed by different researchers. Later the comparisons of the methods will be presented and finally, we present the conclusions of this paper in the last section.

2 ESD Protection Methods

Various ESD protection designs and methods have been proposed by many researchers. For a better discussion of these methods that are developed or designed for ESD protection, brief descriptions of some selected methods are explained as follows.

2.1 Diode-Triggered SCR (DTSCR)

A diode-chain triggered silicon-controlled rectifier (DTSCR) for ESD protection [3] was presented for applications having less signal or supply voltage less than or equal to 1.8 V and very constricted ESD design limits. Advanced technologies to solve two general problems for ESD protection design: 1. SiGe HBT RF-ESD protection in BiCMOS, 2. Ultra-thin Gate oxide protection in CMOS-0.09u was also discussed.

For latching the SCR in ESD difficulty condition, the DTSCR triggered by a chain of the diode is used at the adequate insertion of trigger current into the gates of the SCR. Figure 1 shows various triggering methods of DTSCR.

2.2 Active-Source Pump (ASP) Technique

In advanced sub-90 nm, CMOS technologies for the significant reduction in effect of ESD for ultra-thin gate inputs, an active-source pump (ASP) circuit technique [4]

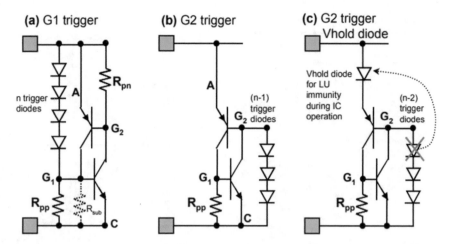

Fig. 1 **a** G1-triggered DTSCR forward bias of SCR G1-Cathode junction. In CMOS-SCR an intrinsic connection to the substrate is present, **b** G2-triggered DTSCR forward bias of G2-Anode junction, **c** G2-triggered DTSCR, Vh diode for LU immunity replaces trigger diode

was introduced. At the certain stage where ESD causes stress, the MOS transistors source potential was managed using ASP technique to bound the effective voltage at the gate by activation of the ASP circuit at less than the greatest acceptable voltage across the thin gate oxide, i.e., V_{max}.

Figure 2 represents the example considered for the illustration of general ASP idea (left) for an uncomplicated input inverter for primary I/O ESD protection and (right) ASP circuit using a chain of diode in combination with a poly source resistor. And for noise control of resistor for instance in an RF-LNA, it used a parallel source capacitance C_s (optional).

2.3 Combination of DTSCR and ASP Scheme

An ESD protection approach for ultrasensitive input–output applications containing thin GOX [5] was presented, merging low-voltage DTSCR clamps [3] that let a competent voltage clamping with an ASP circuits technique applied for ultra-thin gate oxide safety [4]. By injection of a small ESD current in the source impedance, a significant improvement was achieved, which efficiently dropped the most significant gate-source ESD voltage.

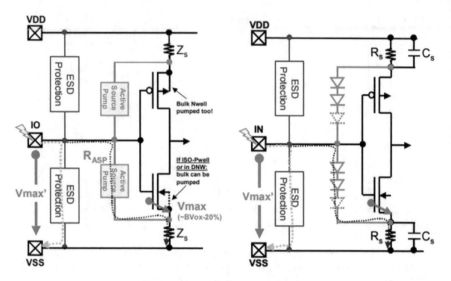

Fig. 2 An scheme for Active-Source-Pump technique with ASP circuit and source impedance for input inverter stage (left). Specific ASP implementation with diode-chain ASP and source resistor (right)

2.4 Stacked Diodes with Embedded Silicon-Controlled Rectifier (SDSCR)

With the purpose of reducing the parasitic capacitance existing in ESD protection diodes excluding the sacrifice of robustness in ESD, the diodes arranged in stacked are used for ESD protection. The stacked diodes embedded with SCR (SDSCR) have already designed and it was informed to be helpful for ESD protection providing high turn-on speed, low parasitic capacitance, and good robustness for ESD [6].

In CMOS technology for I/O applications with high speed, this SDSCR-based electrostatic discharge (ESD) protection circuit [7] was proposed in which a couple of primary SDSCR (SD$_{P1}$ and SD$_{N1}$), a couple of secondary SDSCR (SD$_{P2}$ and SD$_{N2}$), and a resistor (R_{ESD}) in series were included as represented in Fig. 3. For all ESD-test pin arrangements, the whole chip ESD protection was achieved by this ESD protection design.

2.5 Self-Matched ESD Cell in CMOS Technology for RF Applications

For the RF applications of 60 GHz broadband in CMOS technology, cell library of self-matched ESD protection [8] was presented to decrease the complication of design for RF circuit providing proper security from ESD. For the implementation of

Fig. 3 ESD protection
circuit with SDSCR

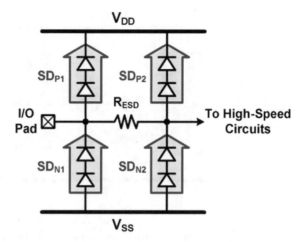

ESD cells, it used the protection scheme of distributed ESD. Four ESD cells having dissimilar ESD robustness were provided by this ESD cell library. These four ESD cells were intended to maintain 1/2, 1, 3/2, and 2 kV human body model (HBM) ESD tests.

With an on-chip spiral inductor, a low C pad and a couple of ESD diodes, the 1-stage ESD protection was intended which is used in 0.5 and 1 kV cells circuit diagram as shown in Fig. 4a. With two on-chip spiral inductors, a low C pad and two couples of ESD diodes, the 2-stage ESD protection was designed which is used in 1.5 and 2 kV cells circuit diagram as shown in Fig. 4b. On above of it, the power-rail ESD clamp circuit is included in all ESD cells to offer ESD current paths from V_{dd} to V_{ss}. 50 Ω of I/O matching was achieved with these ESD cells.

2.6 Reduced Capacitance and Overshoot Voltage ESD Protection Structure

For fast applications, dual diodes embedded SCR (DDSCR) [10] were introduced as a protection element at the input pin which is connected directly to the transistors input gate through an RC circuit. Use of stacked diodes has the drawbacks of increasing complexity in the optimization of the parasitic of interconnect metallization and comparatively high on-state resistance. Figure 5 represents a DDSCR ESD protection circuit.

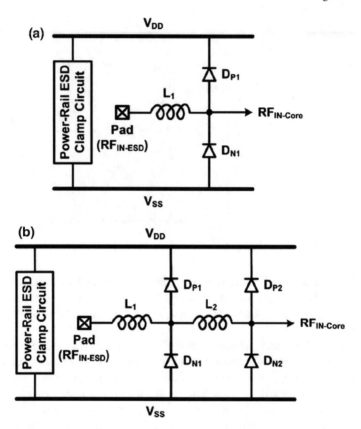

Fig. 4 ESD cells with **a** 1-stage ESD protection and **b** 2-stage ESD protection circuit diagrams

2.7   High Robustness and Low Capacitance SCR for High-Speed I/O ESD Protection

A low capacitance clamp having good robustness for on-chip ESD protection [9] was presented. Doping n-well/p-well junction lightly, it justified the capacitance coupled and hence, the low capacitance was obtained. It also optimized the same structure of an SCR for forward conduction processes and a diode for reverse conduction processes to achieve the high robustness for ESD. It gives an improved ESD performance with much lesser parasitic capacitance than the ordinary SCR and a low capacitance SCR.

Fig. 5 DD-SCR ESD
protection circuit

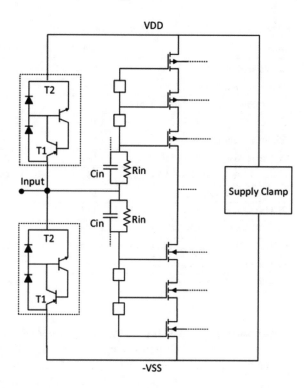

3 Comparisons and Analysis of the Results

As various methods have been proposed by many researchers and now we will see their performance in this session.

Initially, the diodes parasitic capacitance, traditional SCR parasitic capacitance, diodes trigger SCR (DTSCR) parasitic capacitance and stacked diode SCR (SDSCR) parasitic capacitance for some ESD design was observed as in Table 1.

And we observe that stacked diode SCR (SDSCR) is having lowest parasitic capacitance among the diode, traditional SCR, DTSCR, and SDSCR.

Table 2 gives the comparison of clamp1, clamp2, and clamp3, where clamp1 represents the ordinary SCR like low-voltage SCR, clamp 2 represents the improved SCR and clamp 3 for the reference low capacitance clamp.

For the study of robustness under the human body model in ESD protection of Clamp1 and Clamp2, transmission line pulsing (TLP) measurements were observed.

Table 1 Diode, traditional SCR, DTSCR, and SDSCR parasitic capacitance

Device	Diode	LSCR	DTSCR	SDSCR
Capacitance (fF)	550	543	443	278

Table 2 Comparison of Clamp1, Clamp2, and Clamp3

Parameter	Clamp1	Clamp2	Clarnp3[9]
Device area (μm^2)	50 × 10	50 × 10	45 × 20
Failure current-forward (A)	1.2	1.55	1.0
On-state resistance-forward (Ohm)	5.0	4.5	4.8
Trigger voltage-forward (V)	10.5	8.1	8.5
Failure current-reverse (A)	1.2	2.0	1.5
On-state resistance-reverse (Ohm)	13.0	2.0	2.5
VFTLP voltage overshoot (V) @ 1A	25.2	20.5	22
VFTLP turn-on time (ns) 1A	1.36	1.0	1.2
VFTLP voltage overshoot (V) @ 2A	33.6	27.6	NA
VFTLP turn-on time (lis) @ 2A	1.44	0.96	NA
VFTLP voltage overshoot (V) @ 3A	42.8	35.8	NA
VFTLP turn-on time (ns) @ 3A	1.4	1.1	NA
Capacitance @ 0 V bias (fF)	335	94	95

Post-stress leakage currents were found to be 3.6 V which is normal operating voltage (3.3 V). For forward direction, Clamp2 has a less significant trigger voltage and larger I_f than Clamp1 and Clamp3. For the reverse direction, it is observed that Clamp2 got lesser on-state resistance and a larger I_f than Clamp1 and Clamp3 where I_f is the failure current.

Among these three devices, considering its exceptional characteristics in parasitic capacitance, current handling ability, turn-on time, voltage overshoot and trigger voltage, Clamp2 is best for radio-frequency ESD protection.

Device A: STI—bounded DD Silicon-Controlled Rectifier
Device B: Single-diode Silicon-Controlled Rectifier
Device C: Stacked diode Silicon-Controlled Rectifier
Device D: Junction-bounded Dual Diode SCR with $D = 0.4$
Device E: Junction-bounded Dual Diode SCR with $D = 0.2$
Device F: Junction-bounded Dual Diode SCR with $D = 0.09$
Device G: Metal-bounded Dual Diode SCR.

Later it was also observed in Table 3 that the STI-bounded Dual Diode Silicon-Controlled Rectifier has the least capacitance than the stacked diode, single diode, and metal-bounded Dual Diode SCRs. Junction-bounded Dual Diode SCRs found to have the great robustness of ESD protection with less on resistance, R_{ON}, large I_f and less overshoot voltage [10].

Table 3 Measurement results of tested devices

Device	0 V 100 kHz	TLP 600 ps 100 ns			VFTLP 100 ps 5 ns	VHBM/C	$C \times R_{ON}$
	Capacitance (fF)	Trigger V (V)	R_{ON} (Ω)	It_2 (A)	Overshoot (V) @ 1 A	V/fF	fF * Ω
A	12.1	1.88	2.97	1.19	9.9	147.5	35.9
B	40.0	0.99	1.83	1.11	4.49	41.6	73.2
C	14.3	1.70	3.16	1.24	13.2	130.1	45.2
D	19.8	1.89	1.37	1.42	4.98	107.6	27.1
E	17.4	1.89	1.28	1.40	4.55	120.7	22.3
F	20.6	1.92	1.55	1.38	5.2	100.5	31.9
G	22.2	1.81	1.12	1.42	5.06	95.9	24.8

4 Conclusions

With the rapid evolution of ICs in reducing the feature size, high integration capability, less power consumption and improving frequency characteristics, for the implementation of the ICs, CMOS technology has been used. In nanoscale CMOS technology, the robustness of ESD in ICs seriously degrades. And so, the on-chip ESD protections have been discussed in this paper.

Various works on ESD protection have been surveyed and we can conclude that for ESD protection, we must consider all factors like low parasitic capacitance, the robustness of ESD protection in ICs, switching voltage, and transient-induced latch-up, along with reduction in complexity and faster I/O applications. So, we can go for a better ESD protection for ICs in Complementary Metal–Oxide–Semiconductor technology by modifying some of the parameters in already proposed methods and combining ASD with DDSCR for ESD protection.

References

1. M.D. Ker, C.Y. Chang, ESD protection design for CMOS RF integrated circuits using polysilicon diodes. J. Microelectron. Rel. **42**(6), 863–872 (2002)
2. M.D. Ker, K.C. Hsu, Overview of on-chip electrostatic discharge protection design with SCR-based devices in CMOS integrated circuits. IEEE Trans. Device Mat. Rel. **5**(2), 235–249 (2005)
3. M. Mergens et al., Diode-triggered SCR (DTSCR) for RF-ESD protection of BiCMOS SiGe HBTs and CMOS ultra-thin gate oxides, in *IEDM* (2003), p. 515
4. M. Mergens et al., Active-source-pump (ASP) technique for ESD design window expansion and ultra-thin gate oxide protection in sub-90 nm technologies, in *CICC* (2004), p. 251
5. M. Mergens et al., ESD protection circuit design for ultra-sensitive IO applications in advanced sub-90 nm CMOS technologies, in *Proceeding of IEEE International Symposium Circuits Systems* (2005), pp. 1194–1197
6. R. Sun, Z. Wang, M. Klebanov, W. Liang, J. Liou, D. Liu, Silicon-controlled rectifier for electrostatic discharge protection solutions with minimal snapback and reduced overshoot voltage. IEEE Electron Device Lett. **36**(5), 424–426 (2015)

7. J.T. Chen, C.Y. Lin, R.K. Chang, M.D. Ker, T.C. Tzeng, T.C. Lin, ESD protection design for high-speed applications in CMOS technology, in *Proceeding of International Midwest Symposium Circuits Systems* (2016), pp. 1–4

8. C.Y. Lin, L.W. Chu, M.D. Ker, T.H. Lu, P.F. Hung, H.C. Li, Self-matched ESD cell in CMOS technology for 60-GHz broadband RF applications, in *Proceeding of IEEE Radio Frequency Integrated Circuits Symposium* (2010), pp. 573–576

9. Q. Cui, J.A. Salcedo, S. Parthasarathy, Y. Zhou, J.J. Liou, J.J. Hajjar, High-robustness and low-capacitance silicon-controlled rectifier for high-speed I/O ESD protection. IEEE Electron Device Lett. **34**(2), 178–180 (2013)

10. A. Dong et al., ESD protection structure with reduced capacitance and overshoot voltage for high speed interface applications. Microelectron. Rel. **79**, 201–205 (2017)

Efficient VLSI Implementation of CORDIC-Based Multiplier Architecture

Rashmita Baruah, Preetisudha Meher and Ashwini Kumar Pradhan

Abstract Multipliers have always played an important role in the field of VLSI architectures. The speed latency, area, and cost are the main encounters in designing a processor. Multiplication process takes a large amount of time and hence tries to reduce the speed of the processor. In this paper, an effort has been made to reduce the latency of a processor by implementing a multiplier less architecture. Volder introduced the CORDIC algorithm for computing trigonometric relationships. Using CORDIC, multiplier-less architectures can be obtained. The analysis is carried out using Xilinx ISE Design suite 14.2.

Keywords CORDIC · Analog resolver · Vedic

1 Introduction

In 1959, Volder proposed an iterative algorithm for computation trigonometric operations. He named the algorithm as CORDIC algorithm [1]. This algorithm was a digital equivalent for an analog resolver. This iterative algorithm uses a simple shift and adds operation. In the year 1962, J.E. Meggit introduced an effective processor for a faster pseudo multiplication and pseudo division [2]. Later Walther reformulated the CORDIC algorithm for elementary functions in 1971 [3]. Most of the DSP architectures are based on mathematical approaches. In the comparison to addition

R. Baruah · P. Meher (✉)
Department of Electronics and Communication Engineering, National Institute of Technology, Arunachal Pradesh, Yupia 791112, India
e-mail: preetisudha1@gmail.com

R. Baruah
e-mail: rashmita.baruah@gmail.com

A. K. Pradhan
Department of Computer Science and Engineering, College of Engineering, Bhubaneswar, Bhubaneswar, India
e-mail: ashwini.10.pradhan@gmail.com

© Springer Nature Singapore Pte Ltd. 2019   441
S. C. Satapathy et al. (eds.), *Information Systems Design and Intelligent Applications*,
Advances in Intelligent Systems and Computing 862,
https://doi.org/10.1007/978-981-13-3329-3_41

operation, multiplication process takes a larger amount of time. These operations reduce the speed of the processor. Multipliers have always been an important part in a system design. These hardware architectures can be efficiently implemented on the different hardware platform.

This paper attempts to implement a CORDIC algorithm for performing the multiplication operation on FPGA based platform. The design is implemented and the analysis is done based on the efficiency of the implemented system.

2 CORDIC Algorithm

The coordinate rotational digital computer (CORDIC) algorithm is an iterative algorithm. It uses simple shifts and adds operations. The main components and the flow of the information lie in performing the cross-addition with the X and Y factors. It consists of two shift registers and three adder–subtractor along with some special interconnections. The principle of the CORDIC algorithm lies in the simple rotation as shown in Fig. 1. Here, the rotational vector $p_0 = [x_0 y_0]$ is rotated through an angle θ in order to obtain a rotated vector $p_n = [x_n y_n]$. The basic equations for rotation are given by [4],

$$x_{out} = x_{in}\cos\theta - y_{in}\sin\theta,$$
$$y_{out} = x_{in}\sin\theta + y_{in}\cos\theta \qquad (1)$$

where (x_{in}, y_{in}) indicates the initial coordinates of the vector and (x_{out}, y_{out}) gives the final coordinates of the vector.

CORDIC Rotation

Fig. 1 Rotation of a vector in a 2D plane

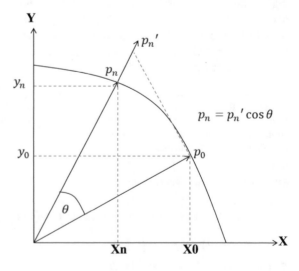

$$\begin{bmatrix} x_m \\ y_m \end{bmatrix} = \begin{bmatrix} \cos\theta & -\sin\theta \\ \sin\theta & \cos\theta \end{bmatrix} \begin{bmatrix} x_0 \\ y_0 \end{bmatrix}$$

$$= \cos\theta \begin{bmatrix} 1 & -\tan\theta \\ \tan\theta & 1 \end{bmatrix} \begin{bmatrix} x_0 \\ y_0 \end{bmatrix}.$$

$$= \text{Rot}(\theta) \begin{bmatrix} x_0 \\ y_0 \end{bmatrix} \tag{2}$$

The matrix product is given by $p_m = Rp_0$, where R gives the rotation matrix. The following equation gives the Rotation matrix:

$$R = \begin{bmatrix} \cos\theta & -\sin\theta \\ \sin\theta & \cos\theta \end{bmatrix}.$$

By factoring the cosine term in (2), the rotation matrix can be written as [4]

$$R = \begin{bmatrix} 1 & -\tan\theta \\ \tan\theta & 1 \end{bmatrix}.$$

The above rotation operation changes its magnitude by a scale factor of $K = \cos\theta$ and after scaling the final output, a pseudo-rotated vector is obtained:

$$p_n = k \cdot R_c p_0.$$

But the scale factor is avoided because it increases the complexity in calculations. Instead of multiplying the scale factor at every iteration, the scale factor is multiplied at the final stage of the output.

3　Conventional Circuit

The Coordinate Rotation Digital Computer is a special-purpose computer which is especially suitable for solving the trigonometric relationships involved in the plane coordinate system and to realize multiplier-less architectures. If $\cos(x)$ is projected along the *x-axis* and $\sin(x)$ is projected along *y-axis*, then the iteration process amounts to the rotation of an initial vector. Hence because of this rotation CORDIC algorithm derives its name as Coordinate Rotation Digital Computer. CORDIC rotation is defined by a set of predefined angles denoted by α_k
where $0° \geq \alpha_k > 90°$

$$\alpha_k = \tan^{-1}(2^{-k}), \quad k = 0, 1, 2, 3, \ldots$$

Table 1 Pre computed angles

i	$\tan \alpha = 2^{-i}$	$\alpha = \tan^{-1} 2^{-i}$
0	1	45
1	1/2	26.565
2	1/4	14.036
3	1/8	7.125
4	1/16	3.576
5	1/32	1.79
6	1/64	0.895
7	1/128	0.448
8	1/256	0.224
9	1/512	0.112
...	...	...

$$\text{For } k = 0, \quad \alpha_k = 45°$$
$$\text{For } k = 1, \quad \alpha_k = 26.56°$$

$$\vdots \qquad \vdots$$

Table 1 shows the precomputed angles α_k. Hence, $\alpha_0 > \alpha_1 > \alpha_2 \ldots > \alpha_{k-1} > \alpha_k \ldots \rightarrow 0°$ with the increasing values of k the value of α_k starts decreasing. These sets of angles are predefined and are stored in a memory. In order to store these predefined angles, I designed a read-only memory (ROM). These angles help in micro-rotations and depending on the sign of new θ_{i+1} (3.11) the next iterations take place.

Now, if we are to rotate $(x_0 y_0)$ to $(x_m y_m)$ through an angle θ then we can write

$$\theta = \sum_{i=0}^{\infty} \delta_i \alpha_i \tag{3}$$

where $\delta_i = \pm 1$ and $\delta_0 = +1$.

The CORDIC arithmetic can be solved using the following set of generalized equations [4].

$$x_{i+1} = x_i - \delta_i 2^{-i} y_i$$
$$y_{i+1} = y_i + \delta_i 2^{-i} x_i.$$
$$\theta_{i+1} = \theta_i - \delta_i \alpha_i. \tag{4}$$

The above equations of CORDIC can be used in two operating modes i.e. rotation mode and vectoring mode. They differ basically on how the directions of the micro-rotations are chosen. Equation (4) is valid for rotation of angles between $-\frac{\Pi}{2} \leq \theta \leq \frac{\Pi}{2}$, i.e., from $-90 \leq \theta \leq 90$.

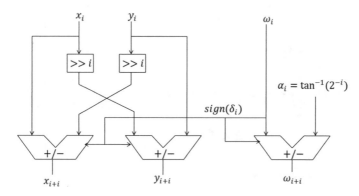

Fig. 2 Folded hardware implementation of CORDIC iteration

Figure 2 shows the folded CORDIC architecture block. This architecture uses two shift blocks, three adder–subtractor, a look up table for storing the predefined angles and some interconnects. This architecture after some iteration is able to make a complete rotation to obtain the coordinate components of the original vector. The complexity of this architecture lies in the latency of the implementation. So, later on, pipelined architectures are proposed for faster applications. Here, instead of using a shifter at every stage, one shift operation is fixed for each CORDIC block and hence there is no requirement for waiting to complete one iteration and start the next hence reducing the critical path. The predefined angles are feed individually to every CORDIC block of the iterative stage. The pipelined architecture is shown in Fig. 3.

4 Multiplier Architecture

In 1971, Walther reformulated the CORDIC algorithm which is suitable for performing rotations in circular, linear and hyperbolic systems. This generalized equation includes a new variable "m" [4].

$$x_{i+1} = x_i - m\delta_i y_i 2^{-i}$$
$$y_{i+1} = y_i + \delta_i x_i 2^{-i}.$$
$$z_{i+1} = z_i - \alpha_i \tag{5}$$

where

$$\delta_i = \begin{cases} \text{sign}(\theta_i), & \text{for rotation mode} \\ -\text{sign}(y_i), & \text{for vectoring mode} \end{cases}.$$

Table 2 summarizes the value of "m" and "α" for different coordinate systems.

$x = x_0 \quad y = y_0$

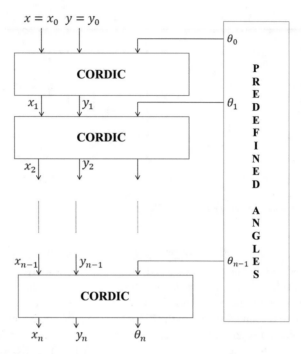

Fig. 3 Pipelined CORDIC unit (PCU)

Table 2 Values of "m" and "α"

System	M	α
Linear	0	2^{-i}
Circular	1	$\tan^{-1} 2^{-i}$
Hyperbolic	-1	$\tanh^{-1} 2^{-i}$

Table 3 Realization of function in linear mode of CORDIC algorithm

m	Mode	Initialization	Output
Linear (0)	Rotation	$x_0 = x_{in}$	$x_n = x_{in}$
		$y_0 = y_{in}$	$y_n = y_{in} + x_{in} \cdot z$
		$z_0 = z$	$z_n = 0$

Multipliers have been a challenging domain. A circuit can perform all the operation of multiplication without the need of multipliers. A CORDIC algorithm is tried to be designed for replacing multipliers. In order to design a multiplication operation using CORDIC, linear system is used. Table 3 shows the realization of CORDIC algorithm in linear mode.

Table 4 Precomputed angles α

i	$\alpha = 2^{-i}$	
0	1	1
1	1/2	0.5
2	1/4	0.25
3	1/8	0.125
4	1/16	0.0625
5	1/32	0.0312
6	1/64	0.0156
7	1/128	0.0078
8	1/256	0.0039
9	1/512	0.0095
…	…	

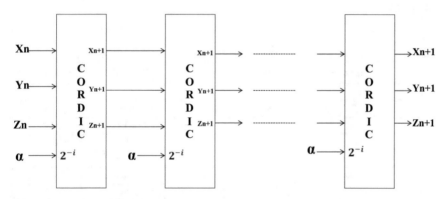

Fig. 4 Pipelined architecture of CORDIC for multiplication

Here, the initialization of x, y, and z registers for the realization of multiplication using CORDIC. Realization of multiplication is done by replacing the value of $y_{in} = 0$, we get $y_{in} = x_{in} \cdot z$.

For implementing the algorithm, the values of α are precomputed and stored in an LUT (Table 4).

Scale factor compensation: The obtained results after the implementation of Eq. (5) is required to be scaled by the factor

$$A = \prod_i A_i,$$

where $A_i = \sqrt{1 + m \cdot d_i^2 \cdot r^{-2i}}$.

The term A_i gives the elementary scaling factor of the ith iteration, and A is the resultant scaling factor after n iteration. The scaling factor "A_i" will be 1 for $m = 0$ in linear coordinate systems. Hence the resultant scaling factor "A" will be 1 (Fig. 4).

5   Results and Discussions

The desired angle of rotation is obtained by performing a series of successively smaller elementary rotation, where $i = 0, \ldots n - 1$ for $\tan \alpha = 2^{-i}$. The desired rotation is assumed for 57° (Table 5).

We start with the desired with the desired angle of rotation $\theta_i = 57°$. For each iteration, if $\theta_i > 0$ then we subtract the current iteration angle from θ_i, otherwise we add the current iteration angle to θ_i and also make our appropriate X and Y.

The results are simulated on XILINX ISE 14.2 and the RTL schematic is shown in Fig. 5.

A.   *Simulations*:

The simulations results of the pipelined architecture are shown in Figs. 5 and 6 (Table 6).

Table 5   Computation of θ_i through the number of iterations

i	$\tan \alpha = 2^{-i}$	$\alpha = \tan^{-1}2^{-i}$	θ_i	Rotation	Final angle
0	1	45	57	−45	12
1	1/2	26.565	12	−26.565	−14.565
2	1/4	14.036	−14.565	14.036	−0.529
3	1/8	7.125	−0.529	7.125	6.596
4	1/16	3.576	6.596	−3.576	3.02
5	1/32	1.789	3.02	−1.789	1.231
6	1/64	0.895	1.231	−0.895	0.336
7	1/128	0.448	0.336	−0.448	−0.112
8	1/256	0.224	−0.112	0.2238	0.112
9	1/512	0.112	0.1117	−0.112	~0
10	1/1024	0.0559	0.0003	–	–

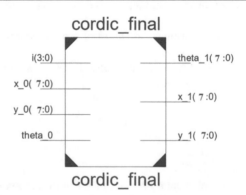

Fig. 5   RTL schematic of CORDIC

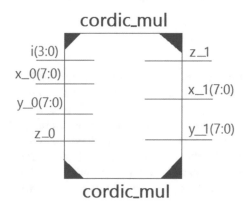

x1[7 :0]			11011011	
y1[7:0]			00110001	
theta1[7:0]			00000000	
theta[7 :0]			00111001	
alpha[7 :0]			00000001	
i[3:0]			1001	
x[7 :0]			00001111	
y[7 :0]			00001011	

Fig. 6 Simulated output of CORDIC

Table 6 Calculated results

Delay (ns)	Number of slices (LUTs)
6.156	140

$$\text{cordic\_mul}$$

i(3:0)

x_0(7:0)

y_0(7:0)

z_0

z_1

x_1(7:0)

y_1(7:0)

$$\text{cordic\_mul}$$

Fig. 7 RTL schematic of a CORDIC multiplier

B. *Simulations*

Calculation of two numbers:

$$x_0 = 15$$
$$y_0 = 0.$$
$$z_0 = 12$$

Simulated output, $y_n = x_0 z_0 = 179$
Theoretically calculated output $y_n = x_0 z_0 = 180$

$$\text{Percentage error} = \frac{(180 - 179) \times 100}{180} = 0.55\%.$$

The simulations results of the CORDIC multiplier architecture are shown in Figs. 7 and 8 (Table 7).

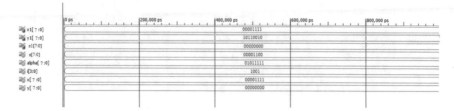

Fig. 8 Simulated results for CORDIC multiplication

Table 7 Calculated results

Multipliers	Delay (ns)	Number of slices (LUTs)
CORDIC	5.256	246

6 Conclusion

Through the works mentioned in this work, an effort has been made to introduce a technique for multiplication. In this proposed architecture, it is seen that multiplication can be done very smoothly with the help of a simple rotation operation. The analysis results are almost accurate with a minimal of 0.55% percentage error. Modifications can be to get accuracy. These implementations help us to achieve efficient and low-cost architectures. These CORDIC based VLSI architecture are very an alternative to the architecture based on conventional multiply and add hardware as comparatively work on to improve the latency of the system.

References

1. J.E. Volder, The CORDIC trigonometric computing technique. IRE Trans. Electron. Comput. **3**(EC-8), 330–334 (1959)
2. J.E. Meggitt, Pseudo division and pseudo multiplication process. IBM J. **6**(2), 210–226 (1962)
3. S. Walther, A unified algorithm for elementary functions, in *Spring Joint Computer Conference* (1971), pp. 379–385, 1–7
4. P.K. Meher, J. Valls, T.B. Juang, K. Sridharan, 50 years of CORDIC: algorithms, architectures and applications. IEEE Trans. Circuits Syst. I Regul. Pap. **56**(9), 1893–1907 (2009)
5. B. Lakshmi, A.S. Dhar, CORDIC architectures: a survey. VLSI Des. **2010**(794891), 1–19. Hindawi Publishing Corporation (2010)
6. S. Sharma, M. Bansal, Designing of CORDIC processor in verilog using XILINX ISE simulator. IJRET: Int. J. Res. Eng. Technol. **03**(05), 342–349. eISSN: 2319-1163 | pISSN: 2321-7308
7. J.E. Volder, The birth of CORDIC. J. VLSI Sig. Proc. **25**(2), 101–105 (2014). http://dx.doi.org/10.1023/A:1008110704586. Accessed on March 2014
8. R. Andraka, *A Survey of CORDIC Algorithms for FPGA based Computer*. Also available on http://www.andraka.com/files/crdcsrvy.pdf. Accessed on March 2014
9. A. Mandal, K.C. Tyagi, B.K. Kaushik, VLSI architecture design and implementation for application specific CORDIC processor, in *2010 International Conference on Advances in Recent Technologies in Communication and Computing* (IEEE Computer Society, 2010), pp. 191–193

Study of Electrical and Magnetic Properties of Multiferroic Composite $(BiFeO_3)_x(Ba_5RTi_3V_7O_{30})_{1-x}$

Hage Doley, Anuradha Panigrahi and Pinaki Chakraborty

Abstract Polycrystalline Samples of the composite $(BiFeO_3)_x(Ba_5RTi_3V_2O_{30})_{1-x}$ (R—Rare Earth Element) is prepared by high-temperature Solid-State Reaction Technique. The formation of the material has been confirmed through XRD (X-ray Diffractometer) to ensure the single-phase formation. Dielectric characterization has been done by Impedance Analyser (HIOKI-IM3536) to study the variation of dielectric constant with frequency and temperature in a wide range to see the dielectric anomaly. Magnetization study is also done by Vibration Magnetometer for different values of x in the composite $(BiFeO_3)_x(Ba_5RTi_3V_2O_{30})_{1-x}$.

Keywords Dielectric · Impedance analyzer · Neel temperature magnetization XRD

1 Introduction

In 1957, Royen and Swares synthesized a material $BiFeO_3$ which is single-phase material exhibiting both ferroelectric and ferromagnetism in the same material popularly known as multiferroic (smart material) [1]. These materials can be used for fabrication of multifunctional electronics and communication devices such as multiple state memories (new-generation RAM), data storage media, [2] transducers, sensors, and spintronic devices. Among many multiferroic materials, $BiFeO_3$ is found to be the most significant because it exhibits its multiferroic properties at and above room temperature. It has a high paraelectric/ferroelectric transition temperature $T_c = 1100$ K [3] and a high Neel temperature $T_N = 643$ K. Unfortunately, this compound has high leakage current and low dielectric constant and high loss tangent. Due to

H. Doley · A. Panigrahi
Department of Physics, Dera Natung Government College, Itanagar 791113, India
e-mail: anu_plasma@yahoo.com

P. Chakraborty (✉)
Department of Physics, National Institute of Technology, Itanagar 791113, India
e-mail: pinakichk@gmail.com

© Springer Nature Singapore Pte Ltd. 2019
S. C. Satapathy et al. (eds.), *Information Systems Design and Intelligent Applications*,
Advances in Intelligent Systems and Computing 862,
https://doi.org/10.1007/978-981-13-3329-3_42

this limitation, it is not very suitable for device application. This problem can be solved by mixing pure ferroelectric material with this multiferroic BFO in a suitable proportion. A proper coupling can be established between the electric and magnetic order making it more suitable for use as a multiple state memories and data storage device. The ferroelectrics ($Ba_5RTi_3V_2O_{30}$) has a TB structure [4] which has many numbers of cationic sites, which provides a scope of interaction between different sites and favors a coupling between magnetic and ferroelectric order. So there is the scope of improving the property of BFO by adding BRTV in different proportion to make a solid solution $(BiFeO_3)_x(Ba_5RTi_3V_2O_{30})_{1-x}$. Some works hav been done by us [5] making the composite $(Ba_5HoTi_3V_2O_{30})_{1-x}$ $(BiFeO_3)_x$, which shows that in a particular composition the magnetic order appears suppressing the electric order in the ferroelectric BHTV. Now, we are planning to study how the magnetic order in BFO will be affected by adding BRTV and in certain composition (i.e., for which value of x) the composite is showing dielectric and magnetic order most effectively.

2 Methodology

2.1 Preparation of $Ba_5RTi_3V_2O_{30}$ (R = Rare Earth Element)

For the synthesis of ceramics sample for our investigation, solid-state reaction technique could be employed.

For the preparation of polycrystalline specimens of $Ba_5RTi_3V_7O_{30}$, (R = Ho, Pr, Tr, etc.). The stoichiometric amount is calculated by the reaction:

$$5BaCO_3 + (1/2)R_2O_3 + TiO_3 + (7/2)V_2O_5 = Ba_5RTi_3V_7O_{30} + 5CO_2$$

These stoichiometric compounds are ground in agate mortar by mixing with methanol then the mixture is calcined in a muffle furnace at a temperature of ~750 °C.

The calcined powder is further ground to fine form so that the size will be in nanoscale.

The powder is die pressed under a pressure of ~7 tons by hydraulic pressure to pallet form of size about 1 cm in diameter. The pallets are sintered at ~800 °C.

2.2 Preparation of $BiFeO_3$ (BFO)

Fe_2O_3 and Bi_2O_3 are mixed in a stoichiometric ratio and calcined at 700 °C for 4 h to form $BiFeO_3$.

2.3 Preparation of Solid Solution $(BFO)_{1-x}(BRTV)_x$

The calcined powder of $BiFeO_3$ and $Ba_5RTi_3V_2O_{30}$ with different proportions are mixed with a small amount of PVB (polyvinyl butyral) as a binder and the powder is die pressed under a pressure of ~7 tons by hydraulic pressure to pallet form of size about 1 cm in diameter. The pallets are sintered at 800 °C for 6 h followed by cooling at 2 °C/min.

Finally, the pallets are painted with silver paste to make it conducting on both sides then the electrical measurement is done.

3 Characterization

3.1 Structural Characterization

The structural characterisation of the calcined powder has been put into X-ray Diffractometer (XRD). The XRD results of BFO, BHTV and $(BHTV)_{1-x} (BFO)_x$ are shown in Fig. 1a–d.

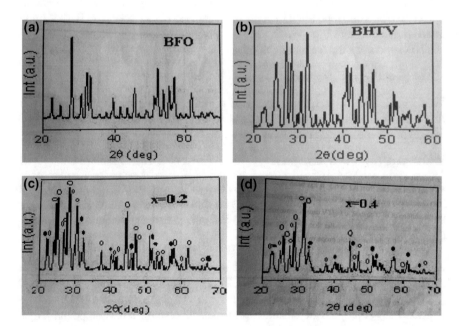

Fig. 1 **a** XRD for BFO. **b** XRD for BHTV. **c** XRD for $(BHTV)_{1-x}(BFO)_x$ for $x = 0.2$. **d** XRD for $(BHTV)_{1-x}(BFO)_x$ for $x = 0.4$

3.2 Electrical Characterization

Using a HIOKI model, dielectric constants and the loss factor (tanδ) could be measured for different frequencies at different temperatures. The transition temperature (T_c) is marked for different variations of x and see how T_c is changing for different x values. Here the result is reported for R = Ho.

3.3 Magnetic Characterization

Magnetic hysteresis is obtained by the Vibration magnetometer, the remnant magnetization is measured for different value of x in the solid solution.

4 Results and Discussions

Result of X-Ray Diffraction study

When the XRD profile is compared then it is seen that Fig. 1c, d contain separate reflection peaks corresponding to both BFO (indicated by symbol) and BHTV (indicated by symbol) thus it proves that the composite indeed a solid solution of BHTV and BFO and not a different compound.

Variation of (ε_r) with frequency

From Fig. 2, it can be seen that the value of relative dielectric constant (ε_r) decreases from ~ 64 to ~ 18 for frequencies increasing from 1 kHz to 1 MHz, which is a general feature of polar dielectric materials.

Variation of (ε_r) with temperature

From Fig. 3, it can be seen that with the rise of temperature the value of ε_r first increases reaches its peak value which is its transition temperature (T_c), and after slight decrease of dielectric constant (ε_r) it further increases (undermentioned frequencies).

From Table 1, it can be seen that for 1 kHz, dielectric constant, ε_{rmax} is 2316 which is relatively very high in comparison to values at 10 and 100 kHz, it may be because it represents the grain boundary region. This dielectric anomaly at $T_c = 240$ °C (513 K) corresponds to the ferroelectric–paraelectric phase transition. The presence of peaks in ε_r–T graph is an indicator for ferroelectric nature of the compound. There is another peak at 378 °C for 10 kHz.

Hence, the dielectric constant may be due to the contribution from both dipole orientation and long-range migration of charge species. The second peak may be considered because of an ionic transition.

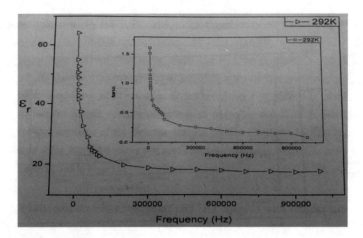

Fig. 2 Frequency variation of ε_r and tanδ of BHTV (at room temp) $(BiFeO_3)_x(Ba_5HoTi_3V_2O_{30})_{1-x}$ for $x = 0$

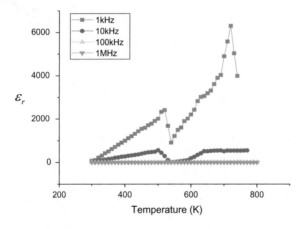

Fig. 3 Variation of ε_r with temperature (K) for BHTV at various frequencies. $(BiFeO_3)_x(Ba_5HoTi_3V_2O_{30})_{1-x}$ for $x = 0$

Dielectric study of $(BHTV)_{1-x}(BFO)_x$

Figure 4 shows the dependency of dielectric constant (ε_r) with the temperature at frequency 100 kHz for $(BHTV)_{1-x}(BFO)_x$ with x ranging from 0 to 0.5 and temperatures ranging 300–800 K.

Table 2 shows the values of T_c and ε_r for $(BHTV)_{1-x}(BFO)_x$ ($x = 0, 0.1, 0.2, 0.3, 0.4, 0.5$). It is clear that T_c significantly increases when x changes from 0 to 0.3 and then decreases for 0.3 to 0.5.

The change in T_c is due to the addition of BFO into BHTV in different proportions, which might have caused a mismatch in the ionic radii of Fe^{3+} and V^{5+} as the V^{5+}

Table 1 Comparison of dielectric properties and $T_c(\mathrm{K})$ of BHTV compounds at different frequencies

Compounds	$Ba_5HoTi_3V_7O_{30}$			
f(kHz)	1	10	100	1000
ε_{RT}	64	41	23	18
ε_{max}	2316	651	163	41
$T_c(\mathrm{K})$	513	501	500	498

Fig. 4 Variation of ε_r with temperature (K) for $(\mathrm{BHTV})_{1-x}(\mathrm{BFO})_x$ at 100 kHz frequency

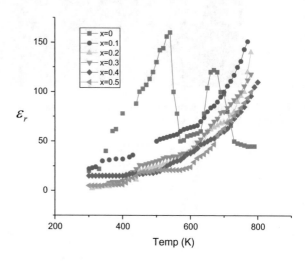

Table 2 Comparision of dielectric properties and $T_c(\mathrm{K})$ of $(\mathrm{BHTV})_{1-x}(\mathrm{BFO})_x$ compound for different values of x at frequency 100 kHz

Frequency	100 kHz	
$(\mathrm{BHTV})_{1-x}(\mathrm{BFO})_x$	$T_c(\mathrm{K})$	ε_r
$x = 0$	500	163
$x = 0.1$	603	48
$x = 0.2$	625	41
$x = 0.3$	631	42
$x = 0.4$	613	43
$x = 0.5$	538	48

ionic radii is smaller than Fe^{3+}. It was found that even with small amount of BFO, i.e., $x = 0.1$, there is considerable change in the ferroelectric property of BHTV.

The decrease of T_c for $x = 0.4$ and 0.5 indicates that there is a modification in ferroelectric property of BHTV by addition of BFO. Up to $x = 0.3$, the compound is predominately ferroelectric in nature.

There is an increase in Curie temperature till $x = 0.3$ and beyond $x = 0.3$ the solid solution is dominated by BFO (i.e., multiferroics), where Neel temperature (T_N) decreases because of the presence of BHTV. This mismatch may be caused by the

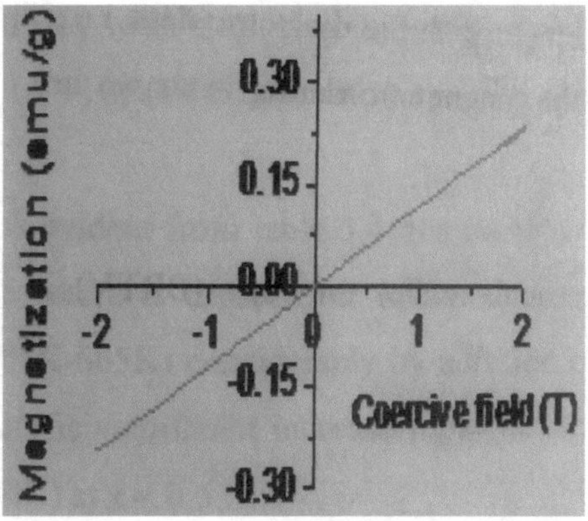

Fig. 5 Coercive field (T) versus Magnetization (M) for $(BHTV)_{1-x}(BFO)_x$ at $x = 0.1$

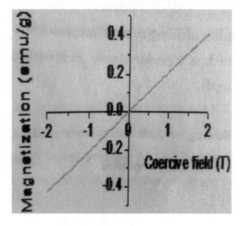

Fig. 6 Coercive field (T) versus Magnetization (M) for $(BHTV)_{1-x}(BFO)_x$ at $x = 0.2$

presence of different concentrations of BHTV and BFO, which introduces distortion in the oxygen octahedron of BHTV showing significant variation in T_c and ε_r. The oxygen octahedron undergoes distortion due to mechanical stresses caused by the increase in the percentage of BFO, resulting in the decrease of transition temperature since T_c depends upon the structural changes. Here, the little amount of BFO can make a complete solid solution which indicates that until $x = 0.3$ in composite falls in the morphotropic phase boundary (MPB) region. Even with the low concentration of BFO in solid solution, BFO dominates the behavior of the compound and it is substantiated by the magnetization behavior.

Magnetization study

The magnetization versus magnetic field has been studied by VSM and Fig. 5 show that for $x = 0.1$, it is perfectly paramagnetic. However, at $x = 0.4$, it starts showing ferromagnetic properties (Fig. 7). These results are in agreement with the dielectric data, i.e., $x > 0.3$ onwards it is predominately multiferroics (Figs. 6 and 8).

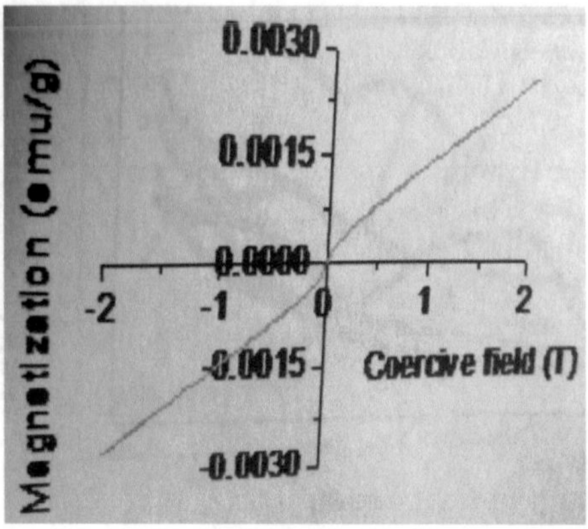

Fig. 7 Coercive field (T) versus Magnetization (M) for $(BHTV)_{1-x}(BFO)_x$ at $x = 0.4$

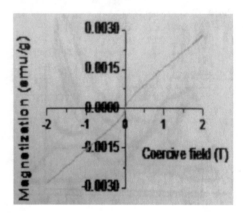

Fig. 8 Coercive field (T) versus Magnetization (M) for $(BHTV)_{1-x}(BFO)_x$ at $x = 0.5$

5 Conclusions

It has been observed that the transition temp T_c of BHTV changes considerably by the addition of BFO. T_c increases up to $x = 0.3$ and then decreases for $x = 0.4$ and $x = 0.5$. It may be due to the interaction of Bismuth ion with different lattice site of BHTV.

It is also seen that when multiferroic $BiFeO_3$ is mixed with ferroelectric materials $Ba_5HoTi_3V_2O_{30}$ to forming a solid solution $(BiFeO_3)_{1-x}(Ba_5HoTi_3V_2O_{30})_x$, then magnetic property starts appearing for certain composition (for $x = 0.4$) in BHTV. That means the composite is predominately paramagnetic at $x = 0.1$ and $x = 0.2$, but for $x = 0.4$ and 0.5, it shows ferromagnetic behavior.

References

1. M. Fiebig, J. Phys. D Appl. Phys. **38**, R123–R152 (2005)
2. M. Kumar, K.L. Yadav, G.D. Varma, Mater. Lett. **62**, 1159 (2008)
3. A.Z. Simoes, E.C. Aguiar, A.H.M. Gonzalez, J. Andres, E. Longo, J.A. Varela, J. App. Phys. **104**, 104115 (2008)
4. A. Magneli, Arkiv Foer Kemi **1**, 213–221 (1949)
5. A. Panigrahi, K. Kathayat, Integr. Ferroelectr. **1180**(01), 8–15 (2010)

Low-Power Process-Insensitive Operational Amplifier Design

D. Anitha, K. Manjunatha Chari and P. Satish Kumar

Abstract A low-power process-insensitive buffered operational amplifier (Opamp) is presented. Current reference circuit of the Opamp is designed by a process invariant current source to achieve better tolerance to process variations without degrading other performance parameters. Simulation results are obtained by using Cadencetools with GPDK 180 nm library and supply voltage of ±0.6 V. The proposed Opamp has shown 42% better process tolerance than the basic CMOS Opamp and it consumes only 54 uW of power, whereas the power consumption is 91 uW for the later.

Keywords Opamp · Process insensitive · Low power · Current reference

1 Introduction

Operational amplifier (Opamp) is the most useful basic component in analog and mixed signal circuits. Huge demand for portable devices leads the requirement of smaller channel length transistors. The decrease of channel length includes many short channel effects which make the design of Opamp with reduced channel length is a challenging task.

An Opamp is described by many performance parameters like gain, unity gain bandwidth (UGB), input impedance, output impedance, output swing, power dissipation, common mode rejection ratio (CMRR), supply voltage rejection ratio (PSRR), slew rate, etc. Opamp was considered as a general purpose analog block till last two decades. So, the designers considered only few performance parameters, i.e., high gain, high input impedance, and low output impedance. But, today's Opamp design is a trade-off among the abovementioned performance parameters based on the application chosen [1–3].

D. Anitha (✉) · K. Manjunatha Chari
Department of ECE, GITAM University, Hyderabad, Telangana, India
e-mail: neetadid@gmail.com

P. Satish Kumar
Department of ECE, ACE College of Engineering, Hyderabad, Telangana, India

© Springer Nature Singapore Pte Ltd. 2019
S. C. Satapathy et al. (eds.), *Information Systems Design and Intelligent Applications*,
Advances in Intelligent Systems and Computing 862,
https://doi.org/10.1007/978-981-13-3329-3_43

Process variations pose more problems on the design, which include smaller channel length transistors. Sources of process variations are: doping differences, mismatch in devices due to lithography, ion implantation and diffusion depth, which leads to threshold voltage (V_{th}) and oxide thickness (T_{ox}) variations [4, 5].

Current equation of the MOSFET is given as

$$I = K(W/L)\left(V_{gs} - V_{th}\right)^2 \tag{1}$$

Here K and V_{th} are process-dependent parameters.

Mobility of electrons (μn), threshold voltage (V_{th}), and resistance (R) are also dependent on temperature variations as shown.

$$\mu_n \alpha\, \mathrm{T}^{-\alpha_k}$$
$$V_{th}(T) = V_{th}(T_0)(1 + \alpha V_{th}(T - T_0))$$
$$R(T) = R(T_0)(1 + \alpha_R(T - T_0)) \tag{2}$$

In this paper, a two-stage buffered Opamp design is proposed, whose current reference circuit is a process, temperature tolerant current source. This current source also helps in reducing the total power consumption with improved stability. Reducing power dissipation of Opamp against process, temperature variations is focused in this paper.

This paper is organized as follows. Basic Opamp design is given in Sect. 2. Proposed Opamp and current source design are presented in Sect. 3. Results of proposed Opamp are discussed in Sect. 4. The conclusion is given in Sect. 5.

2 Opamp Structure

A basic CMOS buffered Opamp structure is shown in Fig. 1. Differential amplifier is used as the first stage to provide high gain. Common source amplifier is used as the second stage to obtain high swing. The second stage must be followed by a common drain amplifier (buffer) to achieve low output resistance with large output swing [2]. Miller compensation capacitor is connected as negative feedback element to obtain the stability [6].

The differential amplifier consists of a differential pair formed by two PMOS transistors M1, M2 with an NMOS load M3, M4. Tail current source M5 is generated by a mirror transistor M10 and current reference I_{Ref}.

Common source amplifier consists of M6 and M7 followed by a buffer consists of M8, M9. Here, the buffer stage is used to drive low resistive load and large capacitive load. The compensating capacitor 'C_c' connected between the two stages plays a vital role in obtaining the dominant pole.

The design of Opamp in Fig. 1 is as follows. The current through transistor M5 is obtained from the Slew Rate (SR) value.

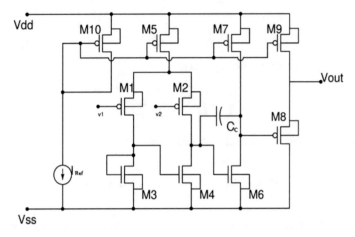

Fig. 1 CMOS Opamp structure

$$I_5 = \text{SR} * C_c \tag{3}$$

where $C_c = 0.22C_L$ to achieve stability of the Opamp.

Specification of M3 and M4 is obtained from Eq. (4).

$$\left(\frac{W}{L}\right)_3 = \frac{I_5}{K_3\left[\left(V_{DD} - V_{in(max)}\right) - V_{t3(max)} + V_{t1(min)}\right]^2} \tag{4}$$

where $V_{in(max)}$ is the maximum input common mode range (ICMR_{max}). K_3 is the process transconductance.

M3 and M4 form NMOS current mirror load. Therefore, they have same (W/L) ratios.

$$\left(\frac{W}{L}\right)_3 = \left(\frac{W}{L}\right)_4 \tag{5}$$

The (W/L) ratio of transistor M1 is

$$\left(\frac{W}{L}\right)_1 = \frac{g_{m1}^2}{I_5 K_3} \tag{6}$$

where

$$g_{m1} = \text{UGB} * C_c * 2\Pi \tag{7}$$

Since M1 and M2 form differential pair, they should have same (W/L) ratio.

$$\left(\frac{W}{L}\right)_1 = \left(\frac{W}{L}\right)_2 \tag{8}$$

Transistor M5's aspect ratio is given as

$$\left(\frac{W}{L}\right)_5 = \frac{2I_5}{K_n(V_{Ds5})^2} \tag{9}$$

where V_{DSat} is obtained from the following equation:

$$V_{\text{DS5}} = V_{\text{in(min)}} - V_{\text{ss}} - \sqrt{\frac{I_5}{\beta_1}} - V_{t1(\text{min})} \tag{10}$$

$V_{\text{in(min)}}$ in the above equation refers to minimum ICMR_{min} and $\beta 1$ is the device transconductance of M1.

To obtain phase margin $= 60°$, $g_{m6} \geq 10\, g_{m1}$

Therefore

$$\left(\frac{W}{L}\right)_7 = \frac{I_7}{I_5}\left(\frac{W}{L}\right)_5 \tag{11}$$

And the current through M6 is

$$I_6 = \frac{g_{m6}^2}{2K_n(W/L)_6} \tag{12}$$

For the balanced condition,

$$I_6 = \frac{(W/L)_6}{(W/L)_3}I_3 \tag{13}$$

To obtain required current ratios, for M7

$$\left(\frac{W}{L}\right)_7 = \frac{I_7}{I_5}\left(\frac{W}{L}\right)_5 \tag{14}$$

From the output resistance specification and current mirror requirement, the aspect ratio of M8 and M10 can be calculated, respectively.

3 Proposed Opamp Structure

Generalized circuit diagram of Opamp is shown in Fig. 1. Current reference (I_{ref}) of this Opamp is generally analyzed by a MOSFET working in saturation region,

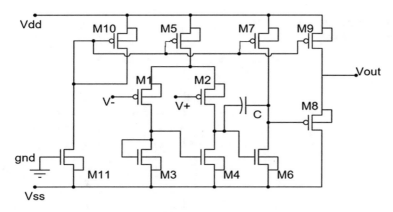

Fig. 2 Schematic of Opamp with MOSFET current source

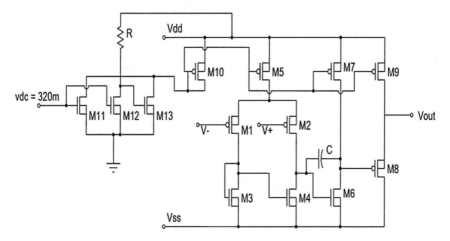

Fig. 3 Schematic of proposed Opamp circuit

providing the required current generated by I_{ref} [7]. The circuit diagram is given Fig. 2.

Proposed Opamp structure is shown in Fig. 3, where the current reference circuit (I_{ref}) is designed by a current source, which is a process, temperature tolerant current source. To ensure the stability and also low power dissipation against the process variations, the MOSFET in Fig. 2 is considered as addition based current source [5].

The circuit diagram of this current source is given Fig. 4. Here, transistors M1 and M3 match each other. Since M1 and M3 are identical transistors, the currents I_1 and I_3 are same. By the current source topology, it can be observed that the currents I_1 and I_2 are negatively correlated, i.e., $I = I_1 + I_2$. If there is any increase in current I_1 due to process variations, I_2 decreases and keeps I stable.

(W/L) ratios of MOS transistors of the Opamp are given in Table 1. The current reference is designed for $I = 30 \, \mu A$ to drive the required load. DC analysis of the

Fig. 4 Circuit of addition based current source

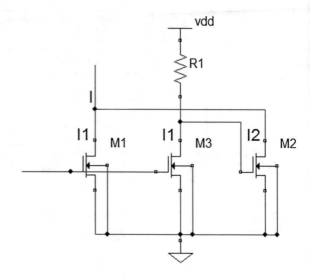

Table 1 Aspect ratios of MOS transistors

MOS transistors	Aspect ratios
PM1, PM2	3
NM3, NM4	1
PM5	13
NM6	4
PM7	19
PM8	135
PM9	75
PM10	75
C_c	1.5 pF

current source circuit is obtained and given in Fig. 5. Current source biasing voltage is at $V_{dc} = 320$ mV to obtain $I_1 + I_2 = 30$ μA.

4 Simulation Results

4.1 AC Analysis

The proposed Opamp circuit is designed by using GPDK 180 nm with a supply voltage of ±0.6 V and simulated using spice tool. The magnitude and phase plots are obtained as in Fig. 6. Gain and phase margin are obtained as 75 db and 55° respectively and the UGB obtained is 4.67 MHz.

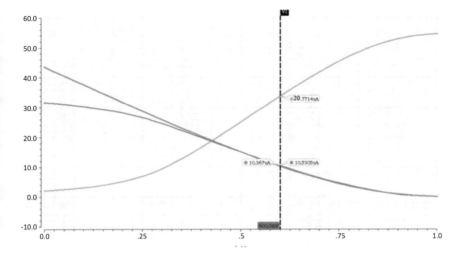

Fig. 5 Dc characteristics of current source

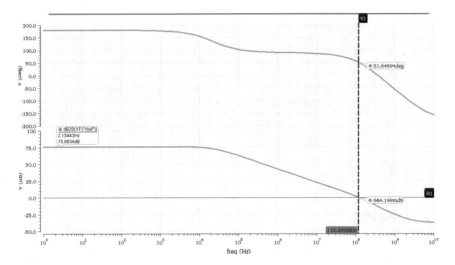

Fig. 6 Gain and phase plot of proposed Opamp

4.2 Common Mode Rejection Ratio (CMRR)

To calculate the CMRR, the two Ac inputs of Opamp are shorted and common mode gain (Acm) is calculated. From the Fig. 7, Acm = −5.7 db.

$$CMRR = Ad(db) - Acm(db) = 80.7 \text{ db}$$

CMRR = Ad(db)-Acm(db) =80.7db

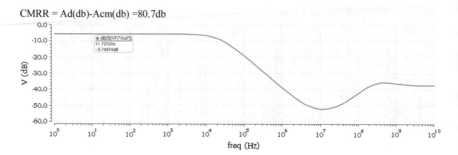

Fig. 7 Common mode gain of proposed Opamp

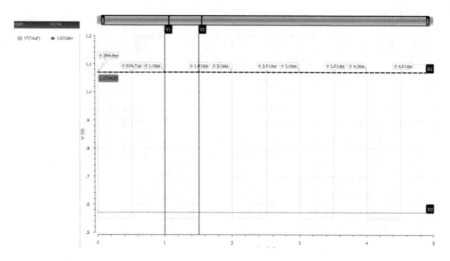

Fig. 8 Slew rate of proposed Opamp

4.3 Slew Rate

To measure the slew rate (dV_o/dt), Opamp is connected in negative feedback mode and applied a square input and obtained the output as shown in Fig. 8. It is observed that the positive slew rate is 40 v/us and negative slew rate is 55 v/us.

4.4 Process Variations

This section deals with the process variations [8, 9], which is the main objective of this paper. Introducing 10% variations in threshold voltage (V_{th}) and oxide thickness (T_{ox}) in the specter file, the Monte Carlo analysis [10] for 100 runs is carried out.

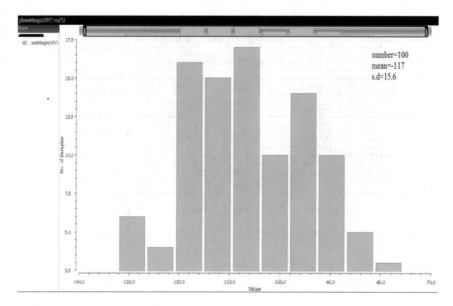

Fig. 9 Phase margin histogram for basic Opamp

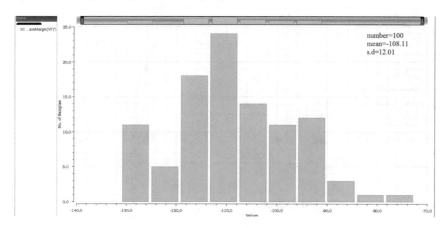

Fig. 10 Phase margin histogram for proposed Opamp

Since the Opamp stability is mainly dependent on phase margin, histograms are plotted and are shown in Figs. 9 and 10. The proposed Opamp shows 42% better immunity to variations.

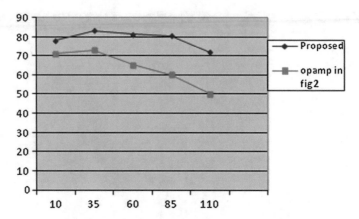

Fig. 11 Gain versus temperature variations

Table 2 Comparison of performance parameters

Performance parameter	Proposed Opamp	Basic Opamp
Gain (dB)	75	75
Voltage supply (mV)	±600	±600
UGB (Mhz)	4.67	4.55
Power dissipation (μW)	54	91
Phase margin	55°	54°
Slew rate (V/us)	40	42
ICMRmax (mV)	263	260
ICMRmin (mV)	−205	−214
CMRR (db)	80.5	78
Phase margin variations σ/μ (%)	0.16	0.28

4.5 Temperature Variations

Figure 11 compares the gain of the two Opamps for a temperature range of 0°–110°. The current source used in the proposed Opamp has tolerance to temperature variations. Therefore, it shows fewer variations in the gain for the given temperature range.

Comparison of performance parameters of proposed Opamp and basic Opamp is given in Table 2. It is also compared with the other existing CMOS Opamps and summarized in Table 3.

Layout of the proposed Opamp is shown in Fig. 12 and post-layout simulation is carried out. The gain obtained for post-layout simulation is 74 dB, which is almost the same as that of simulation results. But, the proposed Opamp occupies more area

Table 3 Performance metrics of Opamp

Parameters	This work	Ref. [11]	Ref. [12]	Ref. [13]	Ref. [14]
C_L	15 pF	17 pF‖1 M	250 fF	20 pF	20 pF‖16 k
Voltage supply (V)	±0.6	±0.5	1.8	±0.5	±0.5
Open loop gain (dB)	75.4	76.2	67.6	72	55
Phase margin (deg)	55	–	65	–	36
UGB (MHz)	4.67	8.1	482	10	1.6
Power dissipation (uW)	54	358	872	100	55

Fig. 12 Layout of the proposed Opamp

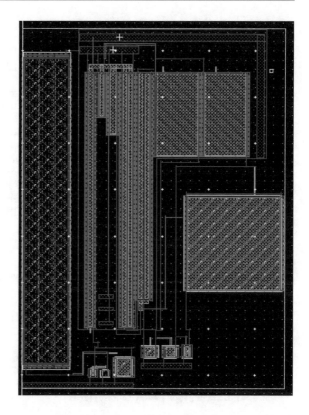

than that of basic Opamp, as it involves more components in the current reference circuit.

5 Conclusion

A novel low-power process-insensitive buffered Opamp is designed using GPDK 180 nm technology. An addition based current source is used to achieve the process insensitivity. The proposed Opamp can drive a 15 pF capacitive load and resistive load of 1 kΩ, with a gain of 75 dB and Unity gain bandwidth of 4.67 MHz. 42% better process tolerance is achieved than the basic Opamp with a power consumption of 54 μW. The CMRR, slew rate of the Opamp is also measured and layout is drawn in virtuoso Layout XL environment and post-layout simulation results are obtained.

References

1. R. Behzad, *Design of Analog CMOS Integrated circuits* (McGraw-Hill, New York, 2010)
2. P.E. Allen, D.R. Holberg, *CMOS Analog Circuit Design*, 3rd edn. (Oxford University Press, Oxford, 2011)
3. R.J. Baker, *CMOS Circuit Design, Layout, and Simulation*, 3rd edn. (IEEE Press, New York, 2010)
4. X. Zhang, A.B. Apsel, A low-power process-and- temperature- compensated ring oscillator with addition-based current source. IEEE Trans. Circ. Syst. I Reg. Pap. **58**(5), 868–878 (2011)
5. M. Pappu, A.V. Harrison, A.B. Apsel, X. Zhang, Process -invariant current source design: methodology and examples. IEEE J. Solid-State Circ. **42**(10), 2293–2302 (2007)
6. G. Prasad, K. Sharma, 170 MHz GBW, two stage CMOS operational amplifier with high slew rate using 180 nm technology, in *Proceeding of IEEE INDICON* (2015)
7. D. Anitha, K.M. Chari, P.S. Kumar, Low voltage and low power PVT compensated Opamp using addition based current source, in *Proceeding of IEEE ICSCTI* (2015)
8. O. Abdelfattah, G.W. Roberts, I. Shih, Y.-C. Shih, An Ultra low-voltage CMOS process-insensitive self-biased OTA with rail-to-rail input range. IEEE transactions on circuits and systems **62**, 2380–2390 (2015)
9. S. Dai, X. Cao, T. Yi, A.E. Hubbard, Z. Hong, 1-V low-power programmable rail-to-rail operational amplifier with improved transconductance feedback technique. IEEE Trans. Very Large Scale Integr. Syst. **21**(10), 1928–1935 (2013)
10. K. Singh, V. Mehta, M. Singh, in *Physical Design of Two Stage Ultra Low Power, High Gain CMOS Op-amp for Portable Device Applications*
11. J.M. Carrillo, G. Torelli, R. Pérez-Aloe, J.F. Duque-Carrillo, 1-V rail-to-rail CMOS opamp with improved bulk-driven input stage. IEEE J. Solid-State Circ. **42**(3), 508–517 (2007)
12. M. Pourabollah, A new gain-enhanced and slew-rate-enhanced folded-cascode opamp. Analog Integr. Circ. Sig. Process. **88**(1), 43–56 (2016)
13. S. Chatterjee, Y. Tsividis, P. Kinget, 0.5-V analog circuit techniques and their application in OTA and filter design. IEEE J. Solid-State Circ. **40**(12), 2373–2387 (2005)
14. G. Raikos, S. Vlassis, Low-voltage bulk-driven input stage with improved transconductance. Int. J. Circ. Theor. Appl. (Wiley Library) (2011)

Static Task Scheduling Heuristic Approach in Cloud Computing Environment

Biswajit Nayak, Sanjay Kumar Padhi and Prasant Kumar Pattnaik

Abstract Scheduling is to assign a resource with a starting and ending time. Mapping refers to the assigning resource to the task without specifying the start time. Mapping can be possible in several conditions. Mapping can be possible when you know what tasks are scheduled or when you do not know what tasks are scheduled. If it is known then it only requires to choose the way so that it can be mapped correctly otherwise it needs to consider varying circumstances. The proposed article focuses on the way to choose an algorithm for mapping when the tasks are scheduled. This article also analyzes experimentally the different algorithms to get out the best of it in different conditions.

Keywords Computing · Datacenter · Task scheduling · Algorithm · Makespan

1 Introduction

Cloud computing can be referred to as a massive computing resource which is deployed through data centers and allocated to tasks and required users through the Internet. Cloud itself consists of huge numbers of servers in different locations, harnessed to provide services. It has massive scalability and reliability, provides efficient fault tolerant, ensures instant infrastructure and application deployment, and provides highly virtualized infrastructure. Due to these reasons cloud computing is being treated as simple, efficient in infrastructure and application management,

B. Nayak (✉) · S. K. Padhi
Computer Science and Engineering, Biju Patnaik Technical University, Rourkela, Odisha, India
e-mail: biswajit.nayak.mail@gmail.com

S. K. Padhi
e-mail: sanjaya2004@yahoo.com

P. K. Pattnaik
School of Computer Science and Engineering, Kalinga Institute of Industrial Technology University, Bhubaneswar, Odisha, India
e-mail: patnaikprasantfcs@kiit.ac.in

© Springer Nature Singapore Pte Ltd. 2019
S. C. Satapathy et al. (eds.), *Information Systems Design and Intelligent Applications*, Advances in Intelligent Systems and Computing 862, https://doi.org/10.1007/978-981-13-3329-3_44

473

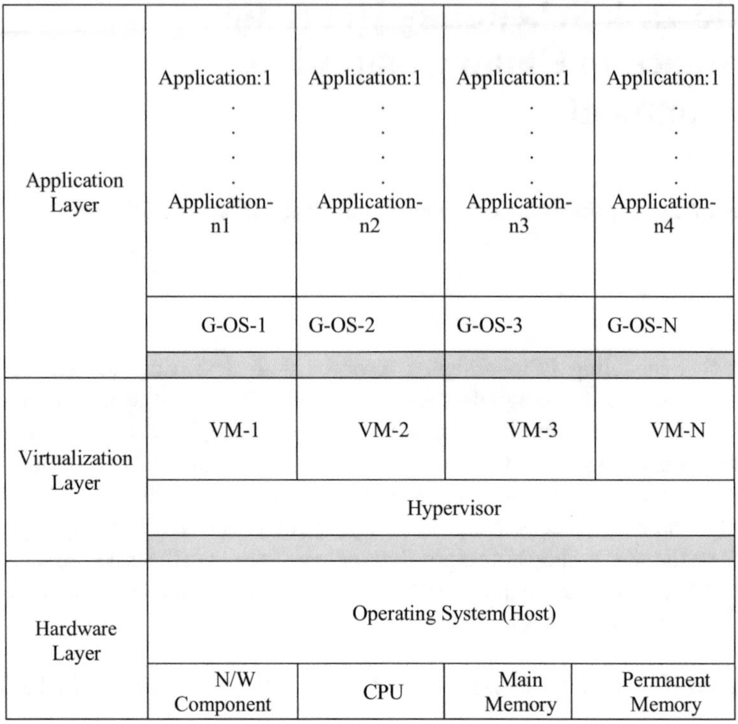

Application Layer	Application:1 . . . . Application-n1	Application:1 . . . . Application-n2	Application:1 . . . . Application-n3	Application:1 . . . . Application-n4
	G-OS-1	G-OS-2	G-OS-3	G-OS-N
Virtualization Layer	VM-1	VM-2	VM-3	VM-N
	Hypervisor			
Hardware Layer	Operating System(Host)			
	N/W Component	CPU	Main Memory	Permanent Memory

Fig. 1 Different components of computing environment

provides high service quality with reduced cost and proper time. Figure 1 shows different components that can be used to schedule task efficiently [1, 2].

The subsequent part of this article consists of different performance metrics both makespan and throughput in Sect. 2; Sect. 3 illustrates the concept of different scheduling algorithms related to task and resource scheduling. Section 4 describes the performance analysis based on metrics of load balancing including several basic techniques as well as some other technique based on basic techniques. Section 5 describes the conclusion of the article.

2 Performance Metrics

A. Makespan

It is defined as the time required completing all the tasks. One of the major characteristics of good task scheduling algorithm is to diminish makespan. In other words, it can be defined as the maximum time required completing a job among group of task [3].

$$\text{Makespan} = \text{maximum}\{CT_i | i \in I \}$$

whereas

CT Completion Time
I Set of jobs
i Particular job from a set of jobs.

B. Throughput

It is defined as the number of tasks executed within a specified completion time or finishing time, in other words, tasks those are finished with respect to deadline. It assumes a heard deadline [4, 5].

$$\text{Throughput} = \sum_{i=1}^{I} X_i$$

whereas

$X_i =$ 1, if task satisfies the deadline. Otherwise $X_i = 0$.
I Set of jobs
i Particular job from a set of jobs.

3 Scheduling Algorithms

Traditional Min-Min algorithm is based on the minimum completion time as a concept. It works on the basis of two stages. In one stage, it calculates the expected completion time and in the other stage, tasks are assigned on the basis of overall minimum completion time. A major advantage of this model is that the processor does not sit idle or wait for small tasks but there will be a starvation for tasks with larger size. It does not work properly when the numbers of small tasks are more in numbers or when small tasks schedule initially [6, 7].

Traditional Max–Min algorithm is based on maximum completion time as a concept. It works on the basis of two stages. In one stage, it calculates the expected completion time and in the other stage, tasks are assigned on the basis of overall maximum completion time. A major advantage of this model is that the processor does not sit idle or wait for large tasks but there will be a starvation for tasks with smaller size. It does not work properly when the number of tasks with large size increases. But it ensures an efficient load balancing when number of small size tasks are more than number of large size tasks [8, 9].

The intention of OLE algorithm is to make each and every node busy whatever it may be the work load of all nodes. It allocates task in random way to the specified available nodes. The OLE algorithm maps the task onto the machine that becomes ready next for the execution. It does not consider the execution time of the task. If multiple machine available then the resultant machine is randomly chosen.

The minimum completion time algorithm uses the earliest completion time to assign each task. It is treated as a benchmark for immediate mode. This algorithm is used as both static and dynamic mapping. But the difference is that the dynamic algorithm has to react to changing situation instead of fixed list of tasks [10, 11].

The algorithm minimum execution time uses least execution time only without considering the machine ready times. Like MCT it is also used as both static mapping and dynamic mapping. This algorithm provides an opportunity to load each task onto machine which has least completion time. But as it is not using machine ready time, there may be a possibility of imbalance in load balance [12, 13].

The Duplex algorithm is a combination of the concepts of Min–Min and Max–Min. It uses the better result after performing both Min–Min algorithm and Max–Min algorithm. It performs better with negligible overhead [14, 15].

4 Performance Analysis

Performance of the above mentioned algorithms is measured by number of tasks and certain available resources. This analysis takes makespan as performance parameter. It finds the makespan on given data by using different algorithms.

Table 1 specifies different task with resources. The diagrams (Figs. 2, 3, 4, 5, 6 and 7) below show scheduling of different algorithms mentioned in this article. Each diagram specifies the available resources in "X-axis" and time required to complete the task can be observed in "Y-axis" [14–16].

Based on the investigation and examination the results can be represented between resources at X-axis and time required at Y-axis. Figure 2 shows the result of the opportunistic load balancing scheduling. Figure 3 shows the result of the minimum execution time scheduling. Figure 4 shows the result of the minimum completion time. Figure 5 shows the result of the Min–Min Heuristics scheduling. Figure 6 shows the result of the Max–Min Heuristics scheduling. Figure 7 shows the result of the duplex heuristics scheduling.

Table 1 Represents tasks and available resources with required time

Tasks → Resources ↓	T1	T2	T3	T4	T5	T6
R1	4	2	1	3	4	4
R2	2	4	2	4	1	5
R3	3	1	3	2	5	5

Fig. 2 Opportunistic load balancing

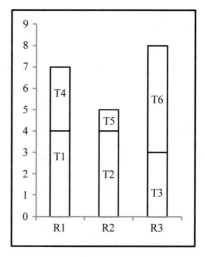

Fig. 3 Minimum execution time

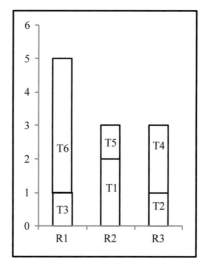

5 Conclusion

In this article, different scheduling algorithms behavior have been investigated and analyzed. The algorithms are described with innovative idea and exposed to rigorous testing using various benchmark datasets and its performance is evaluated. Efficiency evolution is also important to realize large-scale system. The scheduling algorithms have been extensively analyzed on the basis of major characteristics like makespan and throughput. Analysis says some of the tasks are beneficial and there is no single algorithm which provides optimum result. It signifies that the algorithm needs to be selected base on the required quality of service.

Fig. 4 Minimum
completion time

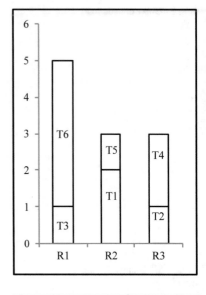

Fig. 5 Min–Min heuristics

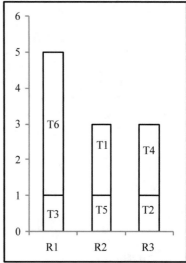

Fig. 6 Max–Min heuristics

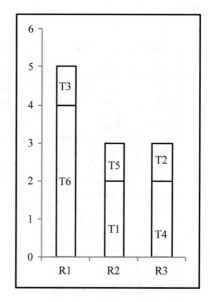

Fig. 7 Duplex heuristics

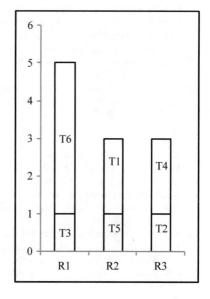

References

1. I.A. Mohialdeen, Comparative study of scheduling algorithms in cloud computing environment. J. Comput. Sci. **9**(2), 252–263 (2013)
2. B. Nayak, S.K. Padhi, P.K. Pattnaik, Impact of cloud accountability on clinical architecture and acceptance of health care system, in *6th International Conference on Frontiers of Intelligent Computing: Theory and Applications* (Springer, 2018), pp. 149–157. https://doi.org/10.1007/978-981-10-7563-6_16
3. P.K. Suri, S. Rani, Design of task scheduling model for cloud applications in multi cloud environment, in *ICICCT 2017, CCIS 750* (2017), pp. 11–24. https://doi.org/10.1007/978-981-10-6544-6_2
4. D.W. Brinkerhoff, Accountability and health systems: toward conceptual clarity and policy relevance. Health Policy Plann. **19**(6), 371–379 ©Oxford University Press, 2004; all rights reserved https://doi.org/10.1093/heapol/czh052
5. P. Banga, S.P. Rana, Heuristic based independent task scheduling techniques in cloud computing: a review. Int. J. Comput. Appl. (0975–8887) **166**(1) (2017)
6. B. Nayak, S.K. Padhi, P.K. Pattnaik, Understanding the mass storage and bringing accountability, in *National Conference on Recent Trends in Soft Computing & It's Applications (RTSCA)* (2017), pp. 28–35, ISSN: 2319 – 6734
7. D. Le, V. Bhateja, G.N. Nguyen, A parallel max-min ant system algorithm for dynamic resource allocation to support QoS requirements, in *2017 4th IEEE Uttar Pradesh Section International Conference on Electrical, Computer and Electronics (UPCON)* (2017), pp. 697–700
8. L. Bao, D.-N. Le, G.N. Nguyen, V. Bhateja, S.C. Satapathy, Optimizing feature selection in video-based recognition using Max-Min Ant System for the online video contextual advertisement user-oriented system. J. Comput. Sci. **21**, 361–370 (2017)
9. S. Singh, M. Kalra, Task scheduling optimization of independent tasks in cloud computing using enhanced genetic algorithm. Int. J. Appl. Innov. Eng. Manage. (IJAIEM) **3**(7), 286–291 (2014). ISSN 2319 – 4847
10. N.M. Reda, An improved sufferage meta-task scheduling algorithm in grid computing systems. Int. J. Adv. Res. **3**(10), 123–129 (2015). ISSN 2320-5407
11. E.K. Tabak, B.B. Cambazoglu, C. Aykanat, Improving the performance of independent task assignment heuristics minmin, maxmin and sufferage. IEEE Trans. Parallel Distrib. Syst. **25**(5), 1244–1256 (2014)
12. E. Kumari, A. Monika, Review on task scheduling algorithms in cloud computing. Int. J. Sci. Environ. Technol. **4**(2), 433–439 (2015). ISSN 2278-3687
13. T. Mathew, K.C. Sekaran, J. Jose, Study and analysis of various task scheduling algorithms in the cloud computing environment, in *International Conference on Advances in Computing, Communications and Informatics (ICACCI)* (IEEE, 2014), pp. 658–664. 978-1-4799-3080-7/14/$31.00_©2014
14. N.S. Jain, Task scheduling in cloud computing using genetic algorithm. Int. J. Comput. Sci. Eng. Inf. Technol. Res. (IJCSEITR) **6**(4) (2016). SSN(P): 2249-6831; ISSN(E): 2249-7943
15. R.K. Devil, K.V. Devi, S. Arumugam, Dynamic batch mode cost-efficient independent task scheduling scheme in cloud computing. Int. J. Adv. Soft Comput. Appl. **8**(2) (2016) (ISSN 2074-8523)
16. R.M. Singh, S. Paul, A. Kumar, Task scheduling in cloud computing: review. Int. J. Comput. Sci. Inf. Technol. **5**(6), 7940–7944 (2014)

Impact of Outlier Detection on Neural Networks Based Property Value Prediction

Sayali Sandbhor and N. B. Chaphalkar

Abstract Detecting outliers is an important step in data mining. Outliers not only hamper data quality but also affect the output in case of prediction models. Prediction tools like Neural Networks (NN) need outlier free dataset in order to achieve better generalization of the network as errors in the dataset hinder the modelling process and produce misleading results. Thus, range of the dataset needs to be curbed in order to make it fit for generating better prediction results. However, outlier detection faces one difficulty. There is no standard framework for the treatment of outliers found in the literature. The present study is an effort to identify the most suited outlier detection method for a specific problem, which deals with the use of NN for prediction of real property value. 3094 cases of property sale instances are presented to various univariate outlier detection methods like Tukey's method, Standard Deviation (SD) method, median method, Z score method, MAD method, modified Z score method, etc. The datasets prepared after removing outliers marked for respective methods are used for prediction using NN. Comparison of results show that the median method is the best-suited outlier detection method for the present study.

Keywords Outliers · Data mining · Neural networks · Valuation · Real property
Data preprocessing

1 Introduction

Outlier is a part of dataset which is not in agreement with the available dataset [1]. The process of outlier detection refers to finding patterns in data that do not substantiate with the expected usual behaviour. These anomalous patterns are often referred

S. Sandbhor (✉)
SIT, Symbiosis International (Deemed University), Pune, India
e-mail: sayali.sandbhor@sitpune.edu.in

N. B. Chaphalkar
JSPM's RSCOE, Pune, India
e-mail: nitin_chaphalkar@yahoo.co.in

© Springer Nature Singapore Pte Ltd. 2019
S. C. Satapathy et al. (eds.), *Information Systems Design and Intelligent Applications*,
Advances in Intelligent Systems and Computing 862,
https://doi.org/10.1007/978-981-13-3329-3_45

to as outliers, also known as anomalies or discordant observations or exceptions or aberrations or noise or errors or novelty or peculiarities or contaminants in various application domains [2]. Outlier is an observation that diverges from other observations in the same dataset leading to suspicion about the origin and the originality of the data [3]. A study [4] indicates that an outlying observation is the one that seems to diverge distinctly from other observations of the sample [5]. Outliers, widely separated from the main body of data points in a sample, are a common characteristic of real data sets. Moreover, the occurrence of outliers can have an adverse effect on any further processing or analysis of the data. It may lead to deceptive results if the dataset with outliers is used [6]. A huge number of observations are recorded in any data analysis related task. Due to large data of variables, consistency of the data gets hampered. Hence, the first step towards attaining a coherent analysis is the detection of outlaying observations. Identified outliers are candidates for anomalous data that may adversely lead to biased parameter estimation and erroneous results. It is, thus, important to spot them prior to modelling and analysis [5–7]. However, outliers may have oddly large or small values in comparison with other values of data set and can be caused by inaccurate measurements that either contains data entry errors or change of source of information than the rest of the data [8]. A number of outlier labelling methods have been used in the literature. However, some methods, like the SD method, are responsive to extreme values and some methods, like Tukey's method, are resistant to extreme values. Although these methods are quite influential with huge traditional information, it is challenging to use them to nontraditional information or small sample sizes devoid of knowing their features in such circumstances. Each labelling method uses diverse measures to detect outliers and their percentages change according to the sample size and detection method used [8]. A major concern in dealing out outliers is that there is no universally accepted theoretical basis for detecting outliers [9]. In the present study, an effort is made to implement several methods of outlier detection to final dataset, to obtain and compare the output of NN for better predictability in order to select the most suitable outlier detection method.

2 Problem Statement

Real property value is the current worth of a property in the market and is required to be computed for varied reasons including buying and selling decision making, renting of property, legal issues, stamp duty, mortgage and many more. Traditional methods make it cumbersome to compute the value. This study, on a broader scale, is an attempt to prepare a value prediction model using neural network. In an attempt to do so, a total of 3094 sale instances from Pune city of India have been collected from professional valuers and converted in identified variable formats. It is observed that application of neural network to the raw dataset does not produce favourable result, and thus requires data preprocessing in order to obtain homogeneous dataset. This is achieved by selection of the most suited outlier detection method and processing the outliers, in order to arrive at better predictability of the model. Following is the methodology used for the study to achieve defined objectives (Fig. 1).

Based on discussions with experts, a total of 19 variables pertaining to property characteristics and five city-specific parameters have been identified prior to this study. Information from the valuation reports is then converted into pre-identified variable format. To reduce the dimensionality of data, PCA has been implemented and 19 variables are reduced to 7 variables for simplification of further processes. This gives two datasets, one with a total of 24 variables and one with a total of 12 variables with five common city-specific parameters. The dataset thus contains categorical and numerical values. Hence, it is preprocessed to convert it into an equivalent numeric dataset. The values of the categorical attributes are substituted with numeric values. Data obtained in the form of valuation reports are converted to quantitative format based on standard conversion designed and normalization is carried out with min-max normalization method. All the other variables have a specific range of observations except saleable area of property which is the direct value of area in Square Feet (sq. ft.). Hence it is used to identify outliers in the dataset. Univariate data analysis for outlier removal is carried out with available methods. Both datasets have been considered for outlier detection, application of neural network and further comparison.

3 Outlier Detection

Outlier detection is a complicated task to execute because of the uncertainty involved in it [10]. Outlier represents semantically correct but infrequent situation in a database. Detecting outliers allow extracting useful and actionable knowledge to the domain experts [11]. As a potential outlier is an extreme data point, it should be labelled as unlikely either by observation or by some descriptive criterion [12]. A broad assessment of various outlier detection techniques has been presented in Markou and Singh [13, 14]. Hodge and Austin [15] have provided an extensive survey of outlier detection techniques developed in domains such as machine learning and statistics. A broad review of outlier detection techniques for numeric as well as symbolic data is presented in another study by Agyemang et al. [16]. The studies have implemented neural networks and statistical approaches, respectively. A survey of

Fig. 1 Methodology of the study

outlier detection techniques that are used specifically for cyber-intrusion detection is presented by Patcha and Park [17]. The detection of outliers has gained considerable interest in data mining with the realization that outliers can be the key discovery to be made from very large databases and a comparison of distance-based, density-based and other techniques for outlier detection is presented in the study by Gogoi et al [18].

There are formal as well as informal methods of outlier detection, called as tests of disordancy and outlier labelling methods, respectively. For parametric data, univariate outlier detection methods are most suited. Outlier labelling techniques are usually very simple to use and they give extreme values that lie away from the majority of observations in the dataset. There are several statistical univariate methods of outlier detection. For the present study, labelling methods have been considered and the methods used are Tukey's or boxplot 1.5 IQR, Tukey's or boxplot 3 IQR, median, 2SD, 3SD, 2 MAD, 3MAD, Z score and modified Z score method. It is observed that the sample mean is a reasonably robust approximation of the sample central tendency so that one outlier among a large sample will have a limited impact [19]. If standard deviations or coefficients of variation are compared across conditions, the measures will be considerably more influenced by the presence of outliers than the mean. The quartile deviation may be a better statistic in such cases [9]. Suitability of outlier detection method would vary as per the characteristics of the available dataset. It is thus necessary to check the suitability of any particular method to selected application. In order to implement univariate methods, it is imperative to study them in detail. Following subsections attempt to give brief introduction of selected methods as observed from available literature.

3.1   Methods of Outlier Detection

Tukey's method or boxplot method [20] is a graphical means of presenting information about the median, upper and lower quartiles and upper and lower extremes of a data set. Quartiles are resistant to farthest values of the dataset. Hence, Tukey's method is less sensitive to extreme values of the data than other methods that use sample mean and standard variance [9]. In this method, it is required to compute quartiles where the Inter Quartile Range (IQR) is the distance between the lower (Q1) and upper (Q3) quartiles. In 1.5 Tukey's method, the boundaries are located at a distance of value equal to 1.5 times IQR below Q1 and above Q3. Similarly, in 3 Tukey's method, boundaries of suitable dataset are located at a distance 3 times IQR below Q1 and above Q3 [8].

The classical approach of detecting outliers is Standard Deviation (SD) method that uses sample mean and standard deviation with certain coefficient. Based on value of coefficient of SD, there are two methods namely 2SD and 3SD which use the following relation as shown in Eqs. (1, 2), respectively to get the boundary limits of outlier labelling. Plus sign in the equation gives outer boundary and use of negative sign gives inner boundary of the range.

$$2 \text{ SD Method: } x \pm 2 \text{ SD} \tag{1}$$

$$3 \text{ SD Method: } x \pm 3 \text{ SD} \tag{2}$$

where x represents the mean of the sample and SD represents standard deviation. The observations not belonging to the identified intervals are considered as outliers [8].

Another method of outlier detection is median method where median is the value exactly in the centre of the data when arranged in order. The median substitutes the quartiles of Tukey's method, and a different scale of the IQR is employed. This method is more resistant and its target outlier percentage is less affected by sample size than Tukey's method. The scale of IQR can be adjusted and 2.3 is chosen as the scale of IQR that lies between Tukey's method of 1.5 IQR and 3 IQR. The inner and outer boundaries are obtained by using Q2 ± 2.3 IQR, where Q2 is the sample median [9].

An alternative method that is used to classify data for detecting outliers is the Z Score that uses the mean and standard deviation [9]. Z score is computed using relation as below:

$$Z = (xi - x)/\text{SD} \tag{3}$$

where xi is a particular reading of a case and x is the sample mean as seen in Eq. (3). Z scores exceeding absolute value of three are usually considered as outliers [21].

Both sample mean and standard deviation can be affected by extreme values and to avoid this issue, median and the median of the Median Absolute Deviation (MAD) are employed in the modified Z Score method. MAD, similar to the standard deviation, is basically an estimator of the spread in a data but has division point like the median [8]. Formulae for computing MAD and modified Z score are as given below in Eqs. (4) and (5).

$$\text{MAD} = \text{Median}\{|xi - x|\} \tag{4}$$

$$\text{The modified } Z \text{ Score } (Mi) = \{0.6745(xi - x)/\text{MAD}\} \tag{5}$$

where xi is the reading for a particular case and x is the sample median.

An alternative method of outlier detection considered for present study i.e. MADe method, uses median and MAD value for outlier computation instead of the mean and standard deviation as in SD method. It is one of the basic robust methods which are mostly unaltered by the existence of extreme values of the data set. The MADe method computations are carried out as shown below in Eq. (6) for 2 MAD method and in Eq. (7) for 3MAD method where MADe = 1.483 × MAD for large normal data.

$$2 \text{ MADe Method : Median} \pm 2 \text{ MADe} \tag{6}$$

$$3 \text{ MADe Method : Median} \pm 3 \text{ MADe} \tag{7}$$

Fig. 2 Scatterplot

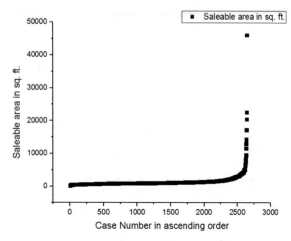

Fig. 3 Number of outliers

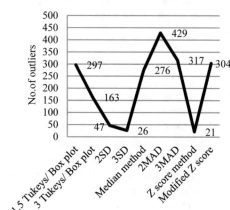

3.2 Comparison of Outlier Detection Methods

The dataset of 3094 cases, in ascending order of saleable area, is presented for outlier detection by the identified methods. Available data ranges from saleable area of 198 to 46,070 sq. ft. Figure 2 illustrates a scatterplot of available data with a clear indication of outlying observations. Y axis of scatterplot represents saleable area in magnitude and the plot is represented by a thick line due to many overlapping data cases. On the basis of equations in earlier section, upper bound and lower bound range of observations is found out for each method. Total 21 outliers are observed for Z score method, 26 for 3SD method and 47 outliers for 2 SD method. Median method and modified Z score method have marked 276 and 304 outliers, respectively. 2MAD and 3MAD methods have nominated a total of 429 and 317 outliers, respectively. It is observed that 2 MAD method exhibits maximum number of outliers whereas Z score method has the least number of outliers (Fig. 3).

4 Neural Network

Neural network, an artificial intelligence technique, works on the lines of computing mechanism of neurons in the human brain. It finds application in classification [22] or prediction related problems [23]. It learns from the historic data of inputs and corresponding outputs provided and predicts future trend for a given unknown input. NN's working is observed to be better than other prediction tools as it can comprehend and identify complicated patterns and it can easily be applied with a slight statistical understanding of the data set [24–26].

4.1 Introduction to NN

The multilayer network comprises of input layer, hidden layer and output layer. The layers are interconnected via links which have random initial weights attached to them. For acquiring the ability to predict the output, the network learns from training where available dataset is presented to the network and the network attempts to identify pattern of relationship between input and output by means of various forward and backward runs and continuous updation of weights of the links in order to reduce the Mean Square Error (MSE) in training. One cycle of forward and backward run make one epoch and it helps in reducing the loss function by improving the weights assigned by better understanding of the pattern. Network overtraining is curtailed by cross-validation MSE. The network is then used for testing where it is presented with new cases, not present in either training set or cross-validation set, with known outputs and network generated output is compared for analyzing the predictive power of the trained network.

4.2 Application of NN to Processed Data

NN is applied on new dataset obtained after outlier removal based on information from Table 1. Neurosolutions 7.0 has been used to carry out network runs for various combinations of the dataset. A study [27] of similar type was carried out to examine the influence of outliers on neural network's performance. Present study specifically figures out outliers' influence on the performance of neural network in predicting the value of property. Based on the outliers detected, separate runs of training of neural network have been carried out per method. A standard combination of 70% of total cases for training, 15% for testing and 15% for cross-validation has been considered for all methods and combinations.

Multilayer perceptron network architecture is selected for the study with number of input processing elements as 12 or 24 as the case may be, 1 hidden layer, five hidden layer Processing Elements (PEs) and one output element. The most suited

Table 1 Dataset details for application of neural network

Method	1.5 Tukeys/ Box plot	3 Tukeys/ Box plot	2SD method	3SD method	Median method	2MAD method	3MAD method	Z score method	Modified Z score
Cases after outlier removal	2797	2931	3047	3068	2818	2665	2777	3073	2790
Training cases (70%)	1959	2053	2133	2148	1974	1867	1945	2153	1954
CV cases (15%)	419	439	457	460	422	399	416	460	418
Testing cases (15%)	419	439	457	460	422	399	416	460	418

learning rule is Levenberg–Marquardt and transfer function is Tanh for present type of prediction problem. Number of epochs considered for the network are 1000. Network parameters have been kept constant for all the runs in order to maintain consistency for comparison. A total of 10 cases are kept aside for prediction purpose, output for which is available and thus can be compared with predicted value to find % prediction error.

4.3 Comparison of Results

NN runs have been taken on dataset with 24 variables and PCA reduced 12 variables [28]. Total three combinations, Combination 1: Dataset of 24 variables with normal sequence, Combination 2: Dataset of 24 variables with randomized sequence and Combination 3: Dataset of 12 variables with randomized sequence, have been considered. As the results of dataset of 12 variables with normal sequence were similar to those of combination 1, it is not mentioned here. Evaluation of performance of the NN model is based on the counts of prediction case records correctly and incorrectly predicted by the model. These counts in terms of % error cases within 20% error are tabulated in Table 2. It shows the performance metrics considered for selection of outlier detection method which is the correlation coefficient and number of cases out of 10 falling within 20% error frame. In case of combination 1, correlation coefficient of 2SD, 3SD, 2MAD, 3 MAD, Z score and modified Z score method is less than 50% and hence can be rejected as the results are not acceptable. Tukey's 1.5 IQR, Tukey's 3 IQR, median methods only, have correlation coefficient more than 0.5. Out of total

Table 2 Performance metrics of neural network and comparison of combinations

Method	Tukey's 1.5 IQR	Tukey's 3 IQR	2SD	3SD	Median	2MAD	3MAD	Z score	Modified Z score
Combination 1	r	0.57	0.55	0.32	0.12	0.52	0.12	0.01	0.41	0.46
	Error cases <20%	4	5	5	5	7	3	1	6	6
Combination 2	r	0.56	0.44	0.69	0.84	0.84	0.43	0.55	0.84	0.85
	Error cases <20%	3	5	0	3	7	4	2	5	4
Combination 3	r	0.83	0.81	0.83	0.78	0.82	0.52	0.58	0.52	0.79
	Error cases <20%	3	6	2	4	6	7	1	4	4

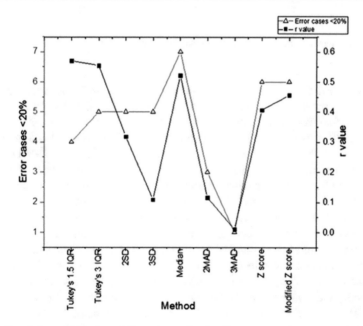

Fig. 4 Correlation coefficients and % error cases for combination 1

10 cases given for prediction, 7 cases in median method have % error less than 20% (Fig. 4).

Further, for combination 2, correlation coefficient of Tukey's 3 IQR and 2MAD is less than 0.5 and hence can be rejected as the results are not acceptable. Tukey's

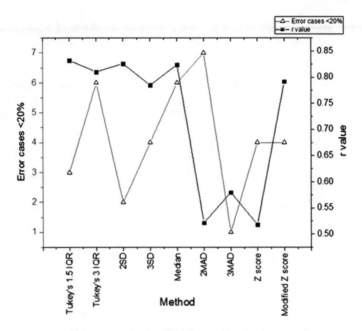

Fig. 5 Correlation coefficients and % error cases for combination 2

1.5 IQR, 2SD, 3MAD methods have r values within 0.7. 3SD, median, Z score and modified Z score methods have r value more than 0.8. Out of total 10 cases given for prediction, four cases in median method have % error less than 10% and three cases within 10–20% error (Fig. 5) which is more than any other method with more than 0.8 r value. Similarly, for combination 3, correlation coefficient of 2MAD, 3MAD and Z score method is very less and hence can be rejected. Out of total 10 cases given for prediction, three cases in median method and two cases in Tukey's 3 IQR method have % error less than 10%. Three cases in median method and four cases in Tukey's 3 IQR method have % error within 10–20% (Fig. 6). However, median method has more r value as compared with Tukey's 3 IQR method.

Figures 4, 5 and 6 are graphical representations of the performance metrics for visual classification and selection of most suited method of outlier detection. It is seen that Tukey's 1.5 IQR, Tukey's 3 IQR, median, modified Z score methods have consistently high values of correlation coefficients as compared with other methods in case of combination 1 and 2. However, median method outperforms all the methods in case of all the three types of sequences of dataset used. Based on the statistical parameters as well as accuracy of the corresponding NN model, in terms of % prediction error, it is observed that median method has a balanced performance (Table 3). Parameters like RMSE, MAE, r and overall score are generated by application of dataset to NN model. However, the number of cases out of 10 which are within 20% error are maximum in case of median method.

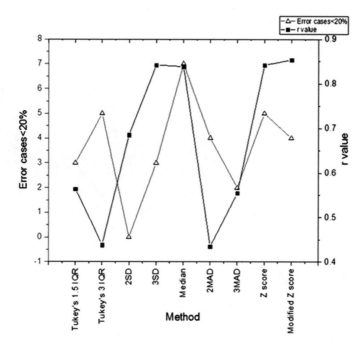

Fig. 6 Correlation coefficients and % error cases for combination 3

Table 3 Summary statistics of performance of network for combination 2

Performance	RMSE	MAE	r	Score
Tukey's 1.5 IQR	0.110	0.074	0.563	69.855
Tukey's 3 IQR	0.150	0.103	0.438	63.575
2SD	0.068	0.055	0.685	80.506
3SD	0.019	0.010	0.842	90.155
Median	0.057	0.028	0.838	88.594
2MAD	0.121	0.077	0.435	66.830
3MAD	0.121	0.082	0.554	72.732
Z score	0.018	0.009	0.842	88.754
Modified Z score	0.037	0.022	0.853	89.968

Visual comparison of desired output (value) and predicted output (value output) as generated for testing cases is shown in Figs. 7 and 8. The figures represent the distinction of actual and predicted value for testing cases which are represented on the x axis and nominated as exemplars. Figure 7 shows comparison of the two measures for 2SD method which shows the least number of cases within 20% error. Figure 8 shows comparison of desired and predicted value for median method that shows the maximum number of cases within 20% error.

Fig. 7 Comparison of desired and actual output of NN model with 2SD method

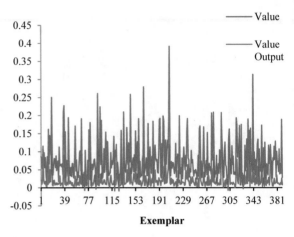

Fig. 8 Comparison of desired and actual output of NN model with median method

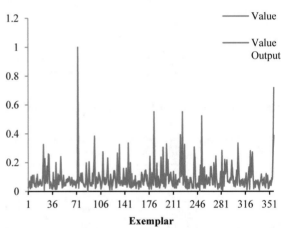

Table 4 represents the neural network's performance metrics for each combination for median method. The network's mean square error for training drops to the minimum within 1000 epochs in case of training and the cross-validation MSE avoids network overtraining for 5 hidden processing elements. Combination 2 and combination 3 comparatively have better minimum and final MSE as compared with combination 1. Five hidden PE's have been observed to be a good choice. It can thus be said that median method of outlier detection with either combination 2 or combination 3 data is most suited for the present study of value prediction for real properties in Pune city, India.

Table 4 Combination wise network performance for median method

Best networks	Combination 1		Combination 2		Combination 3	
	Network training	Cross-validation	Network training	Cross-validation	Network training	Cross-validation
Hidden PEs	5	5	5	5	5	5
Epoch #	1000	25	999	11	1000	14
Minimum MSE	0.0052	0.0356	0.0015	0.0037	0.0027	0.0041
Final MSE	0.0052	0.0372	0.0015	0.0044	0.0024	0.0078

5 Conclusion

Recognizing outliers in the dataset is an important step towards data analysis. Application of NN insists a high-quality data that can form a fine training set for the network for improved comprehension of the indefinite relations between the study-measurements under consideration. From the results of application of various univariate outlier detection methods and implementation of neural network technique to improved dataset, for real property value prediction model, best-suited outlier detection method is median method. This method is suitable in case of the presented data and selected variables of the study. Range of outliers as observed from applied methods varies from 21 to 429. It is observed that four methods namely Tukey's 1.5 IQR, Tukey's 3 IQR, median, modified Z score give number of outliers within a range of 150–305. From results of NN predictions, it is seen that the same four methods give consistent results as far as r value and number of error cases within 20% are considered. In remaining methods, in some cases, even though r value is high, number of error cases with <20% error in prediction are very less. Further, it is observed that median method of outlier detection gives good result for all the three combinations with good r value and better rate of prediction. The present study offers a guideline for selection of best suited univariate outlier detection method in case of implementation of NN in any type of prediction problem. This methodology not only helps in homogenizing the dataset but also in improving the performance of the neural network prediction model.

References

1. R. Johnson, D.W. Wichern, *Applied Multivariate Statistical Analysis*, 6th edn. (Prentice Hall, Englewood Cliffs, 1992)
2. V. Chandola, A. Banerjee, V. Kumar, Anomaly detection: a survey. ACM Comput. Surv. **41**(3), Article 15 (2009)
3. V. Barnett, T. Lewis, *Outliers in Statistical Data*, 3rd edn. (Wiley, New York, 1994). (ISBN 978-0-471-93094-5)

4. I. Ben Gal, Outlier detection, in *Data Mining and Knowledge Discovery Handbook: A Complete Guide for Practitioners and Researchers*, ed. by O. Maimon, L. Rockach (Kluwer Academic Publishers, Dordrecht, 2005)

5. R.J. Bullen, D. Cornforda, I.T. Nabneya, Outlier detection in scatterometer data: neural network approaches. Neural Netw. Spec. Issue **16**, 419–426 (2003)

6. G.J. Williams, R.A. Baxter, H.X. He, S. Hawkins, L. Gu, A comparative study of RNN for outlier detection in data mining, in *IEEE International Conference on Data-Mining* (ICDM'02), Maebashi City, Japan, CSIRO Technical Report CMIS-02/102 (2002)

7. H. Liu, S. Shah, W. Jiang, On-line outlier detection and data cleaning. Comput. Chem. Eng. **28**, 1635–1647 (2004)

8. S. Seo, A review and comparison of methods for detecting outliers in univariate data sets, Dissertation, Graduate School of Public Health, Kyunghee University, University of Pittsburgh (2006). Accessed at: http://d-scholarship.pitt.edu/7948/. Accessed on 20 Apr 2017

9. D. Cousineau, S. Chartier, Outliers detection and treatment: a review. Int. J. Psychol. Res. **3**(1), 58–67 (2010)

10. N. Upasania, H. Omb, Evolving fuzzy min-max neural network for outlier detection, in *International Conference on Advanced Computing Technologies and Applications* (ICACTA-2015) Proceedia Computer Science, vol. 45, pp. 753–761 (2015)

11. A.M. Rajeswari, M. Sridevi, C. Deisy, Outliers detection on educational data using fuzzy association rule mining, in *International Conference on Advance in Computer Communication and Information Science* (ACCIS-14), pp. 1–9 (2014)

12. M. Nkurunziza, L. Vermeire, A comparison of outlier labelling criteria in univariate measurements, Sustainability in statistics education, in *Proceedings of the Ninth International Conference on Teaching Statistics* (ICOTS9, July, 2014), Flagstaff, Arizona, USA (International Statistical Institute, Voorburg, The Netherlands), pp. 1–4 (2014)

13. M. Markou, S. Singh, Novelty detection: a review-part 1: statistical approaches. Sig. Process. **83**(12), 2481–2497 (2003)

14. M. Markou, S. Singh, Novelty detection: a review-part 2: neural network based approaches. Sig. Process. **83**(12), 2499–2521 (2003)

15. V. Hodge, J. Austin, A survey of outlier detection methodologies. Artif. Intell. Rev. **22**(2), 85–126 (2004)

16. M. Agyemang, K. Barker, R. Alhajj, A comprehensive survey of numeric and symbolic outlier mining techniques. Intell. Data Anal. **10**(6), 521–538 (2006)

17. A. Patcha, J.M. Park, An overview of anomaly detection techniques: existing solutions and latest technological trends. Comput. Netw. **51**(12), 3448–3470 (2007)

18. P. Gogoi, D.K. Bhattacharyya, B. Borah, J.K. Kalita, A survey of outlier detection methods in network anomaly identification. Comput. J. **54**(4), 570–588 (2011)

19. M. Daszykowski, K. Kaczmarek, Y. Vander Heyden, B. Walczak, Robust statistics in data analysis-a review basic concepts. Chemometr. Intell. Lab. Syst. **85**, 203–219 (2007)

20. J.W. Tukey, *Exploratory Data Analysis*, 1st edn. (Addison-Wesely Publishers, Reading, MA, 1977). (ISBN-13: 978-0201076165)

21. S.S. Tripathy, R.K. Saxena, P.K. Gupta, Comparison of statistical methods for outlier detection in proficiency testing data on analysis of lead in aqueous solution. Am. J. Theor. Appl. Stat. **2**(6), 233–242 (2013)

22. A. Gautam, V. Bhateja, A. Tiwari, S.C. Satapathy, An improved mammogram classification approach using back propagation neural network, in *Data Engineering and Intelligent Computing* (Springer, Singapore, 2018), pp. 369–376

23. S. Sandbhor, N.B. Chaphalkar, Effect of training sample and network characteristics in neural network based real property value prediction, *Proceedings of 2nd International Conference on Data Engineering and Communication Technology* (ICDECT) 2017, 15th–16th December 2017, 'Advances in Intelligent Systems and Computing (AISC) Series of Springer' (SCOPUS), 303–313 (2018)

24. D. Tay, D. Ho, Artificial intelligence and the mass appraisal of residential apartments. J. Prop. Valuat. Invest. **10**, 525–539 (1991)

25. A. Do, G. Grudnitski, A neural network approach to residential property appraisal. R. Estate Apprais. **58**, 38–45 (1992)
26. N. Nghiep, Al Cripps, Predicting housing value: a comparison of multiple regression analysis and artificial neural networks. J. R. Estate Res. **22**(3), 313–336 (2001)
27. A. Khamis, Z. Ismail, K. Haron, A. Muhammed, The effects of outliers data on neural network performance. J. Appl. Sci. **5**(8), 1394–1398 (2005)
28. S. Sandbhor, N.B. Chaphalkar, Determining attributes of Indian real property valuation using principal component analysis. J. Eng. Technol. **6**(2), 483–495 (2017)

ADAS^DL—Innovative Approach for ADAS Application Using Deep Learning

Ramachandra Guda, V. Mohanraj, J. V. Kameshwar Rao and N. A. Chandan kumar

Abstract The main goal of this paper is to develop an application using deep learning to enhance advanced driver assistance system (ADAS) features based on internal, external and environmental factors from a driver perspective (henceforth called **ADAS^DL**). It covers features like drowsiness detection, de-raining and traffic sign detection models. In this approach, we will analyse driver eye blinking to detect drowsiness of driver and also removing rain streaks from individual images based on the deep convolutional neural network (CNN). This proposed approach is implemented with deep learning models. The results including inference time are discussed.

Keywords ADAS^DL · Deep learning model · SSD · EAR · Resnet · Drowsiness

1 Introduction

The main motivation for the project is the survey results of the US National Highway Traffic Safety Administration. It estimated approximately 100,000 crashes each year caused mainly due to driver fatigue or lack of sleep and also no clear vision due to rain and fog. The effects of rain can also severely affect the performance of outdoor vision systems [1]. To minimize such accidents and also to improve ADAS systems we are providing a deep-learning-based hybrid solution.

R. Guda · V. Mohanraj (✉) · J. V. Kameshwar Rao · N. A. Chandan kumar
HCL Technologies Limited, Noida, India
e-mail: mohanraj4072@gmail.com

R. Guda
e-mail: grcksrgm@gmail.com

J. V. Kameshwar Rao
e-mail: kameshjvkr@gmail.com

N. A. Chandan kumar
e-mail: chandankumar.na@gmail.com

© Springer Nature Singapore Pte Ltd. 2019
S. C. Satapathy et al. (eds.), *Information Systems Design and Intelligent Applications*,
Advances in Intelligent Systems and Computing 862,
https://doi.org/10.1007/978-981-13-3329-3_46

As we know that due to drowsiness and carelessness of drivers while driving and violation of traffic rules, a large number of accidents occur everyday. So, intelligent transport systems (ITS) play a great role in safe driving and in saving lives of pedestrians as well as in saving time and money. These systems are interconnected to the emerging technologies such as the Internet, general packet radio service (GPRS), artificial intelligence, smart sensors, geographical information systems and many more. ITS gives great importance to the field of road sign detection and recognition as it is a part of driving assistance system and autonomous navigation system. These systems must be fast and robust to detect traffic sign in real time [2].

The other core important issues in most of the studies are preventing car accidents which happen each year because of driver's drowsiness and bad weather like fog and rain. So, it is necessary to have a real-time system that supervises driver's drowsiness environment continuously. In some studies, researchers gave more attention to video and image processing; they have used driver's eye and face videos for drowsiness detection. Ueno et al. [3] presented a driver drowsiness detection model to improve the driver assistance system.

Various techniques and methods are developed in last few years about ADAS [4, 5]. In this paper, we have added few works related to ADAS. Tzu-Chi Kao et al. proposed advanced driver assistance system based on head pose recognition. Aleksandra Simić et al. introduced driver monitoring algorithm for advanced driver assistance systems based on the tiredness of driver and raise an alert. Andreas Mogelmose et al. have discussed vision-based traffic sign detection and analysis for intelligent driver assistance systems. In this paper, the author discussed various techniques and datasets for traffic sign recognition.

So, in this paper, we proposed a hybrid approach which can detect driver's drowsiness, traffic sign detection and removing rain/fog strikes from input video frames continually also issue the respective alerts to driver and commands to the controlling unit.

2 Hybrid ADAS^{DL} Proposed Approach

As we stated in last section, our proposed method has three basic submodels, namely, traffic signs, drowsy detection and rain/fog removal. For a prototype implementation, we used a subset of LISA traffic sign dataset (US traffic signs) [6], closed eyes in the wild (CEW) dataset, rainy image dataset—Pretrained model weights [7]. Figure 1 outlines the architecture of the proposed method.

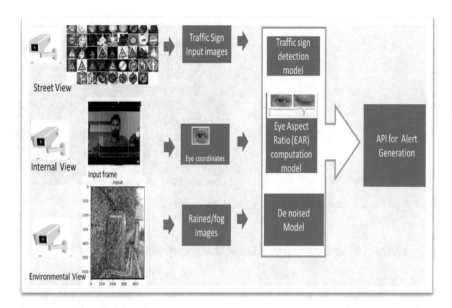

Fig. 1 Proposed method

Input Image | Traffic Sign Detected Image

Fig. 2 Steps in traffic sign detection

2.1 Traffic Sign Detection Model/Street View

In this paper, single shot multibox detector (SSD) [8] is used to detect the traffic sign and the model was trained on LISA traffic sign dataset. Figure 2 shows the steps involved in traffic sign detection.

SSD architecture builds on top of VGG-16 architecture by removing the fully connected layers. A set of auxiliary convolutional layers are added for extracting the features at multiple scales and progressively decrease the size of the input to each subsequent layer. Non-maximum suppression is applied to get the final bounding detections along with the class labels.

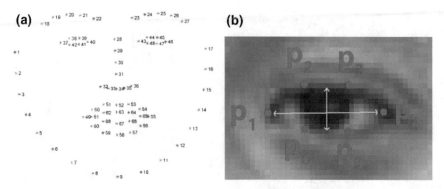

Fig. 3 **a** 68 facial landmark coordinates. **b** Facial eye landmark locations

2.2 Drowsiness Detection Model/Internal View

The existing method calculates only the eye aspect ratio (EAR) value for number of times the person blinked his eyes. The proposed method considers the sequences of EAR value to detect the drowsiness of driver. It has three steps such as face localization, detecting facial structures in the detected face and estimating the eye aspect ratio. Based on reference [9], we have used histogram of oriented gradients (HOG) and linear SVM-based classifier for face localization. Ensemble of regression trees technique [10] is used for facial landmark localization on each detected face image and it detects the $68(x, y)$-coordinates that map to facial structures on the face. Figure 3a shows the 68 facial landmark coordinates.

The facial landmark localization technique detects various regions of the face, namely, eyes, eyebrows, nose, mouth and jawline. We are considering only left and right eyes to calculate eye aspect ratio. Each eye is represented by $6(x, y)$-coordinates, starting at the left corner of the eye, and then working clockwise around the remainder of the region. Figure 3b shows the facial eye landmark locations.

Equation (1) represents the eye aspect ratio calculation for drowsiness detection, where $p1, \ldots, p6$ are 2D facial eye landmark locations.

$$EAR = \frac{||p2 - p6|| + ||p3 - p5||}{2 * ||p1 - p4||} \tag{1}$$

Eye aspect ratio (EAR) [9] value is used to compute the drowsiness of the driver by considering the continuous frame EAR values. Figure 3 shows the EAR value computation for N frames.

From Fig. 4, it is shown that if eye closes, the EAR values became close to zero and if the eye is open, EAR values are above certain threshold. The drowsiness alert will be generated when N frames of EAR values are close to zero. Figure 5 describes the model flow of the drowsiness detection system.

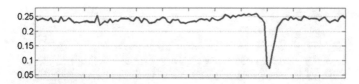

Fig. 4 EAR values *N* frames

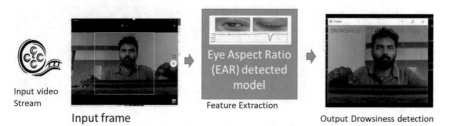

Fig. 5 Steps involved in drowsiness detection model

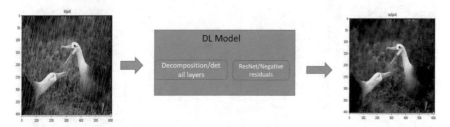

Fig. 6 Steps involved in de-rain model

2.3 De-raining Model/Environmental View

Illumination, noise and severe weather conditions (i.e. rain, snow and fog) adversely affect the performance of many computer vision algorithms such as detection, classification and tracking. In this paper, deep detailed network architecture [7] is used for removing rain streaks. This architecture basically has two submodels such as decomposition and residual network. Decomposition layer is used to get the object boundaries in a frame and residual network is used to identify the region of rain streaks. Figure 6 shows the steps involved in de-rain model.

Negative residual mapping is used to find the difference between clean and rainy images and to learn the network. They created and used a synthetic dataset of 14,000 rainy/clean image pairs. Structural similarity index (SSIM) is used to evaluate the performance of detailed network.

Fig. 7 Sample images of LISA traffic sign dataset

Fig. 8 Overview of ADAS^{DL} application

3   Overview of ADAS^{DL} Application

LISA traffic sign dataset has 6610 sample images for 49 classes of traffic sign and each image size is 640 * 480. Figure 7 shows the sample images of LISA traffic sign dataset.

The application has been developed using Python and related UI stacks. Figure 8 outlines the overall ADAS^{DL} user interface which has three camera views street view, environmental view camera mounted outside of the vehicle and one camera placed inside of the vehicle for internal view. From the street view camera feed, we are detecting the traffic signs followed by internal view camera for drowsiness detection and environment view camera feed for removing rain streaks.

Table 1 Performance of ADAS^{DL}

S. No	Submodule name	Application input	Inference timings on ADAS^{DL} (sec)
1	Street view (Traffic sign)	One frame	4.07
2	Environment view (De-raining)	One frame	3.9
3	Internal view (Drowsiness)	54 s video (810 frames)	1.245[a]

[a] AVG value for response per frame

4 Results and Summary

The overall performance of the ADAS^{DL} is summarized in Table 1. The traffic sign detection model took average of 4 s/frame, de-raining took average of 3.5 s/frame and drowsiness detection took average of 1.2 s/frame.

5 Conclusion and Future Works

In this paper, we proposed hybrid deep-learning-based ADAS^{DL} application which consists of traffic sign detection, drowsiness detection and rain/fog removal. Areas for future work include edge-level deployment, performance improvement and new use case integrations. In future to improve the performance of drowsiness detection along with EAR value, the EEG signal data may be used.

References

1. https://www.its.dot.gov/communications/media/15cv_future.htm
2. H.N. Dean, K.V.T. Jabir, Real time detection and recognition of Indian traffic signs using Matlab. Int. J. Sci. Eng. Res. **4**(5), 684 (2013)
3. H. Ueno, M. Kaneda, M. Tsukino, Development of drowsiness detection system, in *Proceedings of 1994 Vehicle Navigation and Information Systems Conference*, Yokohama, Japan (IEEE, New York, 1994), pp. 15–20
4. T.C. Kao, T.Y. Sun, Head pose recognition in advanced driver assistance system, in *2017 IEEE 6th Global Conference on Consumer Electronics (GCCE)*, Nagoya (2017), pp. 1–3. https://doi.org/10.1109/gcce.2017.8229416
5. A. Simić, O. Kocić, M.Z. Bjelica, M. Milošević, Driver monitoring algorithm for advanced driver assistance systems, in *2016 24th Telecommunications Forum (TELFOR)*, Belgrade (2016), pp. 1–4. https://doi.org/10.1109/telfor.2016.7818908
6. A. Møgelmose, M.M. Trivedi, T.B. Moeslund, Vision based traffic sign detection and analysis for intelligent driver assistance systems: perspectives and survey. IEEE Trans. Intell. Transp. Syst. (2012)

7. X. Fu, J. Huang, D. Zeng, Y. Huang, X. Ding, J. Paisley, Removing rain from single images via a deep detail network, in *IEEE Conference on Computer Vision and Pattern Recognition* (CVPR) (2017)
8. W. Liu, D. Anguelov, D. Erhan, C. Szegedy, S. Reed, C.-Y. Fu, A.C. Berg, SSD: single shot multibox detector, in *ECCV* (2016)
9. T. Soukupová, J. Čech, Real-time eye blink detection using facial landmarks, in *21st Computer Vision Winter Workshop* (2016)
10. V. Kazemi, J. Sullivan, One millisecond face alignment with an ensemble of regression trees, in *2014 IEEE Conference on Computer Vision and Pattern Recognition*, Columbus (2014), pp. 1867–1874

Vowel Onset Point Detection in Hindi Language Using Long Short-Term Memory

Arpan Jain, Amandeep Singh and Anupam Shukla

Abstract In this paper, we have discussed about the Vowel Onset Point (VOP) for the Hindi language and its significance in the speech recognition. We have defined the vowel onset point and how it can be calculated. Alphabets in Hindi language are the combination of the vowel and consonant part. In Hindi, we cannot pronounce a consonant without a vowel. There is a very small region between consonant and vowel where transition happens from consonant to vowel. We have used characteristics of the sound files to get the vowel onset point. To calculate Vowel Onset Point, we have applied filtration process, and after that, we can use energy of the signal and different formants combined with epoch interval and Itakura distance. Filtered energy and filtered formants can be used as cues for accurately detecting VOP within the range of $+/-30$ ms. In order to further increase the effectiveness of the proposed method, we have used Recurrent Neural Network variants to detect VOP which uses speech features and reference point calculated by filtered formants.

Keywords VOP · Filter formants · Filtered energy · RNN · LSTM

1 Introduction

In this paper, we have discussed about the VOP detection in Hindi language. Our proposed method is for single alphabet of Hindi language. In Hindi language, single alphabet is made of consonant and vowel region (vyanjan+swar). Our research found that in Hindi alphabet, consonant and vowel region overlap with each for very small

A. Jain (✉) · A. Singh · A. Shukla
Indian Institute of Information Technology and Management Gwalior, Gwalior, India
e-mail: arpan.jain1405@gmail.com

A. Singh
e-mail: amandeepsinghrattan@gmail.com

A. Shukla
e-mail: dranupamshukla@gmail.com

© Springer Nature Singapore Pte Ltd. 2019 505
S. C. Satapathy et al. (eds.), *Information Systems Design and Intelligent Applications*,
Advances in Intelligent Systems and Computing 862,
https://doi.org/10.1007/978-981-13-3329-3_47

time duration, therefore, we propose VOP is very small region of range 20–50 ms which depends on the type of Hindi alphabet. The applications of VOP are:

1. VOP can be used for detection of Consonant Vowel (CV) units in the Hindi language.
2. VOP can be used for increasing speech recognition accuracy.
3. VOP can be used for calculating speech rate.
4. Determining consonant and vowel region duration.
5. Expressive speech processing.
6. Speaker authentication and identification.

Various methods are proposed for calculating VOP. In [1], they have proposed method based on analyzing the effect of speech coding on the detection of vowel onset point. VOP detection method uses the spectral energy of the sound in the glottal closure region. In [2], the method is based on detecting VOP by marking the points at which there is a sudden increase in the vowel strength. The vowel strength is measured using the change in the energy of each of the peaks and its valleys in the amplitude spectrum. VOP detection method presented in [3] uses product function which can be calculated from the wavelet and scaling coefficients of input speech signal. The values of the product function during vowel region are much higher than the consonant region. Therefore, the product function can be used to find the VOP. In [4], they have calculated VOP with a deviation of 10 ms and with 85% accuracy. They have calculated VOP in two levels; in the first level, they have used excitation source, spectral peaks, and modulation spectrum. In the second level, VOPs at the first level are verified and corrected using the uniform epoch intervals present in the vowel region. In [5], we use methods such as Hierarchical Neural Network (HNN), Multilayer Feed-Forward Neural Network (MLFFNN), and Auto-Associated Neural Network (AANN) models to detect the VOPs. In [6], VOP is calculated using Hilbert envelope of excitation. Our proposed method consists of three stages. The first stage calculates the voice region of the sound so that we can determine the Consonant + Vowel duration through energy and zero crossing rate. On the extracted part, strength of instants, epoch interval, and Itakura distance is calculated to get the idea about VOP location. In the second stage, sound files are filtrated using low-pass and high-pass filters. On filtered files, spectral energy and formants are calculated and graphs are made, change point analysis is done on the graphs. In the third stage, deep learning model is applied.

2 Data Description

We have used our own recorded data set to analyze and test the proposed method for VOP detection for Hindi Language. Sound files are recorded at 44,200 frames per second and in noise-less environment. Each sound file contains one Hindi alphabet pronounced in standard dialect (Khari Boli) by different users. Dataset has 5022 sound files. There are 30 different Hindi alphabets. For each alphabet, there are 165–170 sound files.

3 Methodology

Our proposed method for VOP detection consists of three stages. The first stage calculates the voice region of the sound through spectral energy and zero-crossing rate. On the extracted part, strength of instants, epoch interval, and Itakura distance is calculated to get the idea about VOP location. In the second stage, sound files are filtrated using low-pass and high-pass filters. On filtered files, spectral energy and formants are calculated and graphs are made, change point analysis is done on the graph to get the probable location of the VOP. In stage 3, Recurrent Neural Networks variants are trained and used to detect the VOP.

3.1 Stage 1

3.1.1 Voiced Region Detection in Sound Files

To accurately identify VOP and increase computational efficiency of the proposed method, voiced region is calculated. Voiced region is defined as the region where alphabet is pronounced. In other words, voiced region is the total duration of alphabet in the sound file and it consists of consonant and vowel region. Voiced region calculation is based on energy of the speech signal, autocorrelation, and zero-crossing rate.

Based on short-time energy Amplitude of the speech signal varies with time. Energy of the sound file calculated at small duration demonstrates the variation of the amplitude in the speech signal. Since our sound file contains only one continuous voiced region, we can use Short-Time Energy (STE) for detecting voiced region in the sound file. Speech signal is divided into the intervals of 5 ms. Speech is recorded at the rate of 44,200 frames per second, therefore, each interval will contain approximately 221 frames per interval. After dividing speech signal into intervals, the energy of interval is calculated using below formula.

$$Ex = \sum n|x[n]|^2 \tag{1}$$

Based on zero-crossing rate Zero Crossing is defined as successive samples having different algebraic signs in context of discrete time signals. The rate at which Zero Crossing occurs is known as the Zero-Crossing Rate (ZCR). ZCR is measure of number of times in the given interval or frame the amplitude of the speech signal crosses the axis or change the signs. Spectral properties of ZCR are defined in Rabiner and Schafer [7]. ZCR of unvoiced region is high in comparison to the voiced region therefore this can be used as a cue for calculating voiced region.

$$Z_n = \sum_{m=-\infty}^{\infty} |sgn[x(m)] - sgn[x(m-1)]|w(n-m) \tag{2}$$

where

$$\text{sgn}[x(n)] = \begin{cases} 1, & x(n) >= 0 \\ -1, & x(n) < 0 \end{cases}$$

and

$$w(n) = \begin{cases} 1/2N, & \text{for } 0 <= n < N - 1 \\ 0 \text{ for,} & \text{otherwise} \end{cases}$$

Correlation between the frequencies of speech signal is high for the voiced region in comparison to unvoiced region. Since noise is a random signal, correlation should be zero. Speech signal is divided into the interval of 0.1 ms and correlation for the given interval are calculated with respect to previous interval.

Voiced region detection accuracy by STE and ZCR method is effected by noise, therefore autocorrelation can be used to identify the region where voiced part may occur then ZCR and STE method can be applied to calculate the starting and ending point of voiced region [7].

3.1.2 Epoch Interval

Epoch interval is defined as intervals between two successive significant points of excitation in the speech signal. During the vowel region, epoch intervals remain constant within a range, therefore, epoch interval is used as a reference for detecting the VOP. In some cases, epoch interval becomes constant before VOP, therefore, VOP detected using epochs only might be inaccurate [8].

3.1.3 Itakura Distance

Itakura distance shows the change in the characteristics of speech in the sound file. During the transition from consonant region to vowel region, Itakura distance is generally high therefore this can used as a reference for VOP marking. The points marked by epoch interval and Itakura distance are generally close. Itakura distance fails in marking the VOP in semivowels like Yan [9].

$$d(x, y) = \begin{cases} \sum_j (ln(y_j/x_j) + x_j/y_i - 1), & \text{if } x > 0 \quad \text{and } y > 0 \\ +\infty, & \text{otherwise} \end{cases} \quad (3)$$

3.1.4 Formant Structure

In some alphabets, formants of consonant region are lower in comparison to formants of vowel region, therefore formant structure can be used for marking VOP for reference [10].

Generally, VOP is detected after Itakura distance and epoch interval. Therefore in the second stage of the proposed method, we start VOP detection from the point refer by epoch interval, Itakura distance, and formant distance.

3.2 Stage 2

In the second stage, speech signal is filtered so that any white noise can be reduced in the speech file. We propose two features: filtered energy and filtered formants which are used to find the probable location of the VOP in the stage 2.

3.2.1 Filtration

Two methods are used for filtration one for filtered energy and another for filtered formants. In the first method, time domain audio signal is converted into frequency domain by using Fourier transform. All the frequencies that are not between 250 and 750 Hz are filtered out. The resultant signal is converted into time domain by inverse Fourier transformation. In the second method, low-pass filter and high-pass filter is used for filtered formants. At implementation level, butter low-pass filter and butter high-pass filter of order 5 is used. Cutoff frequency for butter high-pass filter is 250 Hz and for butter low-pass filter varies from 500 to 700 Hz according to the audio file.

3.2.2 Filtered Energy

After stage 1, filtration method one is used on the audio file to remove any unwanted signal so that energy calculated is of alphabet. Energy of vowel region of the Hindi alphabet is generally higher than consonant region part. Therefore, filtered energy is used as a reference so that filtered formants can get the VOP accurately. Filtered energy is the last reference of VOP and it checks the accuracy of VOP calculated by epoch interval, Itakura distance, and formant structure. When vowel region starts, there is significant change in energy for stop consonants, affricates, and fricatives.

3.2.3 Reference Point Calculation

Itakura distance, epoch interval, formant Structure, and filtered energy are used for calculating reference point for VOP detection. For Itakura distance, epoch interval, and formant structure, we take the point which is the average of all three features. Sometimes, one of the features gives absurd result, therefore, we neglect that feature while calculating mean. Absurd results are obtained due to noise or particular feature cannot be used to determine VOP for that alphabet. Mean is compared with the point given by the filtered energy, smallest value of these two is selected as the reference point for the VOP detection.

3.2.4 Filtered Formants

After stage 1, audio files are filtered by the method two explained above. Cutoff frequency for butter low-pass filter is selected. Formant is estimated using Linear Predictive Coding (LPC) coefficients method [11]. Before applying LPC coefficients, two common preprocessing steps are applied to the audio file that is windowing and preemphasis filtering. Hamming window of size equal to the size of audio length is used. After preprocessing, LPC method is applied with model equals to two more than frames per milliseconds. Roots of the prediction polynomial are calculated and roots having imaginary part less than zero are filtered out. Angles of roots are calculated and converted into Hz. Frequency less than 90 Hz or having bandwidth greater than 400 Hz are filtered out. Formants are calculated by dividing audio file into interval of 10 ms and window length of 25 ms. Formants plotted in vowel region formant changes randomly as compared to consonant region. Probable location for VOP is marked by filtered formant generally after the VOP is marked by filtered energy and other reference features (Fig. 1).

3.3 Stage 3

In this stage, variants of Recurrent Neural Network (RNN), Long Short-Term Memory (LSTM), and Gated Recurrent Unit (GRU) are trained on the training dataset.

3.3.1 Data Prepossessing

As discussed in stage 1, voiced region is extracted from the sound files. Since each millisecond plays an important role in the detection of the VOP, therefore, Mel Frequency Cepstrum Coefficients (MFCC) features are calculated with window length of 1 ms and window step of 1 ms. Since the length of alphabets is not constant therefore number of MFCC features for each sound file will be different. Padding is used to tackle this problem.

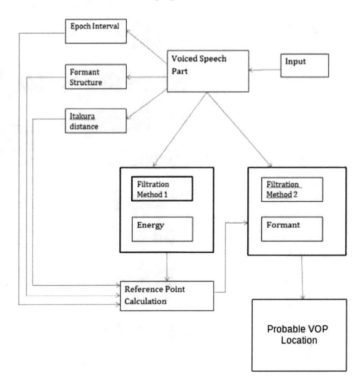

Fig. 1 Block diagram for stage 1 and 2

3.3.2 Architecture

In the architecture, three RNN layers along with two dense layers are used. Probable location of the VOP can be incorporated in the deep learning model using two ways as follows:

- **Method 1**: It is observed that actual VOP always lies within the maximum range of 170 ms from the probable location of the VOP, therefore, 340 ms part can be extracted from the sound file. This makes the sound files of equal length and in most of the cases elevates the need for the padding.
- **Method 2**: Instead of extracting the region around the probable location of the VOP which is done in the method 1, probable location of the VOP calculated in stage 2 can be passed as a feature directly to last layer of the deep learning model. In this method, padding is required for each and every sound file (Figs. 2 and 3).

Fig. 2 Method 1 general
architecture

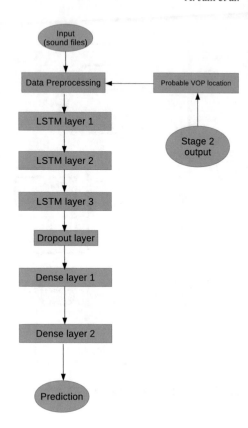

4 Experiment

In order to evaluate the accuracy of the proposed method, dataset is divided into the
training and testing set in the ratio of the 80 and 20%. Two variants of the RNN are
used in different models, i.e., LSTM and GRU. First of all, a base model is trained
and tested without using the probable location of the VOP. There are two base models
one using LSTM and other one using GRU, and they have same three architecture
layers of the respective RNN variant and two dense layers. Then, two LSTM models
are trained and tested by using Method 1 and Method 2 described above. In the same
way, GRU models are trained and tested by using Method 1 and Method 2. Therefore,
a total of six models are trained and tested. All methods are implemented by using the
Python language and for developing the deep learning model, Keras library is used.
System configurations are i7-6850K CPU @ 3.60GHz * 12 and GTX-1080Ti with 11
GB video card memory. It took around 10 mins to train the model. Table 1 shows the
result of the experiments. In order to evaluate the models, average absolute testing
error (in milliseconds) is calculated and testing files are divided into an interval of

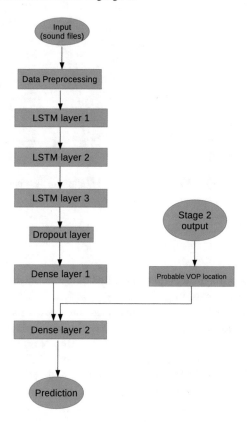

Fig. 3 Method 2 general architecture

Table 1 Results

Model	Average absolute testing error	80%	85%	90%	95%
LSTM Base Model	23.6674	38	42	49	61
LSTM Method 1	22.6156	36	40	47	61
LSTM Method 2	21.6051	32	37	45	55
GRU Base Model	27.05814	42	47	55	68
GRU Method 1	26.8947	41	46	54	67
GRU Method 2	25.5143	39	44	50	62

5 ms based on the error and then, error interval which have 80, 85, 90, and 95% of testing files are calculated and given in the results table (Fig. 4).

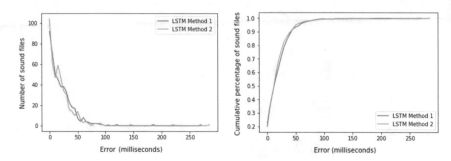

Fig. 4 Two best models (LSTM Method 1 and LSTM Method 2) distribution of sound files with respect to error

5 Conclusion

In our experiments, our hybrid proposed method outperforms the base deep learning model by a significant margin. In our study, it is found that LSTM works better in comparison to the GRU. It was evident from the results that probable location of the VOP calculated in stage 2 of the proposed method plays an important role and is helpful in increasing the accuracy of the deep learning model. Method 2 of incorporating the probable location of the VOP with LSTM model yielded the best result which has achieved the accuracy of 80% with $+/-30$ ms and 95% with $+/-55$ ms with the average absolute error of 21.6051 ms.

6 Future Scope

We have proposed and tested a method for VOP detection for Hindi alphabets having "a" as vowel. Our proposed method can also be modified and tested for VOP detection for Hindi alphabets having other vowels like "e", "u", "ae", etc. We have used LPC method for calculating formants, Burg method for calculating formants can also be used and it may further increase the accuracy of our proposed method.

References

1. A.K. Vuppala, J. Yadav, S. Chakrabarti, K.S. Rao, Vowel onset point detection for low bit rate coded speech. IEEE Trans. Audio Speech Lang. Process. **20**(6), 1894–1903 (2012)
2. D.J. Hermes, Vowel-onset detection. J. Acoust. Soc. Am. **87**(2), 866–873 (1990)
3. J.F. Wang, S.-H. Chen, Ac/v segmentation algorithm for mandarin speech signal based on wavelet transforms, in *1999 IEEE International Conference on Acoustics, Speech, and Signal Processing, 1999. Proceedings.*, vol. 1, March 1999, pp. 417–420

4. A.K. Vuppala, K.S. Rao, S. Chakrabarti, Improved vowel onset point detection using epoch intervals. AEU - Int. J. Electron. Commun. **66**(8), 697–700 (2012) [Online]. Available http://www.sciencedirect.com/science/article/pii/S1434841111003207

5. S.V. Gangashetty, C.C. Sekhar, B. Yegnanarayana, Detection of vowel onset points in continuous speech using autoassociative neural network models, in *Proceedings of the International Conference on Spoken Language Processing*, pp. 401–410 (2004)

6. S.M. Prasanna, B.S. Reddy, P. Krishnamoorthy, Vowel onset point detection using source, spectral peaks, and modulation spectrum energies. IEEE Trans. Audio Speech Lang. Process. **17**(4), 556–565 (2009)

7. L. Rabiner, R. Schafer, Digital speech processing. Froehlich/Kent Encycl. Telecommun. **6**, 237–258 (2011)

8. K.S.R. Murty, B. Yegnanarayana, Epoch extraction from speech signals. IEEE Trans. Audio Speech Lang. Process. **16**(8), 1602–1613 (2008)

9. H.H. Bauschke, R.S. Burachik, P.L. Combettes, V. Elser, D.R. Luke, H. Wolkowicz, Fixed-point algorithms for inverse problems in science and engineering. Springer Optimization and Its Applications (2011)

10. B. Yegnanarayana, R.N. Veldhuis, Extraction of vocal-tract system characteristics from speech signals. IEEE Trans. Speech Audio Process. **6**(4), 313–327 (1998)

11. T. Mellahi, R. Hamdi, Lpc-based formant enhancement method in Kalman filtering for speech enhancement, AEU - Int. J. Electron. Commun. **69**(2), 545–554 (2015) [Online]. Available http://www.sciencedirect.com/science/article/pii/S1434841114003112

An Assessment of Interaction Protocols for Multi-agent Systems

Dimple Juneja, Jasmine, Shefali, Pratibha Kumari and Anjali Arya

Abstract The paper begins by instigating the sketch of agents from different perspectives and expanding to the assessment of various protocols for agent communication and provides an analysis of various available protocols. The paper majorly contributes a concise review of all agent protocols and also presents the issues that can be addressed for further improvement. The article settles by presenting the open challenges in the said domain and future research directions in the said domain.

Keywords Software agents · Intelligent agents · Intelligent computing
Multi-agent systems · Agent interaction protocols

1 Introduction

An agent is a software entity able to observe its environs through devices and reflect through effectors [1]. The word agent is derived from the Latin word "*agere*" (to do) which states an agreement to work on one's behalf [2]. Another definition states that agent is an independent program which is proficient in controlling its decisions and actions based on its perception of its environment [3]. According to researchers, an agent is an entity that is able to carry out goals and belongs to multi-agent sys-

D. Juneja (✉) · Jasmine · Shefali · P. Kumari · A. Arya
National Institute of Technology, Kurukshetra, India
e-mail: dimplejunejagupta@gmail.com

Jasmine
e-mail: jas.031408@gmail.com

Shefali
e-mail: shefali25.1994@gmail.com

P. Kumari
e-mail: pratibhakumari521@gmail.com

A. Arya
e-mail: anjaliarya184@gmail.com

© Springer Nature Singapore Pte Ltd. 2019   517
S. C. Satapathy et al. (eds.), *Information Systems Design and Intelligent Applications*,
Advances in Intelligent Systems and Computing 862,
https://doi.org/10.1007/978-981-13-3329-3_48

tem having joint impact on each other [4]. In other words, a software agent is an intelligent, self-directed program that senses and reacts to its environment without direct human supervision to complete some task for an end user. Agents can perhaps be considered as the next step in a gradual progress of development starting from imperative programming to agent-oriented programming in the field of software engineering. Agents are designed to be autonomous, proactive, reactive, social, etc. Further, agents are categorized according to one or more properties that they possess. There are numerous factors such as mobility, behavior, and role being played that are responsible for classification of agents [5]. In fact, multiple heterogeneous agents form a society of agents to collaboratively achieve an objective forming a multi-agent system [6]. In other words, "A multi-agent system is a loosely coupled network of problem-solving entities that work together to find answers to problems that are beyond the individual capabilities or knowledge of each entity" [1]. Each agent has a limited viewpoint in the sense that it has incomplete information or competence for resolving the issues. Agents in MAS have no control over the system globally and moreover the data is available in decentralized mode leading to asynchronous computation.

The paper is organized into five sections. Section 1 introduced the concept of agents and also provided an insight into classification of agents. Section 2 acknowledges the related work and Sect. 3 assesses various interaction protocols in MAS. Section 4 highlights the unfolded challenges and also provide a vision of future research directions. Section 5 finally concludes.

2 The Related Work

The multi-agent systems have been finding space in wide variety of domains and the growth seems to be exponential. Recent works by Jenabzadeh and his team [7] have addressed the problem of distributed tracking control problem by considering Lipschitz multi-agent systems. Cunha et al. [8] redefine MASs as a community of heterogeneous and independent agents which work toward similar or different aims. Cohen and his coworkers [9] have considered deploying agents in rescue operations offering flexibility in terms of agents entering and exiting the system anytime during the process. Literature also enlists a commitment-based interaction protocol in JaCaMo+ [10] that facilitates reuse of the interaction code and composition of interaction protocols.

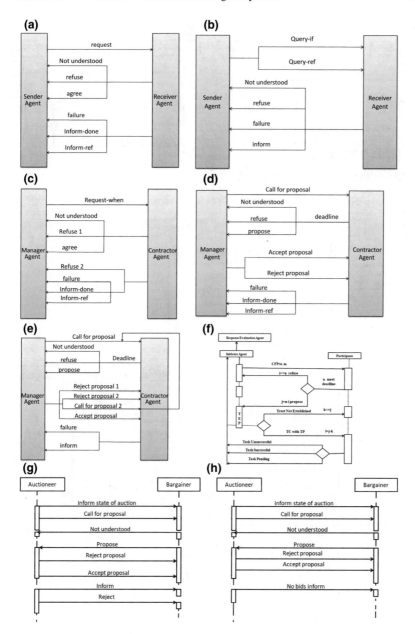

◀**Fig. 1 a** Request interaction protocol. **b** Query interaction protocol. **c** Request when interaction protocol. **d** Contract net protocol. **e** Iterated contract net protocol. **f** CNTEP protocol. **g** Auction English interaction protocol. **h** Auction Dutch interaction protocol. **i** Subscribe interaction protocol. **j** Propose interaction protocol. **k** FIPA recruiting interaction protocol. **l** FIPA brokering interaction protocol

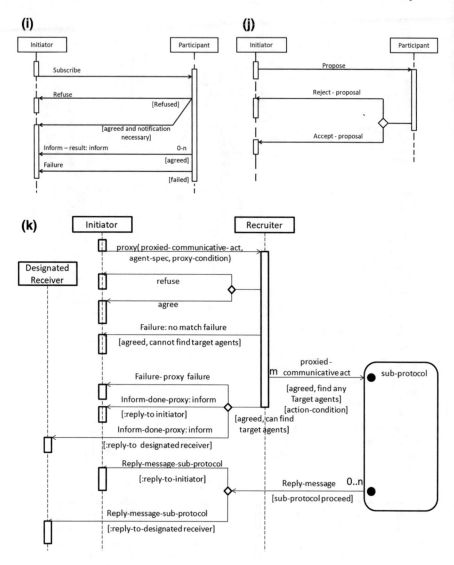

◄**Fig. 1** (continued)

(I)

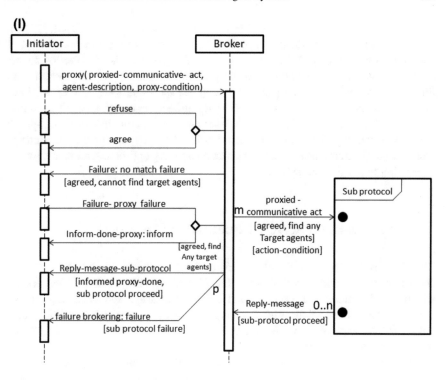

◄**Fig. 1** (continued)

In 2017, Lillis proposed Agent Conversation Reasoning Engine (ACRE), with the intention to facilitate the practical development, debugging, and deployment of communication-heavy MASs. Many recent works [11–13] also favors that agents are champions in offering solution to complex and intelligent problems. Further, an illustrated review pertaining to agent-based systems [14] indicates that agents have been in existence for more than three decades and will continue to strive in future. Recent AI-based applications such as Alexa[1] are another contemporary example of latest developments. The available literature indicates that agents will continue to interact and more emphasis should be given on defining optimizing protocols for the interaction in multi-agent systems.

3 The Assessment of Agent Interaction Protocols

The work by Wajid and Mehandjiev [15] mentions that individual agents interact, cooperate, and negotiate with other agents to achieve some desired objectives for which certain interaction structure and expected message sequence needs to be defined. This pattern of message exchange is known as Agent Interaction Protocol (AIP). As explored from literature, Fig. 1a–l outlines the working of various interaction protocols [16].

The analysis of the listed protocols indicates that communicative acts and proposals are inevitable for all protocols. CNP is one of the most popular protocols and is being used most prominently for establishing communication among agents in MAS. It mainly focuses on the interaction between agents for the specified amount of time period. ICNP is the extended version of CNP offering the iterative function until either the requests are completed or the time period has exceeded. On the other hand, query protocol answers the queries of different agents. Request protocol is based on the requesting factor and responds to the requests. Request when protocol has somehow similar working as that of the request protocol but the only thing is that the precondition should always be true for the request to be processed. Auction English Protocol and the Auction Dutch Protocols are based on the auction property. But, in contrast to Auction Dutch Protocol which chooses from low to high bid, Auction English Protocol proposes a CFP and the bidder is chosen by the auctioneer. Failure-to-understand protocol basically provides the facility to communicate by sending not understood message.

4 The Unfolded Challenges and Research Directions

The literature strongly advocates the use of CNP and its variations [17–21] to solve a problem cooperatively by achieving better communication. However, there are

---

[1]https://www.technologyreview.com/s/608430/growing-up-with-alexa.

many unfolded challenges that shall be addressed for optimal performance. Of the various challenges, the effective allocation of design agents to distributed and cooperative design tasks is the most crucial issue. There is a need to extend the protocols for decentralized task allocation in cooperative engineering design. For instance, although CNP is the most popular and favored protocol; however, the same does not prescribe any specific coordination policies for cooperative engineering design. It requires adding a set of agent selection and task selection policies, to address the issues in agent-based cooperative engineering design such as, how an agent is selected to carry out a design task, how can an agent select from among a list of design tasks which to bid for, and so on. Further, the creation of open agent-mediated electronic marketplaces requires new architectures, capable of coping with unreliable computational and network infrastructures, limited trust among independently developed agents, and the possibility of systemic failures. So some methodology is needed to help guarantee stability and efficiency during communication of agents. In any of the FIPA standardized interaction protocols, the requester agents may get several proposals (bids) from the responders (agents). Now there is a chance that the responders seem to be reliable but they are actually not. Hence, the reliability of the contractor agent should be checked before granting the bids. Although RCNTEP is one of such protocols which advocate computing the reliability of agents; however, the same has not been standardized by FIPA. To the best of our understanding, most of these protocols assume that all agents are approachable and compassionate which leads to non-detection of conflicts (not even in negotiation protocol) and additionally these are unresolvable. Further, most of these agents are egotistic, and hence in most cases these turn out to be practical and unfriendly. Few more prevailing issues are unreliable communication infrastructure, unpredictable life of agents, requirement of quick response and delivery times, and timebound framework. The abovementioned issues once resolved would raise the level of intelligence possessed by multi-agent systems. Since there are very few clustering based protocols [22] available in multi-agent system till date and none of them supports intelligent and autonomous clustering of agents, so a clustering based interaction protocol can be a promising solution. Further, it is evident from the functionality of CNP, the initiator agent establishes contract with contractors on the basis of bids but it nowhere considers validation of agents. A variation in the same is CNTEP that considers computing trust percentile among the agents that are already admitted into the system which again is a loophole as the agents should be authenticated before giving an entry into the system. The in-depth grilling of literature made us understand that agents have been playing a vital role and the field has been witnessing exponential growth. To the best of our understanding and knowledge, the field has ample scope of improvement and the research is still open. Hence, the motivation for future work shall consider extension of CNTEP by authenticating agents before giving admission to a particular system.

5 Conclusion

The paper presented a detailed study about distributed multi-agent systems and the thorough analysis of literature suggested that incorporating clustering technique in one of the agent interaction protocol (preferably cooperation) is a promising solution. It shall lead to better and efficient performance of the protocol. Although the idea of inclusion of clustering is not novel; however, not much work is available to support. Owing to the advantages that clustering based protocols offer in other fields, the performances and the glitches offered by current set of protocols can be removed by introducing a clustering based protocols in the domain of MAS. Further, a protocol that considers authenticating and validating agents before sharing the CFP should offer an edge over current versions of CNP.

References

1. M. Glavic, Agents and multi-agent systems: a short introduction for power engineers, a technical report, University of Liege Electrical Engineering and Computer Science Department, pp. 1–21 (2006)
2. S. Wettig, E. Zehender, A legal analysis of human and electronic agents. Artif. Intell. Law **12**(1–2), 111–135 (2004)
3. N. Jennings, M. Wooldridge, Software agents. IEE Rev. **42**(1), 17–20 (1996)
4. C.E. Georgakarakou, A.A. Economides, Software agent technology: an overview application to virtual enterprises, in *Agent and Web Service Technologies in Virtual Enterprises*, ed. by P.F. Tiako (ed.) IGI-Global, ISBN: 978-1-60566-060-8 (2003)
5. H.S. Nwana, D.T. Ndumu, A brief introduction to software agent technology, in *Agent Technology* (Springer, Berlin, 1998), pp. 29–47
6. S.A. DeLoach, M.F. Wood, C.H. Sparkman, Multiagent systems engineering. Int. J. Software Eng. Knowl. Eng. **11**(03), 231–258 (2001)
7. A. Jenabzadeh, B. Safarinejadian, Distributed tracking control problem of Lipschitz multi-agent systems with external disturbances and input delay. Syst. Sci. Control. Eng. **6**(1), 268–278 (2018)
8. F.J.P. da Cunha, T.F.M. Sirqueira, M.L. Viana, C.J. Pereira, Extending BDI multiagent systems with agent norms. World Acad. Sci. Eng. Technol. Int. J. Comput. Inf. Eng. **12**(5) (2018)
9. J. Cohen, J.S. Dibangoye, A.I. Mouaddib, Open decentralized POMDPs, in *2017 IEEE 29th International Conference on Tools with Artificial Intelligence (ICTAI)*. (IEEE, New York, 2017 November), pp. 977–984
10. M. Baldoni, C. Baroglio, F. Capuzzimati, R. Micalizio, Commitment-based agent interaction in JaCaMo+. FundamentaInformaticae **159**(1–2), 1–33 (2018)
11. V. Bhateja, B. Le Nguyen, N.G. Nguyen, S.C. Satapathy, D.N. Le (eds.), Information systems design and intelligent applications, in *Proceedings of Fourth International Conference India 2017*, vol 672 (Springer, Berlin, 2018)
12. D.J. Lillis, Internalising interaction protocols as first-class programming elements in multi agent systems. arXiv preprint arXiv:1711.02634 (2017)
13. J.K. Mandal, S.C. Satapathy, M.K. Sanyal, P.P. Sarkar, A. Mukhopadhyay (eds.) Information systems design and intelligent applications, in *Proceedings of Second International Conference INDIA 2015*, vol 2 (Springer, Berlin, 2015)
14. D. Juneja, A. Singh, R. Singh, S. Mukherjee, A thorough insight into theoretical and practical developments in multiagent systems. Int. J. Ambient. Comput. Intell. (IJACI) **8**(1), 23–49 (2017)

15. U. Wajid, N. Mehandjiev, Agent interaction protocols and flexible agent interaction in dynamic environments, In *15th IEEE International Workshops on Enabling Technologies: Infrastructure for Collaborative Enterprises, 2006. WETICE'06* (IEEE, New York, 2006 June), pp. 23–28
16. M.P. Huget, J. Odell, Ø. Haugen, M.M. Nodine, S. Cranefield, R. Levy, L. Padgham, FIPA modeling: interaction diagrams. Work. Draft Vers. pp. 07–02 (2003)
17. A. Singh, D. Juneja, An improved design of contract net trust establishment protocol. Int. J. Commun. **4**(1), 19 (2013)
18. A. Singh, D. Juneja, A.K. Sharma, Introducing trust establishment protocol in contract net protocol, in *2010 International Conference on Advances in Computer Engineering (ACE)* (IEEE, New York, 2010 June), pp. 59–63
19. R. Singh, A. Singh, S. Mukherjee, A critical investigation of agent interaction protocols in multiagent systems. Int. J. Adv. Technol. **5**(2), 72–81 (2014)
20. J. Vokřínek, J. Bíba, J. Hodík, J. Vybíhal, M. Pěchouček, Competitive contract net protocol, in *International Conference on Current Trends in Theory and Practice of Computer Science* (Springer, Berlin, 2007 January), pp. 656–668
21. J. Wu, Contract net protocol for coordination in multi-agent system, in *Second International Symposium on Intelligent Information Technology Application, 2008. IITA'08*, vol 2 (IEEE, New York, 2008 December), pp. 1052–1058
22. D. Juneja, R. Singh, A. Singh, S. Mukherjee, A clustering-based generic interaction protocol for multiagent systems, in *Information Systems Design and Intelligent Applications* (Springer, New Delhi, 2016), pp. 563–572

An Analytical Approach for Asthma Attack Prevention

Aakash Gupta

Abstract Asthma patients sometimes immediately need medical attention due to sudden breath loss or asthma attack. An analytical approach is proposed in this paper to prevent a patient from asthma attack. Weather, air quality and pollen data are used to predict the probability of asthma attack for taking preemptive action using logistic regression. The proposed solution implemented on environmental datasets and it achieves 84.2% accuracy.

Keywords Logistic regression · Asthma patient alert system
Asthma attack prediction

1 Introduction

The main motivation for the paper is the survey results of the US National Asthma New York Summary Report [1], which mentions that 467,000 children between 0 and 17 years had lifetime asthma and 368,000 children had current diagnosed asthma. Among adults 18+ years, 1,480,000 had lifetime asthma and 1,087,000 had current diagnosed asthma.

Due to frequent environmental changes in weather, pollen and air quality asthma patients are suffering from losing breathe or asthma attack. But gathering environmental data with the help of technologies like the Internet of Things (IoT), we can analyse the data and give prediction for the user to keep the inhaler handy with them.

The objective of the paper is to provide an analytical approach to prevent asthma attack on a given day or time using weather, pollen and air quality data. If the prediction score for a given day is above a specified threshold, then alert the patient to take preemptive treatment.

A. Gupta (✉)
HCL Technologies, Pune, India
e-mail: aakash.gupta@outlook.com

© Springer Nature Singapore Pte Ltd. 2019   527
S. C. Satapathy et al. (eds.), *Information Systems Design and Intelligent Applications*,
Advances in Intelligent Systems and Computing 862,
https://doi.org/10.1007/978-981-13-3329-3_49

2 Related Works

Asthma disease once seen only in adults, but this is now affecting children's as well. There is an increase in the number of people getting affected by asthma and various techniques have been proposed to predict and diagnose the asthma attack. In this section, the state-of-the-art methods proposed recently in the field of asthma prediction systems are discussed.

Finkelstein and Jeong [2] proposed an advanced system which improved prophecy of asthma that may show remarkable improvement on living conditions and even lower the expenses on medications. Home-based telemonitoring system helps in providing data for developing enhanced systems which are used for the forecasting of increasing asthma. Predictive modelling algorithms like Classification and Regression Trees (CART) result had specificity 0.971, sensitivity of 0.647 and accuracy of 80%.

Shikalgar et al. [3] introduced a system to forecast the asthma attack on basis of asthma triggers history. Neural networks are used to analyse the everyday circumstances of the person which in turn creates the precise and accurate perspective of how is the person's routine. Table 1 will show the many triggers and how they are classified.

Siddiquee et al. [4] proposed a method which will help an asthma patient to handle crucial situations that can cause breathing problems. Fitbit devices are used to collect the real-time information like body temperature, location, heartbeats, etc. with the help of the app, as soon as the patient takes the inhaler notification is sent via Bluetooth to the app which in turn generates data analysis with the real-time data that helps doctors to diagnose the disease more easily and accurately. Sensors are used to determine locations with more pollution or traffic which are not safe for

Table 1 Classification of possible triggers

Body temperature	Blood pressure	Pulse rate	Questions
Drowsiness	Chest pain	Breathing trouble	Allergic food
Exhaustion	Difficulty in exercising	Chest pain	Smoke
Fainting	Dry or wet cough	Difficulty in exercising	Wood smoke
Fatigue	Rapid respiratory rate	Difficulty in speaking	Pollution exposure
Headache	Shortness of breath	Dry or wet cough	Trouble in sleeping
Low alertness level	Sneezing	Increased heart rate	Pattern of symptoms
		Rapid respiratory rate	Family history
		Shortness of breath	Proper medications
		Sighing	
			Allergens and irritants
		Sneezing	Animal dander

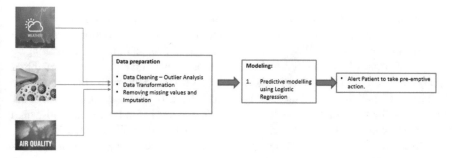

Fig. 1 Proposed asthma alert system

asthma patients and will generate a warning signal suggesting the current situation of the place like too much traffic and smoke is produced in this area and should take this travel route or not.

Al-khassaweneh et al. [5] proposed a method put forward which help in distinguishing and categorising different features of asthma in children's using the sounds of cough. MATLAB, programming language, is used to build an enhanced hardware system for monitoring asthma attacks on the basis of sounds of cough in different children.

3 Proposed Asthma Patient Alert System

The proposed method gathers the environmental data like weather, pollen and air quality from different sources and brings them to a central database. The collected data is preprocessed in which missing values imputation, outlier treatment and data transformation are done. Preprocessed data goes into the modelling part using logistic regression to find the probability score to take the preemptive action for asthma patient on any given day. Once the probability value is greater than the threshold the notification will be sent out to the patient, doctor and caretaker. Figure 1 shows the flow diagram for proposed asthma patient alerting system.

3.1 Data Sampling

Data comprises several data points from our environment such as using weather, pollen and air quality data. The data [6–8] is collected from Hermitage, Nashville, TN and USA. For asthma prediction, totally eight features are computed such as ozone, particle pollution, pollen index, dew point, wind speed, air pressure, humidity and air quality.

Ozone is a gas which is composed of three atoms of oxygen and found both in the Earth's upper atmosphere and at ground level. Particle pollution, also known as particulate matter or PM, is a mixture of liquid droplets and solids floating in the air. Particles less than or equal to 10 micrometres in diameter are so small that they can get into the lungs, potentially causing serious health problems like asthma attacks, etc. Pollen is one of the most common triggers of seasonal allergies and many people know pollen allergy as 'hay fever'. Dew point is the temperature to which air must be cooled to become saturated with water vapour. Wind speed is the rate of the movement of wind in distance per unit of time. It is the rate of the movement of airflow. Air pressure is present in the weather. Humidity is the amount of water vapour present in the air. Air quality pertains to the degree which the air is clean, clear and free from pollutants such as smoke, dust and smog among other gaseous impurities.

3.2 Model Building

To predict asthma attack logistic regression is used for model training as it measures the relationship between the categorical dependent variable and one or more independent variables by estimating probabilities using a logistic/sigmoid function.

The idea here is that in logistic regression, we predict not the actual probability but a transformed version of it, the 'log of odds'. Instead of the probability p, we deal with probability of happening of event/probability of not happening of event ($\log p/(1 - p)$) and find linear regression coefficients for the log odds.

The trained classification model using logistic regression on training data and performance has been evaluated on test data and it achieves accuracy of 84.2 and 79% accuracy in random forest. When compared to random forest, the logistic regression model performs well on asthma attack prediction model.

4 Analytical Results and Discussions

The proposed method is implemented on Windows 10 with 16 GB RAM and I5 processor. For experimental analysis, we have used RStudio software.

The data is divided into train and test based on the train–test split ratio of 70:30 using random sampling. Train set is used for model building and test set is used to evaluate the performance of the model. The dataset comprises 190 rows and 8 attributes, whereas training data is 133 rows and test data is 57 rows for the model building.

Logistic-regression-based classifier is used to predict the probability of preemptive action for asthma attack. Figure 2 represents the actual and predicted results from logistic regression model. The logistic regression model gave accuracy of 84.2% for predicting asthma attack (Table 2).

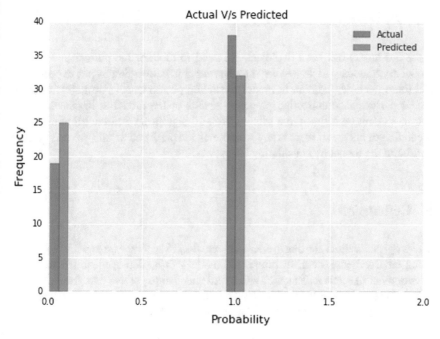

Fig. 2 Asthma attack prediction based on logistic regression model results

Table 2 Actual versus predicted results

Test dataset	Probability of 0	Probability of 1
Actual	19	38
Predicted	25	32

Table 3 Confusion matrix for asthma attack prediction on test data

Prediction	Reference	
	0	1
0	19	9
1	0	29

Figure 2 shows the actual versus predicted score from the logistic regression model, whereas blue bar represents actual values and green bar shows the predicted value. The model predicts the asthma attack on the basis of environmental data points like weather, air quality and pollen. Table 3 represents the confusion matrix for asthma attack prediction model.

4.1 Performance Monitoring

Once the model has been deployed, we need to monitor the performance of the model continuously to make sure that the model is performing as per expectations and there is no degradation in model prediction performance. For maintaining the model performance collecting predicted results of the model at regular intervals and to examine whether the attributes are still significant and contributing to the prediction or not and if there is any variation in the predicted results, we need to refit or rebuild the model by considering latest data.

5 Conclusion

The proposed asthma alerting system achieves 84.2% accuracy on logistic regression using environmental data. In future, the model can be deployed on the real-time environment like Jetson, Raspberry Pi board and also to get live data from different data sources with the help of APIs. The alert is also sent to patient, doctor and caretaker through mail or SMS to take preemptive action.

Acknowledgements I would sincerely like to thank Mr. Kameshwar Rao JV and my colleagues who encouraged me during this journey.

References

1. https://www.health.ny.gov/statistics/ny_asthma/pdf/national_asthma_survey_nys.pdf
2. J. Finkelstein, I.C. Jeong, Using CART for advanced prediction of asthma attacks based on telemonitoring data, in *2016 IEEE 7th Annual Ubiquitous Computing, Electronics & Mobile Communication Conference (UEMCON)*, New York, NY, 2016, pp. 1–5
3. S. Shikalgar, S. Marathe, P. Rai, D. Nadar, Rule extraction for detection and prevention of asthma attacks, in *2015 IEEE 3rd International Conference on MOOCs, Innovation and Technology in Education (MITE)*, Amritsar, 2015, pp. 437–440
4. J. Siddiquee, A. Roy, A. Datta, P. Sarkar, S. Saha, S.S. Biswas, Smart asthma attack prediction system using Internet of Things, in *2016 IEEE 7th Annual Information Technology, Electronics and Mobile Communication Conference (IEMCON)*, Vancouver, BC, 2016, pp. 1–4
5. M. Al-khassaweneh, S.B. Mustafa, F. Abu-Ekteish, Asthma attack monitoring and diagnosis: a proposed system, in *2012 IEEE-EMBS Conference on Biomedical Engineering and Sciences*, Langkawi, 2012, pp. 763–767
6. https://www.airnow.gov/
7. https://www.pollen.com/
8. https://api.darksky.net/

Unusual Crowd Event Detection: An Approach Using Probabilistic Neural Network

B. H. Lohithashva, V. N. Manjunath Aradhya, H. T. Basavaraju and B. S. Harish

Abstract Unusual event detection is a disputing problem in the field of security assets for public places. In this paper, we proposed detection of unusual crowded events in video established on interest point information. The distribution of magnitude and orientation holistic feature descriptor extracted from the histogram of optical flow for detection of usual and unusual events. In the proposed method, the probabilistic neural network approach is employed for unusual event detection in video. The reason for using PNN is more accurate, fast in training process and much faster than support vector machine. The proposed method used three benchmark datasets, UMN, PETS2009, and violent flows datasets, for the experiment. The obtained results compared with state-of-the-art methods, and improved outcomes are depicted to evaluate the outperformance of the proposed method.

Keywords Unusual crowd · Feature descriptor · Videos · HOFO
DiMOGIF · PNN

B. H. Lohithashva (✉) · V. N. Manjunath Aradhya · H. T. Basavaraju
Department of Master of Computer Applications,
JSS Science and Technology University (Sri Jayachamarajendra College of Engineering),
Mysuru, India
e-mail: lohi.bh@gmail.com

V. N. Manjunath Aradhya
e-mail: aradhya@sjce.ac.in

H. T. Basavaraju
e-mail: basavaraju.com@gmail.com

B. S. Harish
Department of Information Science and Engineering,
JSS Science and Technology University (Sri Jayachamarajendra College of Engineering),
Mysuru, India
e-mail: bsharish@sjce.ac.in

© Springer Nature Singapore Pte Ltd. 2019
S. C. Satapathy et al. (eds.), *Information Systems Design and Intelligent Applications*,
Advances in Intelligent Systems and Computing 862,
https://doi.org/10.1007/978-981-13-3329-3_50

1　Introduction

Video surveillance system is a process of analyzing video sequences. It is an active research area in pattern recognition and machine learning. The main arenas of video surveillance are in real-time processing, a machine-controlled system for detection of unusual events and event recrudesce analysis. The unusual event is distinguished as abnormal events from the normal ones. The detection of an unusual event is significant for the social and public security of the video scene. A public sector is an important place which is a necessity for everyday actions. The security of surveillance video in public sector associated with daily lives issues. Therefore, the security guard has to monitor dozens of videos simultaneously; it is quite difficult to detect unusual events due to the quality of information and rather lengthy footage. Therefore, they cannot respond instantly to emergency or unusual activity. The automated surveillance system assists the security guard to identify unusual events in a surveillance video.

In this work, we have researched on unusual event detection problem from the surveillance video. The University of Minnesota (UMN) [1], performance of evaluation of tracking and surveillance (PETS) 2009 [2], and violent flows datasets [3] are the datasets used for the experimentation. The global event detection is a classification problem which can be done on both usual and unusual samples taken for training. So, in this paper, we have used PNN for the classification. The UMN, PETS2009, and violent flows datasets have been tested in our proposed method. The rest of the paper is planned as follows. Section 2 describes related works concisely. The detailed description of the proposed method is presented in Sect. 3. Experimental results are shown in Sect. 4. Finally, we conclude the paper in Sect. 5.

2　Related Works

In the last few years, some of the researchers have mainly focused on unusual event detection and tracking related methods, and they have proposed many algorithms. In [4], local and global unusual event in the video using social force model was introduced. Based on the interaction force, the crowd activity was captured. Cong et al. [5] presented sparse reconstruction cost by a weighted linear construction. Sparse reconstruction cost detects both local and global unusual events. The histogram of the gradient (HOG) was first invented by Dalal and Triggs [6] for human detection. Recently, Wang et al. [7–9] introduced global unusual event detection using a HOFO and mean magnitude based on the movement information. Unusual crowd activity is detected by the distribution of the histogram of optical flow orientation and means magnitude in [10]. Setapathy et al. [11] introduced unsupervised feature selection technique using rough set and teaching based learning optimization. In spite of the significance progression in unusual activity detection, still, some restrictions make cause unmanageable and challenging. This motivated to take unusual event detection problem. In our research work, we have combined histogram of optical flow magnitude and orientation. Probabilistic neural network classifier is used to detect usual and unusual events.

3 Proposed Methodology

In this section, we can characterize usual and unusual events based on both magnitude and orientation of object apparent motion using extraction of distribution optical flow (OF) feature in crowded scenes. The complete model is to detect unusual crowd events using the proposed OF-feature-based method which is shown in Fig. 1. The proposed method comprises the below-mentioned stages.

3.1 Extraction of Spatiotemporal Interest Points (STIP)

In this paper, STIP points have been extracted from interest frames. STIP points vary because of variation of both spatial and temporal directions. Therefore, STIP points are detected when there is an eminent saturation in spatiotemporal arena. The proposed method extracts distribution of magnitude and orientation of global interest frame (DiMOGIF) descriptor. OF can allow for significant selective information around the spatial and variation dispersion of the objects. It is an evident distribution of brightness motion pattern in an image. A global constraint of smoothness in OF is introduced by Horn and Schunck [12]. The distribution of OF is applied to depict unusual events exhibiting in a scene. The unusual activity indicates divergence in

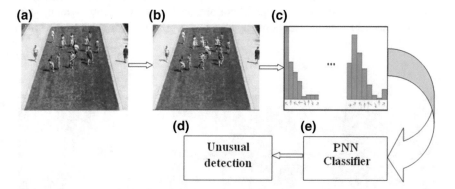

Fig. 1 Architecture of the proposed method; **a** Video sequence, **b** Optical flow for spatiotemporal interest points (STIP) extraction, **c** Distribution of magnitude and orientation of global interest frame (DiMOGIF) extraction, **d** PNN classifier is used for classification of usual and unusual event, **e** Detection of an unusual event

the motion from the usual activity. We find the motion vector $\begin{bmatrix} d_x \\ d_y \end{bmatrix}$ for each pixel $p(x, y)$, where d_x, and d_y are displacements along x and y-axes, respectively. The computations are based on two consecutive frames by taking one instance step. A two-dimensional distribution $[u, v]$ is computed from each one pixel location (x, y). The vector $[u, v]$ extracted magnitude and orientation at each pixel location (x, y).

$$\text{Magnitude} = \sqrt{u_x{}^2 + v_y{}^2} \tag{1}$$

$$\text{Orientation} = \tan^{-1}\left(\frac{v_y}{u_x}\right) \tag{2}$$

3.2 DiMOGIF Feature Extraction

The proposed method describes the extraction of distribution of magnitude and orientation global interest frame (DiMOGIF) descriptor as shown in Fig. 2. First, each sequence of frame is divided into 4×4 non-overlapping window as in [10]. Each window or block comprises a set of STIP points. We interested rich number of STIP points bearing in window. For the selection of STIP points, we have used certain threshold value and discarded the STIP points, if the values are below the threshold value. The next step comprises the model of each block motion changes throughout the time. The classification of spatiotemporal volume (STV) presenting each window by accumulating uninterrupted frames. Each STV is fabricated approximately to the center of all STIP points taken in the window. Indeed, we calculate the distribution of interest points in each pixel within an STV. The combination of magnitude and orientation of the OF is found next.

The window can be equally divided into the separation of nine equal subspaces. Orientation value is $(0°–360°)$ range. Therefore, the orientation information is quantized into one of the nine directions θ_n, $n = 1, 2, \ldots 9$. The histogram of the OF of

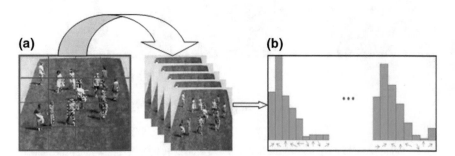

Fig. 2 DiMOGIF feature extraction process; **a** Each window in an interest frame is comprised by an STV built throughout all STIP points in the window, **b** Distribution of magnitude and orientation of optical flow vector is quantized to 9 bins histogram, finally DiMOGIF feature is calculated

nine bins is presented in each window. The combination of the HOFO and magnitude comprising all windows is then added. Histograms present all windows and then added to get the DiMOGIF feature of dimension 144($4 \times 4 \times 9$) for each frame, this constitutes the information of the unusual event that can be well determined by HOFO and magnitude from the OF motion. The proposed feature extraction method is shown in Algorithm 1.

Algorithm 1: Unusual crowd event detection based on DiMOGIF feature

---

Require: Sequence of training images $S_{tr} = (I_1^{N_{tr}})$, Sequence of testing images $S_{te} = (I_1^{N_{te}})$

1: Compute distribution of optical flow of training images, $(I_1^{N_{tr}}) \rightarrow (O_1^{N_{tr}})$

2: Feature extraction is computed established for each sequential n Spatio-Temporal Interest Points (STIPS), $(O_1^{N_{tr}}) \rightarrow (O_1^{N_{Ptr}})$, where $P_a = [\frac{N_{tr}}{n}]$

3: DiMOGIF feature descriptor is computed, $(O_1^{N_{tr}}) \rightarrow (G_1^{N_{Ptr}})$,

4: Sequence of training samples $(G_1^{N_{tr}})$ are determined based on PNN, weighted votes, and corresponding values are derived, $(G_1^{N_{tr}}) \xrightarrow{PNN} (G_1^{P'})$

5: Sequence of testing samples $(G_1^{N_{te}})$ is computed from testing fram $S_{te} = (I_1^{N_{te}})$ for each one sequential n STIPS.

6: Based on the set of weights $W_i \in (W_1^{P'})$ and corresponding values, each test samples $W_j \in (W_1^{N_{te}})$ of sequential of STIPS is classified detection stage. $A(W_i) = 1 \rightarrow Normal, 2 \rightarrow Abnormal$

3.3 Classification

Probabilistic neural network (PNN) is a supervised network, and it is feedforward artificial neural network, which is inspired by the Bayesian network, which can be used for classification and pattern recognition problem. D. F. Specht first introduced PNN approach in the year 1990s [13], which is different from the traditional neural network. Training of PNN is well suited and exigent; it permits addition and removal of data in the network [14, 15], without indigence for retraining, but only a few new data sample vector is added to existing weights when training. PNN classifier maps any input data to a number of classification, and it gives convergence to an optimal classifier. In our research work, global unusual crowd event detection relates to a classification problem. We have trained both usual and unusual samples of training data using PNN. The classification of testing samples is based on the computation of weighted votes and corresponding values during the training stage.

4 Experimentation Results and Discussions

In this section, we have demonstrated the experimentation results conducted to analyze the implementation of the proposed method of global unusual crowd events on three benchmark datasets UMN [1], PETS2009 [2], and violent flows dataset [3].

4.1 UMN and PETS2009 Dataset

The University of Minnesota (UMN) dataset consists of Lawn, Indoor, and Plaza, three different scenes in the video sequences of unusual crowd events. According to the video scenes, we have separated the sequence of frames. Based on the individual frames we have classified the usual and unusual event. In our research work, we have considered both usual and unusual frames for training and testing. Training and testing samples are extracted in performance of evaluation of tracking and surveillance (PETS) dataset based on the usual and unusual events. In experimentation, we have selected sequences as Time 14-16, Time 14-31, Time 14-17, and Time 14-55 in the PETS2009 dataset. Usual and unusual events in the PETS2009 have been selected based on the extraction of training and testing samples. We have selected Time 14-17, Time 14-31, and Time 14-55 for training and Time 14-16, Time 14-17, and Time 14-55 are selected for testing samples.

4.2 Violent Flows Dataset

Violent flows dataset is employed to assess in unusual crowd events. It comprises 246 small video clips in which 123 different clips comprise violent scenes. It includes 9092 frames, which has illumination, variation, shadow, occlusion, and complex background. We have selected all labeled sequences for training and both usual and unusual sequence of frames for testing.

4.3 Details of Implementation on UMN and PETS2009

PETS2009 dataset original resolution is 576×768 and UMN dataset resolution is 240×320. To simplify the procedure of an experiment and analyze the implementation, the PETS2009 resolution is changed to UMN resolution dataset. The dataset samples are initially trained; people who are walking in one direction as a usual event and people who are running toward one direction and all the directions considered as an unusual event. The sequence of testing includes both usual and unusual event samples. In UMN dataset, people walking in different directions are considered as

Table 1 Comparison of proposed method with other state-of-the-art method using area under ROC on UMN dataset

Methods	Lawn	Indoor	Plaza
Mehran et al. [4]	–	0.96	–
Mehran et al. [4]	–	0.84	–
Shi et al. [16]	0.936	0.776	0.966
Cong et al. [5]	–	0.93	–
Cong et al. [5]	0.995	0.975	0.964
Wang et al. [7]	0.985	0.904	0.982
Wang et al. [8]	0.998	0.953	0.978
Patil and Biswas [10]	0.984	0.956	0.968
Wang et al. [9]	0.978	0.923	0.985
Proposed method	1.000	0.987	0.989

a usual event. If there is a sudden change in the sequence of frames or people are running in different directions, it is considered as an unusual event.

4.4 Quantitative Performance Evaluation

To appraise the performance, frame-level measurement is followed. The measurement calculates the corresponding value among predicted result and the ground truth (GT). If the matching GT is unusual, it is considered as true positive otherwise it is considered as false positive. The frame-level evaluation is elementary and justifiable, but it cannot assure the right unusual pixel detection. Because some true positives perhaps induced along co-occurrence of false detection and true unusual. For the measurement at the frame level, we have applied the receiver operating characteristic (ROC) to measure the detection of accuracy. In the ROC curve, true positive rate (TPR) on y-axis measures the correct event and false positive rate (FPR) on the x-axis is not correctly detected. The evaluation measure based on ROC curves is used as the area under the curve (AUC). The proposed method is compared with state-of-the-art methods and has accomplished AUC on UMN scene as provided in (Fig. 3 and Table 1). Frame-level detection of the results is reported. The proposed method has achieved AUC of 100% on UMN Lawn scene, AUC 98.7% on UMN Indoor, and AUC 98.9% on UMN Plaza. Our proposed method outperforms state-of-the-art methods for Lawn, Indoor, and Plaza. Lawn result on UMN indoor and plaza unusual event detection and the ROC curves are shown in Fig. 3. The unusual event detection results on PETS2009 dataset are shown in Figs. 4 and 5. The proposed method has achieved AUC of 99.3% on Time 14-17 scene and AUC of 100% on Time 14-31 scene.

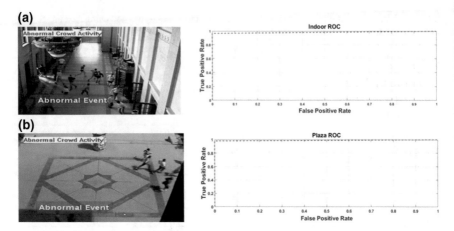

Fig. 3 Results on UMN dataset; **a** Indoor unusual event detection and ROC curve for unusual event detection AUC is 0.987. **b** Plaza unusual event detection and ROC curve for unusual event detection AUC is 0.989

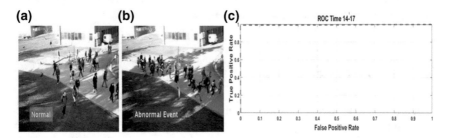

Fig. 4 Time 14-17 scene detection result; **a** Time 14-55 is chosen as training sample which contains people not walking orderly. **b** Time 14-17 is chosen as testing sample which contains crowds running in one direction. **c** The ROC curve for the unusual event detection AUC is 0.993

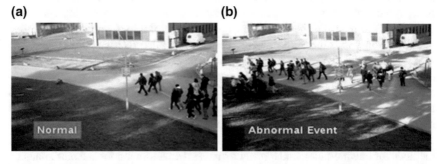

Fig. 5 Result of Time 14-31 scene detection; **a** Time 14-16 is chosen as training sample which contains people walking along the same direction. **b** Time 14-31 is chosen as testing sample which contains one queue split into several flows

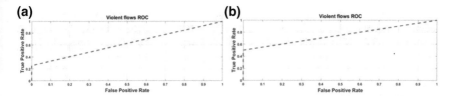

Fig. 6 The result of violent flows scene detection; **a** SVM classifier AUC is 0.25. **b** PNN classifier AUC is 0.50

4.5   Details of Implementation on Violent Flows Dataset

To study the superiority of the proposed method, we used violent flows dataset, which is required to evaluate unusual crowd events. The resolution of violent flows dataset is 240 × 320 which comprises 246 videos and in that 123 clips comprises violent from different scenes. We have combined violent flows dataset with UMN and PETS2009 datasets, which includes 9092 frames, which has illumination, variation, shadow, occlusion, and complex background. Sequences of all frames are selected for training and 15 usual and unusual video clips are given for testing, which contains 200 frames. PNN classifier maps any input data to some classification and it gives convergence to an optimal classifier. From the experimentation, it was clear that the proposed method outperformed the SVM classifier and achieved AUC of 50% against AUC of 25% using SVM classifier as shown in Fig. 6.

5   Discussion and Conclusion

Unusual crowd activity detection is challenging and unmanageable. The selection of features from the specific moving object is a very hard task because it determines substantial description and the analysis of unusual activity. Extraction of relevant and discriminative information from small varies in the visual object is very crucial to identify. Many researchers have used social force model, artificial neural network, optical flow, sparse reconstruction cost, HOG, and HOFO algorithms for unusual event detection using SVM classifier. In this paper, we propose PNN classifier which is used to detect crowd events in the videos. The approach is demonstrated with a distribution of spatiotemporal interest points using a HOFO and magnitude. The experiments conducted on UMN, PETS2009, and violent flows dataset benchmark datasets. Our proposed method when compared with state-of-the-art methods performed robust and improved results are reported. The PNN classifier is more accurate and faster than support vector machine. In the future work, we plan to continue work on some composite and large video applications.

Acknowledgements B. H. Lohithashva has been financially supported by UGC under Rajiv Gandhi National Fellowship (RGNF) Letter no. F1-17.1/2014-15/RGNF-2014-15-SC-KAR-73791/(SA-III/Website), Sri Jayachamarajendra College of Engineering, Mysuru, VTU, Karnataka, India.

References

1. Unusual Crowd Activity Dataset of University of Minnesota (UMN), Department of Computer Science and Engineering (2006), http://mha.cs.umn.edu/movies/crowdactivity-all.avi
2. Performance Evaluation of Tracking and Surveillance (PETS), Benchmark Data. Multisensor Sequences Containing Different Crowd Activities (2009), http://www.cvg.rdg.ac.uk/pets2009/a.html
3. T. Hassner, Y. Itcher, O. Kliper-Gross, Violent flows: real-time detection of violent crowd behavior, in *3rd IEEE International Workshop on Socially Intelligent Surveillance and Monitoring (SISM) at the IEEE Conference on Computer Vision and Pattern Recognition (CVPR)*, Rhode Island (June 2012), https://www.openu.ac.il/home/hassner/data/violentflows
4. R. Mehran, A. Oyama, M. Shah, Abnormal crowd behavior detection using social force model, in *IEEE Conference on Computer Vision and Pattern Recognition*, CVPR 2009, vol 2009 (IEEE, New York, 2009), pp. 935–942
5. Y. Cong, J. Yuan, J. Liu, Sparse reconstruction cost for abnormal event detection in *2011 IEEE Conference on Computer Vision and Pattern Recognition (CVPR)* (IEEE, New York, 2011), pp. 3449–3456
6. N. Dalal, B. Triggs, Histograms of oriented gradients for human detection, in *IEEE Computer Society Conference on Computer Vision and Pattern Recognition* (2005). CVPR 2005, vol 1 (IEEE, New York, 2005), pp. 886–893
7. T. Wang, H. Snoussi, Detection of abnormal visual events via global optical flow orientation histogram. IEEE Trans. Inf. Forensics Secur. **9**(6), 988–998 (2014)
8. T. Wang et al., Abnormal event detection via covariance matrix for optical flow based feature. Multimed. Tools Appl. pp. 1–21 (2017)
9. T. Wang et al., Abnormal event detection based on analysis of movement information of video sequence. Optik-Int. J. Light. Electron Opt. **152**, 50–60 (2018)
10. N. Patil, P.K. Biswas, Detection of global abnormal events in crowded scenes, in *2017 Twenty-third National Conference on Communications (NCC)* (IEEE, New York, 2017), pp. 1–6
11. SCh. Satapathy, A. Naik, K. Parvathi, Unsupervised feature selection using rough set and teaching learning-based optimisation. Int. J. Artif. Intell. Soft Comput. **3**(3), 244–256 (2013)
12. B.K.P. Horn, B.G. Schunck, Determining optical flow. Artif. Intell. **17**(1–3), 185–203 (1981)
13. D.F. Specht, Probabilistic neural networks. Neural Netw. **3**(1), 109–118 (1990)
14. V.N. Manjunath Aradhya, S. K. Niranjan, Analysis of different subspace mixture models in handwriting recognition, in *2012 International Conference on Frontiers in Handwriting Recognition (ICFHR)* (IEEE, New York, 2012), pp. 670–674
15. V.N. Manjunath Aradhya, S.K. Niranjan, G. Hemantha Kumar, Probabilistic neural network based approach for handwritten character recognition. Int. J. Comput. Commun. Technol. **1**(2, 3, 4), 9–13 (2010)
16. Y. Shi, Y. Gao, R. Wang, Real time abnormal event detection in complicated scene, in *Proceedings of International Conference on Pattern Recognition (ICPR)*, vol 13(7), pp. 3653–3656 (2010)

Variance-Based Feature Selection for Enhanced Classification Performance

D. Lakshmi Padmaja and B. Vishnuvardhan

Abstract Irrelevant feature elimination, when used correctly, aids in enhancing the feature selection accuracy which is critical in dimensionality reduction task. The additional intelligence enhances the search for an optimal subset of features by reducing the dataset, based on the previous performance. The search procedures being used are completely probabilistic and heuristic. Although the existing algorithms use various measures to evaluate the best feature subsets, they fail to eliminate irrelevant features. The procedure explained in the current paper focuses on enhanced feature selection process based on random subset feature selection (RSFS). Random subset feature selection (RSFS) uses random forest (RF) algorithm for better feature reduction. Through an extensive testing of this procedure which is carried out on several scientific datasets previously with different geometries, we aim to show in this paper that the optimal subset of features can be derived by eliminating the features which are two standard deviations away from mean. In many real-world applications like scientific data (e.g., cancer detection, diabetes, and medical diagnosis) removing the irrelevant features result in increase in detection accuracy with less cost and time. This helps the domain experts by identifying the reduction of features and saving valuable diagnosis time.

Keywords Random subset feature selection · Random forest
Variance-based selection · Classification accuracy

D. Lakshmi Padmaja
Department of Information Technology, Anurag Group of Institutions (CVSR),
Hyderabad, India
e-mail: glpadmaja@gmail.com

B. Vishnuvardhan (✉)
Department of Computer Science and Engineering, JNTUH, Hyderabad, India
e-mail: mailvishnuvardhan@gmail.com

© Springer Nature Singapore Pte Ltd. 2019
S. C. Satapathy et al. (eds.), *Information Systems Design and Intelligent Applications*,
Advances in Intelligent Systems and Computing 862,
https://doi.org/10.1007/978-981-13-3329-3_51

543

1 Introduction

Dimensionality reduction [1–3] is an important preprocessing task in data mining. This will aid in reducing the dimensionality of dataset thereby addressing curse of dimensionality issue. In general, the datasets, collected from scientific experiments, have more features than instances. This results in an increase in processing time and reduction in classification accuracy. Hence, it is prudent to reduce the dimensionality of the dataset without compromising the intrinsic geometric properties of data. For reducing the dimensionality and removing irrelevant features [4–7], we have proposed modified RSFS algorithm in our earlier work. A set of evaluation measures such as accuracy, information, distance, dependency, and consistency [8–10] are discussed in detail. This paper focuses on consistency of the feature selection as a parameter [11, 12].

Mainly, feature selection algorithms are categorized as filter, wrapper, and hybrid type of algorithms. The filter methods are classifier dependent and wrapper methods are computationally intensive. In view of the same RSFS algorithm is used for selecting the relevant features. This will help in evaluating performance measures like accuracy and consistency of the real-world applications. This will also help in reducing the dimensionality of scientific datasets which are unstructured, sparse, and often contain missing values [13, 14]. The randomization process will eliminate bias and overfitting problems. By eliminating the irrelevant features on scientific data reduces the cost and time (for cancer patients by reducing the tests) with the help of domain experts. This approach is more useful in medical diagnosis.

In this paper, Sect. 1 gives the introduction, Sect. 2 discusses the existing method, Sect. 3 discusses the proposed algorithm, Sect. 4 gives the dataset description, Sect. 5 provides the experimental setup, and Sect. 6 gives the conclusion.

2 Existing Method

We have selected RSFS algorithm along with eight datasets, available publicly, for this study.

2.1 Existing RSFS Algorithm

The aim of the random subset feature selection algorithm (RSFS) [8, 15] is to identify the best possible feature subsets, from a large dataset, in terms of its usefulness in a classification task. The feature selection is carried out iteratively. Traditional feature selection techniques follow to find the best subset of features using different methods such as forward selection algorithm by including new relevant feature for the existing dataset, in backward elimination algorithm, iteratively remove the unwanted features. Another approach is based on the feature correlation and ranking the features according to its usefulness [16]. Using kNN classifier, features are selected in each iteration. Relevancy is calculated as difference between the performance criteria and expected criteria. In RSFS, feature subset selection is based on a statistical comparison against random walk statistics. In RSFS, to improve the classification performance, the random subsets size is not fixed. Iterations are continued until good features are found. The iterations are stopped using stopping criteria which is user-defined threshold or search exhaust whichever is earlier. In RSFS, the random subset classification is performed, as many times as it is necessary in order to identify the good features from features that simply appears useful due to the random components of the process.

2.2 Classification

In random subset feature selection algorithm, kNN classifier is used to classify the features which are generated as subsets by random forest algorithm [17–19]. A kNN classifier is simple to use and easy to select features based on the nearest-neighbor-based measure. Usually, when k value is large, less variance and high bias are observed and for small k values, high variance and low bias are observed.

2.3 Motivation

Since the existing methodology suffers from consistency and selects different subsets of feature in different iterations. This will create suspicion in the view of domain experts and cannot result in a reliable and stable outcome. This has motivated us to enhance the existing methodology with a view to reduce inconsistency and improve the classification accuracy.

3 Proposed Methodology

The proposed methodology is experimented using RSFS algorithm [8, 15, 20]. The procedure is suitably modified to find the consistency, accuracies of the features with criteria as variance of features from mean. It is evident from the table that RSFS +1NN algorithm has better accuracy and reduction of features compared to normal kNN algorithm. The pseudocode for the algorithm is as shown below:

1. Normalize both training and testing datasets.
2. Randomly select the subset from the full feature set.
3. Classify the dataset using kNN classifier.

Table 1 Dataset details

S. No.	Dataset	No. of features	No. of instances	KNN accuracy (%)	KNN + RSFS	
					Features	Accuracy (%)
1	Colon Cancer	62	2000	54.55	20	63.63
2	CTG	20	2126	85.07	10	87.61
3	Lung cancer 32 149	12,533	149	93.75	2	71.88
4	Lung	3312	203	94.20	66	95.65
5	Fisher iris	4	150	92.16	2	96.08
6	ALLAML	7129	72	84.00	2	80.00
7	Carcinom	9182	174	93.22	10	61.02
8	Prostate 102 34	12,600	174	88.14	11	45.76
9	Forest	27	523	86.46	15	82.15
10	Ovarian cancer	4000	216	91.89	27	95.95

4. Record the feature relevance using unweighted average recall (UAR).
5. Select the top 1% of features based on probability scale of performance.
6. Eliminate the bottommost features whose relevancy is less than mean-3 (outliers).
7. Repeat the process until the number of selected features are constant for 1000 iterations then choose the features with best relevancies (whose probability is greater than 0.99).
8. Otherwise, go to step number 2 and repeat all the steps until the selected features are constant.

The algorithm works as shown in Fig. 1.

4 Datasets and Experimental Setup

All the datasets are taken from UCI machine learning repository [16], www. featureselection.asu.edu [21], www.broadinstitute.org [22], and some are from cancer research studies. The dataset properties are mentioned in Table 1.

From the above results, it is evident that the performance of RSFS algorithm is superior when compared to the other algorithms. In order to improve further, the algorithm is modified as mentioned in Fig. 2.

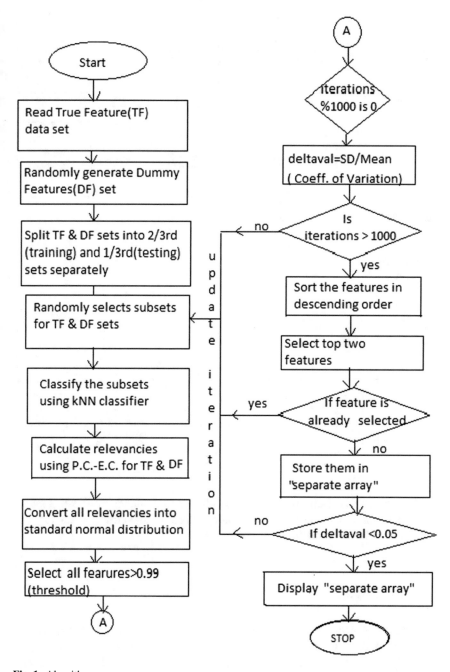

Fig. 1 Algorithm

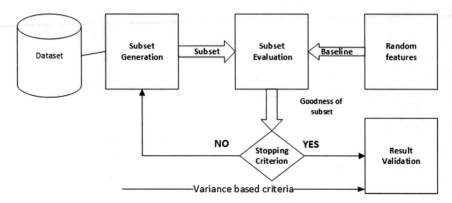

Fig. 2 Modified process for variance-based selection

5 Results

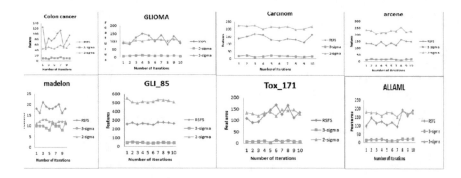

The process, as mentioned in Fig. 2, is followed for reducing the dimensionality while maintaining the consistency of feature selection. The details about variance selection and feature reduction are shown in Fig. 3.

6 Conclusion

Consistency of features is maintained when we have considered the variance of features from mean to three-sigma standard deviation when compared to two-sigma standard deviations away from mean. These selection techniques combined with RSFS algorithm are producing very satisfactory results for feature subset selection. Time complexity is also improved when compared to the original RSFS algorithm. In future, more studies and experiments are required for fine-tuning the algorithm

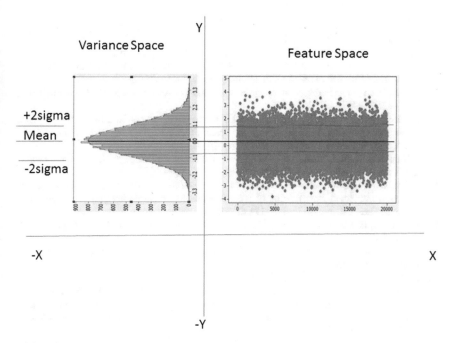

Fig. 3 Mapping of feature space to variance space

and process to apply various types of data such as ordinal, categorical, and mixed along with multi-label and missing values.

References

1. C. Bartenhagen, H.U. Klein, C. Ruckert, X. Jiang, M. Dugas, Comparative study of unsupervised dimension reduction techniques for the visualization of microarray gene expression data. BMC Bioinform. **11**(1), 567 (2010)
2. L. Shen, E.C. Tan, Dimension reduction-based penalized logistic regression for cancer classification using microarray data. IEEE/ACM Trans. Comput. Biol. Bioinform. (TCBB) **2**(2), 166–175 (2005)
3. L. Van Der Maaten, E. Postma, J. Van den Herik, Dimensionality reduction: a comparative. J. Mach. Learn. Res. **10**, 66–71 (2009)
4. C. Ding, H. Peng, Minimum redundancy feature selection from microarray gene expression data. J. Bioinform. Comput. Biol. **3**(02), 185–205 (2005)
5. F. Ding, C. Peng, H. Long, Feature selection based on mutual information: criteria of max-dependency, max-relevance, and min-redundancy **27**(8), 1226–1238 (2005)
6. L. Yu, H. Liu, Efficiently handling feature redundancy in high-dimensional data, in *SIGKDD 03* (Aug 2003)
7. H. Liu, H. Motoda, *Feature Selection for Knowledge Discovery and Data Mining*, vol. 454 (Springer Science & Business Media, Berlin, 2012)
8. J. Pohjalainen, O. Rasanen, S. Kadioglu, Feature selection methods and their combinations in high dimensional classification of speaker likability, intelligibility and personality traits (2013)

9. L. Yu, C. Ding, S. Loscalzo, Stable feature selection via dense feature groups, in *Proceedings of the 14th ACM SIGKDD* (2008)

10. M. Dash, H. Liu, Feature selection for classification **1**, 131–156 (1997)

11. K. Kira, L.A. Rendell, A practical approach to feature selection, in *Proceedings of the Ninth International Workshop on Machine learning*, pp. 249–256 (1992)

12. J. Reunanen, Overfitting in making comparisons between variable selection methods **3**, 1371–1382 (2003)

13. E. Maltseva, C. Pizzuti, D. Talia, Mining high dimensional scientific data sets using singular value decomposition, in *Data Mining for Scientific and Engineering Applications* (Kluwer Academic Publishers, Dordrecht, 2001), pp. 425–438

14. J. Kehrer, H. Hauser, Visualization and visual analysis of multifaceted scientific data: a survey. IEEE Trans. Visual Comput. Graphics **19**(3), 495–513 (2013)

15. J. Pohjalainen, O. Rasanen, S. Kadioglu, Feature selection methods and their combinations in high-dimensional classification of speaker likability, intelligibility and personality traits. Comput. Speech Lang. **29**(1), 145–171 (2015)

16. D. Dheeru, E. Karra Taniskidou, UCI machine learning repository (2017), http://archive.ics.uci.edu/ml

17. S. Li, J. Harner, D. Adjeroh, Random kNN feature selection a fast and stable alternative to random forests. BMC Bioinform. (Dec 2011)

18. L. Breiman, Random forests **3**, 5–32 (2001)

19. I. Guyon, A. Elisseeff, An introduction to variable and feature selection **3**, 1157–1183 (2003)

20. O. Räsänen, J. Pohjalainen, Random subset feature selection in automatic recognition of developmental disorders, affective states, and level of conflict from speech, in *INTERSPEECH*, pp. 210–214 (2013)

21. J. Li, K. Cheng, S. Wang, F. Morstatter, T. Robert, J. Tang, H. Liu, Feature selection: a data perspective arXiv:1601.07996 (2016)

22. https://www.broadinstitute.org/

Attribute Reduction for Defect Prediction Using Random Subset Feature Selection Method

G. N. V. Ramana Rao, V. V. S. S. S. Balaram and B. Vishnuvardhan

Abstract Large software products require high effort to maintain the code base. Most of the time managers face challenging situations for efficient allocation of resources. In this paper, we proposed a novel approach to aid the software engineering managers to predict the software defects using few matrices. In our study, we have used publicly available software engineering repositories concentrating on object-oriented (OO) methodology. Our study suggests that few important matrices are sufficient to predict the defects in the system. We have used kNN classifier for classification and random subset feature selection (RSFS) for dimensionality reduction of the attributes.

Keywords Software defect prediction · Object-oriented metrics
Feature subset selection · Dimensionality reduction · Software engineering

1 Introduction

Software system testing is a time-consuming activity. This will be challenging when the system design is complex with more number of modules to maintain. Due to cost reduction pressure, it is not always possible to allocate equal effort or resources to all modules of the system. Due to uneven distribution of defects in the modules, the resource allocation is a challenging task for the managers. Hence, if the testing

G. N. V. Ramana Rao (✉)
Wipro Ltd, Bengaluru, India
e-mail: ramana.rao@wipro.com

V. V. S. S. S. Balaram
Department of IT, SNIST, Hyderabad, India
e-mail: vbalaram@sreenidhi.edu.in

B. Vishnuvardhan
Department of Computer Science, JNTUH, Hyderabad, India
e-mail: mailvishnuvardhan@gmail.com

© Springer Nature Singapore Pte Ltd. 2019
S. C. Satapathy et al. (eds.), *Information Systems Design and Intelligent Applications*,
Advances in Intelligent Systems and Computing 862,
https://doi.org/10.1007/978-981-13-3329-3_52

managers, responsible for testing resources, can predict the defects using few metrics then the effort for maintaining the code base will be optimal in nature.

Weyuker et al. [14, 15] opined that 80% of the defects are sourced from 20% of the modules in the system. Testers with good defect prediction techniques can detect 80% defects using 20% of effort. Defect prediction studies usually depend on the historical data to predict the defects by building efficient prediction models. But the effort for collection of these matrices increases if the number of metrics is more. So in the interest of testing manager community, small set of metrics for defect prediction will not only reduce the collection cost but also increases the prediction performance of the model as well.

In this paper, we have limited our study on object-oriented (OO) metrics and used datasets from public repositories. Each dataset is with 20 object-oriented (OO) metrics of different instances. Our approach to build and defect prediction model with few metrics is discussed in the conclusion section. This paper is organized as follows: In Sect. 2, the related work is described. In Sect. 3, a detailed description about datasets and OO metrics are discussed and in Sect. 4, the experimental set and results are discussed. The conclusions are discussed in Sect. 5.

2 Current Work and Challenges

In this section, a brief discussion is presented on the dimensionality reduction techniques used in the field of software engineering.

Guyon and Eliseff [4] proposed key approaches for feature selection including feature ranking, feature selection, and efficient search methods. Hall and Holmes presented six attribute selection techniques to produce ranked list of features. This ranking can help to determine the relevancy of each feature.

The application of feature selection in the field of software quality is not extensive till date. Many approaches are toward effort estimation and defect prediction on change history. Very few researches have attempted to develop defect prediction models using dimensionality reduction, the preprocessing stage. Chen and Daoqiang [1] suggested approaches for effort estimation techniques using wrapper-based techniques. Rodriguez [12] concluded in their comparative study that the wrapper methods outperform remaining methods.

Very limited research exists in the dimensionality reduction task of data mining, primarily focusing on software defect prediction. Koru and Liu [6] concluded strangely that generalizable defect prediction cannot be developed as the defect prediction models are not independent of development environment. But, in our study, we have observed that such an attempt is worth considering the amount of advantage it can give to the quality managers. Chen et al. [2] have stated that the features selected in one project cannot be used in another project.

2.1 Existing Feature Subset Selection Methods

The following methods are used for selection of subset of features based on intrinsic property of a large dataset. Primarily, there are three methods called filter, wrapper, and hybrid methods.

Filter methods are the primarily used distance measure for assigning a score to the attributes or features. Usually, filter methods are computationally less intensive but accuracy is low when compared to wrapper methods.

Wrapper methods use predictive model instead of distance measure for ranking the attributes. Each subset of attributes or features is connected to a predictive model and error is accumulated over several iterations. This is computationally intensive but more accurate.

Hybrid methods use the combination of the abovementioned methods to achieve high accuracy with low computational complexity.

2.2 Challenges

From the above studies, it is understood that there is a need for a prediction model which can perform with less attributes and with equal or improved classification accuracy. This will help the managers to reuse the data collected from previous release of the software and predict the defects for future planning. Reducing the attributes will reduce the collection cost and time as well.

In this study, we have selected mutual information (MI) and statistical dependency (SD) methods as feature scoring methods and sequential forward selection (SFS), sequential float forward selection (SFFS), and random subset feature selection (RSFS) are selected as wrapper methods.

We have conducted experiments on nine datasets (refer Table 1), with OO metrics, which are publicly available.

Table 1 Dataset details

S. No.	Dataset name	Features	Instances
1	Prop1	19	18,472
2	Serapion	19	45
3	Skarbonka	19	45
4	Sklebagd	19	20
5	Synapse 1.0	19	157
6	Synapse 1.2	19	256
7	Systemdata	19	65
8	Szybkafucha	19	25
9	Termoproject	19	42

3 Datasets and OO Metrics

The following datasets are collected from PROMISE repository for our study. The datasets are with OO metrics mostly used for Java projects. The metrics are initially developed by Chidamber and Kemerer [3] and further enhanced by Henderson-Sellers [5], Tang et al. [13], and Martin [8]. Size- and complexity-based metrics are used for this study which are as mentioned below:

1. Metrics suggested by Chidambaram and Kemerer [3]

   (a) Weighted methods per class (WMC): Total number of methods in the class. If there are five methods, then the weight of the class would be five.
   (b) Depth of inheritance (DIT): Number of inheritance levels from the object hierarchy.
   (c) Number of children (NOC): Number of immediate descendants of a given class.
   (d) Coupling between objects (CBO): The number of class coupled together though method invocation, system calls, inheritance, and method arguments.
   (e) Response for class (RFC): Number of methods triggered when and an object receives trigger.
   (f) Lack of cohesion in methods (LCOM): Difference between the number of method pairs that share access to that do not share access.

2. Metrics suggested by Henderson-Sellers [5]

   (a) Lack of cohesion in methods (LCOM3):

   $$\frac{\sum_{j=1}^{a} \left\lceil \frac{1}{a}\mu(A_j) \right\rceil - m}{1 - m} \tag{1}$$

   where m is number of methods in a class, a is number of attributes of the class, and $\mu(A)$ is number of attributes that access the same method.

3. Metrics suggested by Bansiy and Davis (5)

   (a) Number of public methods (NPM): Number of methods declared as public also known as class interface size.
   (b) Data access metric (DAM): The ratio of private (public) attributes to total attributes.
   (c) Measure of aggregation (MOA): The number of class fields of user-defined type.
   (d) Measure of functional abstraction (MFA): Ratio of number of methods inherited by a class to total the total number of methods accessible.
   (e) Cohesion among methods of class (CAM): Number of deferent types of method parameters in every method divided by product of number of deferent method parameter type in entire class and number of methods.

4. Metrics suggested by Tang et al. [13]

    (a) Inheritance coupling (IC): Number of parent classes are coupled to a given class.

    (b) Coupling between methods (CBM): Total number of methods to which all inherited methods are coupled.

    (c) Average method complexity (AMC): Average method size of each class.

5. Metrics suggested by Martin [8]

    (a) Afferent couplings (C_a): Number of classes that depends on measured class.

    (b) Efferent couplings (Ce): Number of classes that the measured class depends on.

6. McCabe's metric (1)

    (a) McCabe's metric cyclomatic complexity metric (CC): Number of paths in method plus one.

7. Size metric (1)

    (a) Lines of code (LOC): Number of lines in the program or application under measure.

4 Experimental Setup and Results

The following process is used for conducting the experiments. The original dataset is split into 2/3rd and 1/3rd for as training and testing purpose. The training dataset is used for training the process using random forest algorithm. The training dataset is randomly split into subsets and accuracy is calculated. The process is repeated till the process converges into solution. In parallel, a random walk process is used for selecting the true features from the subset so that the features which are in similar geometry of original dataset are selected as final subsets. This measure will help us to select global maxima always. The stopping criterion is used for stopping process when there are no subsets available for selection. Figure 1 gives a detailed process of the same.

The datasets (refer Table 1) are selected from the publicly available repository PROMISE. The collected datasets are reviewed and cleaned to remove unimportant data from dataset.

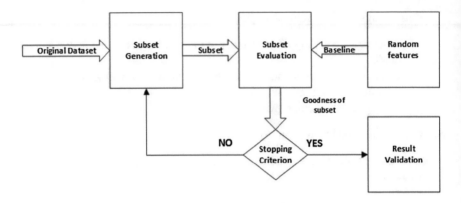

Fig. 1 Random subset feature selection process

4.1 About RSFS Process

In the preprocessing stage, a random subset feature selection (RSFS) algorithm is used for selection of reduced number of features without compromising the prediction accuracy. The RSFS algorithm works on random forest principle to reduce the data. A detailed work has been carried out by Padmaja and Vishnuvardhan in their work [13–15]. From their work, it is concluded that the RSFS algorithm performance is superior to other feature subset selection algorithms. A simple kNN classifier is used as baseline performance measure to validate the results.

4.2 Results

After completing the experiments on nine datasets (refer Table 1) collected from PROMISE repository, the following results are obtained.

From the results (refer Fig. 2), it is evident that RSFS algorithm performance is superior when compared to other algorithms. Accuracy 1 refers to Synapse 1.2 and accuracy 2 refers to szybkafucha dataset. Other datasets are used while conducting the experiments and are not mentioned here to avoid the clutter.

5 Conclusions and Future Work

From the results, it is clear that the RSFS algorithm is highly useful in reducing the dimensionality of the large attribute set to smaller set while retaining the same classification accuracy [7, 9]. The result suggests that with fewer matrices we can perform defect prediction without compromising on the accuracy. However, a detailed

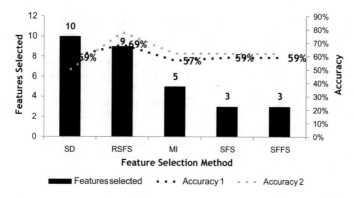

Fig. 2 Attribute selection method results

study is required to cover the remaining technologies as we have concentrated on object-oriented technologies only. Software engineering community will be greatly benefited if the prediction results are improved using less number of attributes. This will reduce the attribute collection cost and time for enhancing the defect prediction accuracy.

The current study is limited to object-oriented (OO) metrics and the same can be expanded to other methodology related software for wider coverage using the same attribute reduction method. A detailed study is carried out in Lakshmipadmaja and Vishnuvardhan study [7] as well as in Painter et al. [10] and Panhalkar et al. [11] on how to implement the same.

References

1. S. Chen, Z. Daoqiang, Semisupervised Dimensionality Reduction With Pairwise Constraints for Hyperspectral Image Classification. IEEE Geosci. Remote Sens. Lett. **8**(2), 369–373 (Mar 2011), https://doi.org/10.1109/LGRS.2010.2076407
2. Z. Chen, T. Menzies, D. Port, D. Boehm, Finding the right data for software cost modeling. IEEE Softw. **22**(6), 38–46 (2005)
3. S.R. Chidamber, C.F. Kemerer, A metrics suite for object oriented design. IEEE Trans. Softw. Eng. **20**(6), 476–493 (1994)
4. I. Guyon, A. Elisseeff, An introduction to variable and feature selection. J. Mach. Learn. Res. **3**(Mar), 1157–1182 (2003)
5. B. Henderson-Sellers, *Object-Oriented Metrics: Measures of Complexity* (Prentice-Hall, Inc., Englewood Cliffs, 1995)
6. A.G. Koru, H. Liu, Building effective defect-prediction models in practice. IEEE Softw. **22**(6), 23–29 (2005)
7. D. Lakshmipadmaja, B. Vishnuvardhan, Classification performance improvement using random subset feature selection algorithm for data mining. Big Data Res. (2018)
8. R. Martin, OO design quality metrics. Anal. Depend. **12**, 151–170 (1994)
9. D.L. Padmaja, B. Vishnuvardhan, Comparative study of feature subset selection methods for dimensionality reduction on scientific data, in *2016 IEEE 6th International Conference on Advanced Computing (IACC)* (IEEE, New York, 2016), pp. 31–34

10. N. Painter, B. Kadhiwala, Comparative analysis of android malware detection techniques, in *Proceedings of the International Conference on Data Engineering and Communication Technology* (Springer, 2017), pp. 131–139

11. A. Panhalkar, D. Doye, An outlook in some aspects of hybrid decision tree classification approach: a survey, in *Proceedings of the International Conference on Data Engineering and Communication Technology* (Springer, 2017), pp. 85–95

12. I. Rodriguez-Lujan, R. Huerta, C. Elkan, C. S. Cruz, Quadratic programming feature selection. J. Mach. Learn. Res.**11**(3), (1532–4435) 1491–1516, (1859900) JMLR.org, (Aug 2010), http://dl.acm.org/citation.cfm?id=1756006.1859900

13. M.H. Tang, M.H. Kao, M.H. Chen, An empirical study on object-oriented metrics, in *Proceedings Sixth International Software Metrics Symposium* 1999 (IEEE, New York, 1999), pp. 242–249

14. D. Wahyudin, R. Ramler, S. Biffl, A framework for defect prediction in specific software project contexts, in *IFIP Central and East European Conference on Software Engineering Techniques* (Springer, Berlin, 2008), pp. 261–274

15. E.J. Weyuker, T.J. Ostrand, R.M. Bell, Do too many cooks spoil the broth? using the number of developers to enhance defect prediction models. Empir. Softw. Eng. **13**(5), 539–559 (2008)

Correction to: Machine Learning on Medical Dataset

M. P. Gopinath, S. L. Aarthy, Aditya Manchanda and Rishabh

Correction to:
Chapter "Machine Learning on Medical Dataset" in:
S. C. Satapathy et al. (eds.), *Information Systems Design and Intelligent Applications*, Advances in Intelligent Systems and Computing 862, https://doi.org/10.1007/978-981-13-3329-3_13

In the original version of the book, the author name "Rishadh" has been changed to "Rishabh" in the Chapter "Machine Learning on Medical Dataset". The chapter and book have been updated with the changes.

The updated version of this chapter can be found at
https://doi.org/10.1007/978-981-13-3329-3_13

Author Index

Printed in the United States
By Bookmasters